Membrane Desalination

Membrane Desalination
From Nanoscale to Real World Applications

Edited by
Andreas Sapalidis

CRC Press
Taylor & Francis Group
Boca Raton London New York

CRC Press is an imprint of the
Taylor & Francis Group, an **informa** business

First edition published 2021
by CRC Press
6000 Broken Sound Parkway NW, Suite 300, Boca Raton, FL 33487-2742

and by CRC Press
2 Park Square, Milton Park, Abingdon, Oxon, OX14 4RN

© 2021 Taylor & Francis Group, LLC

First edition published by CRC Press 2021

CRC Press is an imprint of Taylor
& Francis Group, LLC

International Standard Book Number-13: 978-0-367-03079-7 (Hardback)
International Standard Book Number-13: 978-0-429-02025-4 (eBook)

Reasonable efforts have been made to publish reliable data and information, but the author and publisher cannot assume responsibility for the validity of all materials or the consequences of their use. The authors and publishers have attempted to trace the copyright holders of all material reproduced in this publication and apologize to copyright holders if permission to publish in this form has not been obtained. If any copyright material has not been acknowledged please write and let us know so we may rectify in any future reprint.

Except as permitted under U.S. Copyright Law, no part of this book may be reprinted, reproduced, transmitted, or utilized in any form by any electronic, mechanical, or other means, now known or hereafter invented, including photocopying, microfilming, and recording, or in any information storage or retrieval system, without written permission from the publishers.

For permission to photocopy or use material electronically from this work, access www.copyright.com or contact the Copyright Clearance Center, Inc. (CCC), 222 Rosewood Drive, Danvers, MA 01923, 978-750-8400. For works that are not available on CCC please contact mpkbookspermissions@tandf.co.uk

Trademark notice: Product or corporate names may be trademarks or registered trademarks, and are used only for identification and explanation without intent to infringe.

Contents

Preface vii
Biography ix
Editors xi

Chapter 1 Introduction to Membrane Desalination 1

Andreas A. Sapalidis, Evangelos P. Kouvelos,
George E. Romanos, and Nick K. Kanellopoulos

Chapter 2 Membrane Materials Design Trends: Nanoadditives 17

Shadia R. Tewfik, Mohamed H. Sorour, Abdelghani M.G.
Abulnour, Hayam F. Shaalan, Heba A. Hani, Marwa M.H.
El-Sayed, Yomna O. Abdelrahman, Eman S. Sayed, Aya N.
Mohamed, Amany A. Al-Mansoup, and Nourhan A. Shawky

Chapter 3 Functionalization of Carbon-Based Additives 67

Myrsini Kyriaki Antoniou, Andreas Sapalidis, and
Zili Sideratou

Chapter 4 Nanohybrid Graphene-Based Materials for Advanced Wastewater
Treatment: Adsorption and Membrane Technology..................... 91

George Z. Kyzas, Evangelos P. Favvas, and
Athanasios C. Mitropoulos

Chapter 5 State of the Art and Perspectives in Membranes for Membrane
Distillation/Membrane Crystallization 125

Carmen Meringolo, Gianluca Di Profio, Efrem Curcio, Elena
Tocci, Enrico Drioli, and Enrica Fontananova

Chapter 6 Artificial Water Channels: Toward Next-Generation Reverse
Osmosis Membranes 211

Maria Di Vincenzo, Sophie Cerneaux, Istvan Kocsis, and Mihail
Barboiu

v

vi Contents

Chapter 7 Desalination Membranes: Characterization Techniques 229

Rund Abu-Zurayk, Mohammed R. Qtaishat, and Abeer Al Bawab

Chapter 8 Carbon Molecular Models for Desalination.............................. 251

Georgia Karataraki, Anastasios Gotzias, and Elena Tocci

Chapter 9 Virtual Material Design: A Powerful Tool for the Development of High-Efficiency Porous Media 269

Stéphan Barbe and Aron Kneer

Chapter 10 Estimation of Process Parameters in Industrial Membrane Manufacture Using Computational Fluid Dynamics 285

Stéphan Barbe, Fei Wang, Aron Kneer, Michael Metze, Christian Wenning, Patrick Altschuh, and Britta Nestler

Chapter 11 Towards Energy-Efficient Reverse Osmosis 303

Mohamed T. Mito and Philip Davies

Chapter 12 Membrane Fouling and Scaling in Reverse Osmosis 325

Nirajan Dhakal, Almotasembellah Abushaban, Nasir Mangal, Mohanad Abunada, Jan C. Schippers, and Maria D Kennedy

Chapter 13 Sustainable Development and Future Trends in Desalination Technology... 345

Mattheus Goosen and Hacene Mahmoudi

Terminology 369

Index 383

Preface

Countless pages have been written on the importance of water quality to humanity. All of them include one very specific phrase: *water=life*. This equation leaves no room for interpretation; there are no *buts* or *ifs*. Western societies have found themselves isolated from the natural environment, which they take for granted, with their only concern being what will be the next technological development in terms of the television's resolution, the electric vehicle's mileage, and the smartphone's camera capabilities.

Our natural resources are deteriorating in terms of both quality and quantity. Solutions need to be included in our everyday life that will provide a heathy and sustainable environment for the generations to come.

Looking at the Earth from a distance, no one can assume that this habitat is lacking water resources. The technology to introduce all this water in our everyday life is not widely established, however. The easy answer to why we don't use all the seawater that surrounds us is that associated energy costs and environmental impacts make it unattractive to do so.

Desalination has had major scientific and technological breakthroughs in the last 50 years, reducing the costs for every m^3 of potable water produced massively by introducing better performing membranes, energy recovery systems, effective pretreatment, fouling control, and process optimization.

The aforementioned have been boosted by nanotechnology, which, for the last 30 years has provided ingenious solutions in desalination, and particularly, in membrane synthesis.

In this book, the reader will be introduced to membrane-based desalination methods, from the nanoscale to real-world applications, as well as to future developments.

Biography

Dr. Andreas Sapalidis is a researcher at the Institute of Nanoscience and Nanotechnology of the National Centre for Scientific Research Demokritos, Greece. He holds a bachelor's degree in petroleum engineering from the Technological Institute of Kavala, Greece. In 2004, he obtained his master of science from the Department of Chemistry in the field of polymer science of the National and Kapodistrian University of Athens, and in 2009, a PhD from the same department. He is a member of Membranes and Materials for Environmental Separations Laboratory (MESL) for more than 15 years, with his research interests focusing on novel preparation and characterization techniques of nanostructured materials for environmental and energy applications. Indicatively, Dr. Sapalidis focuses on preparation and characterization of nanocomposite structures, development of porous hybrid polymeric and inorganic membranes, and novel sorbents for liquid and gas applications.

He is the co-author of 40 research papers and has given more than 50 conference presentations.

Editors

Andreas A. Sapalidis
Institute of Nanoscience and
  Nanotechnology
National Centre for Scientific
  Research Demokritos
Agia Paraskevi Attikis, Greece

Evangelos P. Kouvelos
Institute of Nanoscience and
  Nanotechnology
National Centre for Scientific
  Research Demokritos
Agia Paraskevi Attikis, Greece

George E. Romanos
Institute of Nanoscience and
  Nanotechnology
National Centre for Scientific
  Research Demokritos
Agia Paraskevi Attikis, Greece

Nick K. Kanellopoulos
Institute of Nanoscience and
  Nanotechnology
National Centre for Scientific
  Research Demokritos
Agia Paraskevi Attikis, Greece

Shadia R. Tewfik
National Research Centre
Dokki, Giza, Egypt

Mohamed H. Sorour
National Research Centre
Dokki, Giza, Egypt

Abdelghani M.G. Abulnour
National Research Centre
Dokki, Giza, Egypt

Hayam F. Shaalan
National Research Centre
Dokki, Giza, Egypt

Heba A. Hani
National Research Centre
Dokki, Giza, Egypt

Marwa M.H. El-Sayed
National Research Centre
Dokki, Giza, Egypt

Yomna O.Abdelrahman
National Research Centre
Dokki, Giza, Egypt

Eman S. Sayed
National Research Centre
Dokki, Giza, Egypt

Aya N. Mohamed
National Research Centre
Dokki, Giza, Egypt

Amany A. Al-Mansoup
National Research Centre
Dokki, Giza, Egypt

Nourhan A. Shawky
National Research Centre
Dokki, Giza, Egypt

Myrsini Kyriaki Antoniou
Institute of Nanoscience and
  Nanotechnology
National Centre for Scientific
  Research Demokritos
Agia Paraskevi Attikis,
  Greece

Zili Sideratou
Institute of Nanoscience and
 Nanotechnology
National Centre for Scientific
 Research Demokritos
Agia Paraskevi Attikis, Greece

George Z. Kyzas
International Hellenic University
Department of Chemistry
Kavala, Greece

Athanasios C. Mitropoulos
International Hellenic University
Department of Chemistry
Kavala, Greece

Evangelos P. Favvas
National Centre for Scientific
 Research Demokritos
Agia Paraskevi Attikis, Greece

Carmen Meringolo
Institute on Membrane Technology
 of the
National Research Council
 (ITM-CNR)
at University of Calabria (UNICAL)
Rende (CS), Italy

Gianluca Di Profio
Institute on Membrane Technology
 of the
National Research Council
 (ITM-CNR)
at University of Calabria (UNICAL)
Rende (CS), Italy

Efrem Curcio
Department of Environmental and
 Chemical Engineering
(DIATIC)
University of Calabria (UNICAL)
Rende (CS), Italy

Elena Tocci
Institute on Membrane Technology
 of the National Research Council
 (ITM-CNR)
at University of Calabria (UNICAL)
Rende (CS), Italy

Enrico Drioli
Institute on Membrane Technology
 of the National Research Council
 (ITM-CNR)
at University of Calabria (UNICAL)
Rende (CS), Italy

Enrica Fontananova
Institute on Membrane Technology
 of the National Research Council
 (ITM-CNR)
at University of Calabria (UNICAL)
Rende (CS), Italy

Maria Di Vincenzo
Institut Européen des Membranes
University of Montpellier
ENSCM-CNRS
Montpellier (France)

Sophie Cerneaux
Institut Européen des Membranes
University of Montpellier
ENSCM-CNRS
Montpellier (France)

Istvan Kocsis
Institut Européen des Membranes
University of Montpellier
ENSCM-CNRS
Montpellier (France)

Mihail Barboiu
Institut Européen des Membranes
University of Montpellier
ENSCM-CNRS
Montpellier (France)

Editors

Rund Abu-Zurayk
Nanotechnology Center
The University of Jordan
Amman, Jordan

Hamdi Mango Center for Scientific
Research
The University of Jordan
Amman, Jordan

Mohammed R. Qtaishat
Department of Chemical Engineering
The University of Jordan
Amman, Jordan

Abeer Al Bawab
Hamdi Mango Center for Scientific
Research
The University of Jordan
Amman, Jordan

Department of Chemistry
The University of Jordan
Amman, Jordan

Georgia Karataraki
NCSR – Demokritos Institute of
Nanoscience and
Nanotechnology (INN)
Agia Paraskevi
Greece

Anastasios Gotzias
NCSR – Demokritos Institute of
Nanoscience and
Nanotechnology (INN)
Agia Paraskevi, Greece

Elena Tocci
CNR Institute on Membrane Technology (IT)
Rende, Italy

Stéphan Barbe
Faculty of Applied Natural Sciences
Cologne University of applied Sciences
Leverkusen, Germany

Aron Kneer
TinniT Technologies GmbH
Karlsruhe, Germany

Fei Wang
Institute for Applied Materials-
Computational Materials Science
Karlsruhe Institute of Technology
Karlsruhe, Germany

Michael Metze
Sartorius Bioprocess Solutions
Göttingen, Germany

Christian Wenning
Polyplast Müller GmbH
Straelen, Germany

Patrick Altschuh
Institute for Applied Materials-
Computational Materials Science
Karlsruhe Institute of Technology
Karlsruhe, Germany

Britta Nestler
Institute for Applied Materials-
Computational Materials Science
Karlsruhe Institute of Technology
Karlsruhe, Germany

Mohamed T. Mito
School of Engineering
University of Birmingham
Edgbaston, Birmingham, UK

Philip Davies
School of Engineering
University of Birmingham
Edgbaston, Birmingham, UK

Nirajan Dhakal
IHE Delft Institute for Water
Education
Water Supply Sanitation and Environmental Engineering Department
Westvest 7, 2611 AX Delft,
Netherlands

Almotasembellah Abushaban
IHE Delft Institute for Water
  Education
Water Supply Sanitation and
  Environmental Engineering
  Department
Westvest 7, 2611 AX Delft,
  Netherlands

Nasir Mangal
IHE Delft Institute for Water
  Education
Water Supply Sanitation and Envir-
  onmental Engineering Department
Westvest 7, 2611 AX Delft,
  Netherlands

Mohanad Abunada
IHE Delft Institute for Water
  Education
Water Supply Sanitation and
  Environmental Engineering
  Department
Westvest 7, 2611 AX Delft,
  Netherlands

Jan C. Schippers
IHE Delft Institute for Water
  Education
Water Supply Sanitation and Envir-
  onmental Engineering Department
Westvest 7, 2611 AX Delft, Netherlands

Maria D Kennedy
IHE Delft Institute for Water
  Education
Water Supply Sanitation and Envir-
  onmental Engineering Department
Westvest 7, 2611 AX Delft
  Netherlands

Mattheus Goosen
Office of Research & Graduate
  Studies
Alfaisal University
Riyadh, Saudi Arabia

Hacene Mahmoudi
Faculty of Technology
Hassiba Ben Bouali University
Chlef, Algeria

1 Introduction to Membrane Desalination

Andreas A. Sapalidis, Evangelos P. Kouvelos,
George E. Romanos, and Nick K. Kanellopoulos
Institute of Nanoscience and Nanotechnology
National Centre for Scientific Research Demokritos
Agia Paraskevi Attikis, Greece

1.1 DESALINATION NECESSITY

The idea of producing potable water from seawater was conceived in ancient times, when the crews of merchant ships, cruising across the oceans for long periods, met with formidable challenges of water shortage, while oddly enough, being surrounded by huge amounts of nondrinkable water. This contradiction constituted an incentive for conducting the first experiments that proved that distilled seawater does not carry any salt and is drinkable, and therefore, desalinated.

Desalination can therefore be defined as the process by which salts and minerals are removed from water, producing usable water (potable or for other uses such as agriculture).

The worldwide amount of water is limited and sustained through the water cycle process by which water that comes down as rain finally evaporates and recondenses.

With the amount of available water constant and the world's population increasing, it is apparent that some parts of the world will become water-stressed.

World population growth, expected to exceed 10 billion by year 2040 (United Nations, DESA Population Division), introduces a major obstacle towards an effective management of water resources. Climate change is another stumbling block, which affects the Earth in ways yet to be discovered, and creates natural hazards – from severe draughts to hurricanes and monsoons – that seem to increase in severity and occurrence.

Moreover, 66% of the global population (approximately 4 billion people) currently live in conditions of severe water scarcity for at least one month per year.[1]

Additionally according to the UN, nearly 2.4 billion people, who constitute almost one third of the world's current population, live within 100 km (60 miles) of coastlines, and therefore, seawater desalination seems to be a reasonable solution for supplying fresh water to water-stressed coastal regions.

Water use has been increasing worldwide by about 1% per year since the 1980s, driven by a combination of population growth, socioeconomic development, and changing consumption patterns. Global water demand is expected to continue increasing at a similar rate until 2050, accounting for an increase of 20% to 30% above the current level of water use, mainly due to rising demand in the industrial and domestic sectors. Water-stress levels will continue to increase as demand for water grows and the effects of climate change escalate.[2]

Studies analyzing 14 years of NASA's satellite data (Gravity Recovery and Climate Experiment, GRACE) from freshwater resources showed that areas with limited access to water will likely become more stressed than before (Figure 1.1).[3]

Water deficiency often leads to the use of degraded or low-quality deposits that result in public health risks. It is distressing that even nowadays, waterborne diseases such as cholera continue to menace large numbers of people. Estimates show that each year there are 1.3 million to 4.0 million cases of cholera and 21.000 to 143.000 deaths worldwide due to cholera.[4]Access to clean water will be able to greatly attenuate these numbers. According to the WHO *"cholera transmission is closely linked to inadequate access to clean water and sanitation facilities. Typical at-risk areas include peri-urban slums, and camps for internally displaced persons or refugees, where minimum requirements of clean water and sanitation are not met."*[5]

Under the critical conditions of global water scarcity, desalination processes have the potential to become a plentiful and feasible source of fresh water, effectively rising to the challenge of augmented water demands in the near future.

Nowadays, an estimated 15.906 desalination plants are fully operational (Figure 1.2), and are located in 177 countries and territories across all major world regions, producing around 95 million m³/day of desalinated water for human use, of which 48% is produced in the Middle East and North Africa.[6]

1.2 DESALINATION PROCESSES

Desalination processes can be classified in two main categories: thermal systems and membrane systems.

In the next two sections, an outline of the most common methods in each category is given.

1.2.1 THERMAL DESALINATION PROCESSES

The main industrially used thermal desalination processes are the following:

- Multistage flash distillation (MSF)
- Multiple effect distillation (MED), and
- Thermal vapor compression (TVC).

Introduction to Membrane Desalination

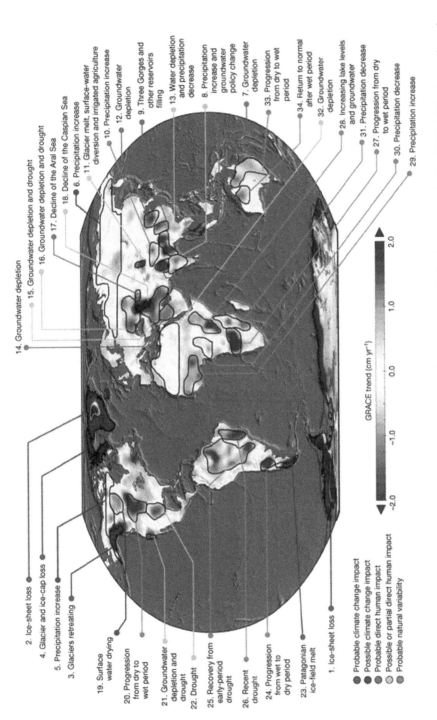

FIGURE 1.1 Annotated map of TWS trends. Trends in TWS (in centimeters per year) obtained on the basis of GRACE observations from April 2002 to March 2016. The cause of the trend in each outlined study region is briefly explained and color-coded by category. The trend map was smoothed with a 150-km-radius Gaussian filter for the purpose of visualization; however, all calculations were performed at the native 3° resolution of the data product.

[Source: Data from M. Rodell et al. Emerging trends in global freshwater availability, *Nature*, Vol. 557, p. 651, 2018].

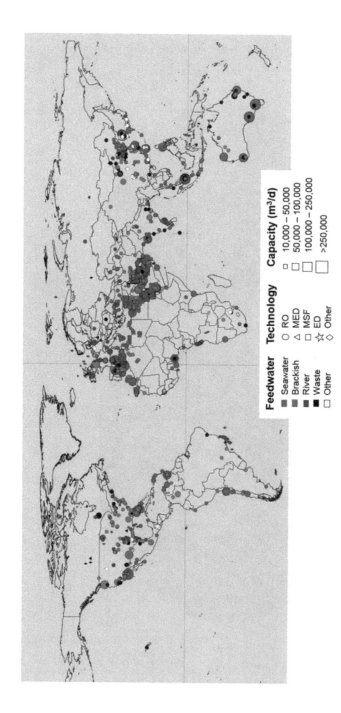

FIGURE 1.2 Global distribution of large desalination plants by capacity, feedwater type, and desalination technology. (reproduced by permission from E. Jones et al. *Science of the Total Environment* 657 (2019) 1343–1356).

Introduction to Membrane Desalination

1.2.1.1 Distillation

Water that is heated enough eventually evaporates, leaving any salt content behind. Condensing this vapor via cooling results in distilled – freshwater production.

A simple still is sufficient to perform the aforementioned process. The heat energy required is the latent heat of evaporation, which is around 627 kW$_h$/m^3, plus losses, making distillation a very energy demanding process.

1.2.2 MULTISTAGE FLASH (MSF)

MSF desalination is a thermally driven process that consists of an array of elements, called stages. Each stage is divided into two parts: the heat exchanger and distiller. In each stage, the seawater that flows in the pipes of the heat exchanger is heated by the steam that condenses on the pipes' surface. At the end, the preheated feed is driven in a brine heater to raise its temperature to nearly the saturation temperature at the maximum system pressure. Afterwards, the feedwater enters the first stage distiller, through an orifice, where there is low pressure. Since the water was at the saturation temperature for a higher pressure, it becomes superheated and flashes into steam. Only a small part of this feed is transformed into water vapor, based on the level of pressure at this stage. The produced vapor passes through a wire mesh (demister) to remove any entrained brine droplets and then through the heat exchanger, where it condenses and drips into a distillate tray. Due to partial evaporation, the temperature of the remaining feed is decreased, so that in the evaporation process to be continued, the pressure of the subsequent stages would be successively lower.

Current commercial installations are designed with 10–30 stages with a typical 2°C -temperature drop per stage. The principle of operation is shown in Figure 1.3.

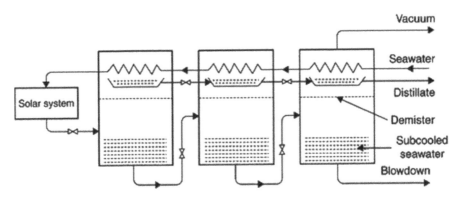

FIGURE 1.3 Principle of operation of the multistage flash (MSF) system.
(reproduced with permission from Soteris A. Kalogirou, in Solar Energy Engineering Elsevier 2014).

1.2.3 MULTIEFFECT DISTILLATION (MED)

Similar to the MSF process, desalination takes place in a series of vessels (known as effects), each one operating at a successively lower pressure. The main difference between MED and MSF is that in MED, as seen in Figure 1.4, the preheated seawater is sprayed onto the hot surface of the heat exchanger's bank of tubes (feed with steam). In each effect, a portion of the feedwater evaporates where the rest of the concentrated seawater is collected at the bottom of the effect. The vapor produced enters the next effect, which operates at a lower pressure than the previous pressure, generating secondary steam. The steam condenses in the tubes and is withdrawn as a product. This operation is repeated along the plant from stage to stage. The primary steam condensate is usually returned to the boiler of the power station since it is of extremely high quality. Up to 8 or 16 effects can be used in this way. MED plants are typically built in units of 2,000 to 10,000 m^3/day capacity, with a top temperature (first effect) of about 70°C.

1.2.4 VAPOR COMPRESSION (VC)

Vapor compression systems are divided into two main categories: mechanical vapor compression (MVC) and thermal vapor compression (TVC) systems. In these plants, the necessary heat for the evaporation of water is supplied by compression of vapor, rather than by direct heat transfer from the steam produced in a boiler. A simplified outline of a VC is depicted in Figure 1.5. The water vapors produced in the distillation chamber (distiller) are drawn by the vacuum line to the suction of the compressor in order to be compressed.

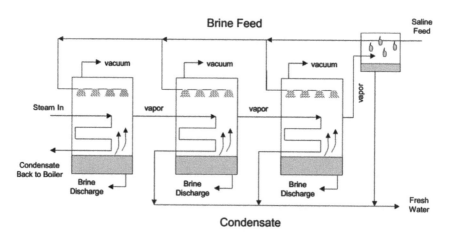

FIGURE 1.4 Principle of operation of the MED system.
(Reproduced from: A. H. M. Saadat. "Desalination Technologies for Developing Countries: A Review", *J. Sci. Res.* **10 (1), 77–97, 2018).**

Introduction to Membrane Desalination

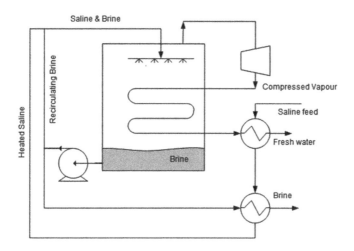

FIGURE 1.5 Basic layout of MVC.

Thereafter, these vapors (as steam) enter the internal of a tube buddle where preheated saline water is sprinkled at the outside surface of these tubes, resulting in a partial evaporation. The partial condensed steam exiting the distiller buddle passes through a heat exchanger preheating the cold feed seawater. As a result, the steam fully condenses and is collected as fresh water. The warm saline feed is further heated by passing through another heat exchanger where it receives energy from the relatively hot discharged brine. Finally, the hot saline feed is mixed with a portion of the brine collected in the bottom of the distillation chamber, while the remaining brine is directed to the saline preheater and finally discharged (brine level in the evaporation chamber must be monitored carefully).

A number of MVC-based desalination plants have been installed worldwide that produce fresh water for industrial and municipal purposes. These plants, however, have the disadvantage of restricted capacity due to scale limitations for large size vapor compressors. Another imperative issue is compressor maintenance resulting from the high possibility of corrosion; consequently, many MVC plants must operate at relatively low temperatures in order to minimize the formation of scaling and corrosion of materials.

1.2.5 MEMBRANE DESALINATION

It is generally recognized that the work entitled "Sea Water Demineralization by Means of an Osmotic Membrane" by Sidney Loeb and Srinivasa Sourirajan, presented in 1963[7] was a major starting point for the commercialization of membranes for desalination purposes in the 1970s, which eventually resulted in the dominance of membranes over thermal processes.

Depending on the applied driving force (chemical potential), membrane desalination processes can be grouped as:

- Pressure driven, such as reverse osmosis (RO) and forward osmosis (FO);
- Electrical potential driven, such as electrodialysis (ED); and
- Temperature driven, such as membrane distillation (MD).

1.2.5.1 Reverse Osmosis (RO)

RO is a membrane filtration process and, in contrast to the previously described thermal desalination processes, does not involve vaporization of water, which in turn, is much more energy-efficient than thermal methods.

RO was introduced in the early 1950s as an alternative process for seawater desalination based on cellulose acetate membranes.[8] In 2000, the volumes of desalinated water produced by thermal technologies (dominated by MSF) and RO were approximately equal to 11.6 million m^3/day and 11.4 m^3/day respectively, together accounting for 93% of the globally produced volume of desalinated water (Figure 1.6a). Since then, both the number and capacity of RO plants has risen exponentially, while thermal technologies have only experienced marginal increases (Figure 1.6b). The current production of desalinated water from reverse osmosis stands at 65.5 million m^3/day, accounting for 69% of the volume of desalinated water.

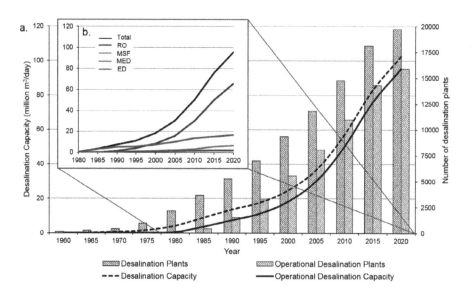

FIGURE 1.6 Trends in global desalination by (a) number and capacity of total and operational desalination facilities and (b) operational capacity by desalination technology through the period 1960-2020.

(reproduced by permission from E. Jones et al. *Science of the Total Environment* 657 (2019) 1343–1356).

Introduction to Membrane Desalination

Osmosis is a surprisingly powerful phenomenon; the osmotic pressure of typical seawater is around 26 bar, and this is the pressure that the pump must overcome in order to reverse the direction of freshwater flux. In practice, a significantly higher pressure is used, typically 50–70 bar, in order to achieve a generous flow of freshwater, which is the product (permeate), while the rejected seawater is known as concentrate or brine.

The ratio of product flow to that of the feed is defined as the recovery ratio. In seawater RO, a recovery ratio of 30% is typical, meaning that the remaining 70% is the concentrate, which is commonly returned to the sea.

The desalination industry makes a distinction between seawater and brackish water. Seawater typically has a salt concentration in the order of 36,000 mg/L total dissolved solids (TDS), while brackish water, usually from underground, might be between 3,000 and 10,000 mg/L TDS.

1.2.6 Forward Osmosis (FO)

The basic principle of FO is the osmotic pressure gradient between seawater and a *draw solution*. This solution has greater osmotic pressure than seawater and therefore, water molecules permeate through a membrane to its side. By this process, a seawater concentrate and a diluted draw solution are created using low (if any) hydraulic pressure.

In order to recover fresh water from the *draw* solution, a posttreatment step such as membrane distillation is necessary. As a whole, FO requires less energy than any RO process.

1.2.7 Electrodialysis (ED)

ED is a process in which solute ions move across membranes by application of an electric field. This electrical potential drives the electrolytes into a concentrated solution, leaving behind a diluted solution. Unlike other desalination technologies, in electrodialysis the salts are removed from the feedwater. The feed water should be free of suspended solids, organic matter, and nonionic contaminants that accumulate in the product. A typical electrodialysis module consists of anion and cation selective membranes that are stacked alternately with spacers interposed between them, as depicted in Figure 1.7. The saltwater is fed into the spacer layers on one side of the stack, and a DC voltage is applied to the stack as a whole. Electrodialysis was introduced as a desalination process during the 1960s and is widely used today for desalination of brackish water. The energy consumption depends very much on the concentration of the feedwater, and for this reason, electrodialysis is rarely used for seawater desalination.

1.2.7.1 Membrane Distillation (MD)

MD is a fairly new concept that is steadily gaining popularity in the field of desalination research; as the term *membrane distillation* indicates, MD combines established desalination methods of membrane filtration and heating.

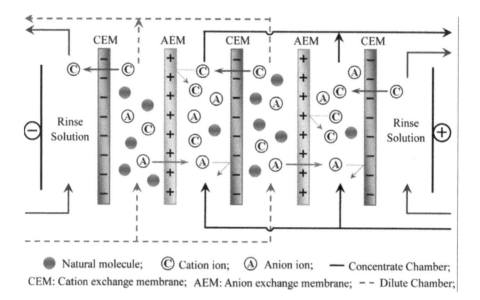

FIGURE 1.7 Representative schematic illustration of a conventional electrodialysis stack (Reproduced with permission from J. Ran et al. *Journal of Membrane Science* **522**, pp 267–291, 2017).

A porous hydrophobic membrane is used as a nonselective interface placed between an aqueous heated solution on the one hand (feed or retentate) and a condensing phase (permeate or distillate) on the other. The hydrophobic nature of the membrane, generally polymeric, prevents penetration of the pores by aqueous solutions due to surface tension, and allows the establishment of a vapor – liquid interface at the entrance of each pore. The temperature gradient between the two streams leads to a vapor pressure difference that causes volatile compounds (most commonly water) to evaporate on the hot feed solution – membrane interface, transfer of the vapor phase through the membrane pores, and condensation on the cold side membrane – permeate solution interface.

1.3 MEMBRANE PREPARATION TECHNIQUES

Asymmetric membranes are mostly fabricated by a process called *phase inversion*, which can be achieved through four principal methods: immersion precipitation (wet-casting), vapor-induced phase separation, thermally-induced phase separation, and dry-casting. In all of these techniques, an initially homogeneous polymer solution becomes thermodynamically unstable due to different external effects, and the subsequent phase separation concludes to a polymer-lean and a polymer- rich phase. The polymer-rich phase forms the matrix of the membrane, while the polymer-lean phase, rich in solvents and nonsolvents, specifies the area where the pores are generated (reproduced with permission from:

Sacide Alsoy Altinkaya & Bulent Ozbas, Journal of Membrane Science Volume 230, Issues 1–2, 15 February 2004, Pages 71–89).[9]

1.3.1 THIN FILM COMPOSITE (TFC)

Thin film nanocomposite (TFN) membranes are a new type of composite membranes prepared via interfacial polymerization (IP) processes. Nanoparticles are incorporated within the thin polyamide (PA) dense layer of the TFC membrane with the aim of improving the characteristics of the interfacially polymerized layer as in Figure 1.8.

In the 1970s, composite membranes comprising ultrathin polyamide films – formed via in situ polycondensation on porous polysulfone supports – were developed to replace integrally skinned, asymmetric RO membranes – formed by phase inversion of cellulose acetate. A great advantage of TFC technology is that it allows development and successful handling of extremely thin layers of barrier materials formed from almost any conceivable chemical combination . In addition, the ultrathin barrier layer and the porous support can be independently optimized with respect to structure, stability, and performance.

Over the last 30 years, water flux and solute rejection by polyamide TFC membranes have continually improved, but reverse osmosis processes remain relatively energy-intensive and fouling-prone. A lack of significant innovations in RO membrane materials persists despite the pressing needs for desalination membranes with (1) increased water permeability for energy savings, (2) improved control of selectivity in membrane design, and (3) more fouling-resistant surfaces. These constraints remain in the face of rising worldwide demand for clean water and the sustainability imperatives to control energy consumption.[10]

1.4 DESALINATION MEMBRANE SHAPES AND MODULES

In industrial membrane plants, an active membrane area of 100,000 m² is required to perform effective separations and achieve sufficient water recovery. There are several ways to economically and efficiently package membranes to provide a large surface area for effective separation. From the perspective of

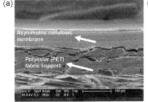 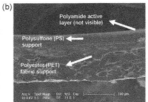 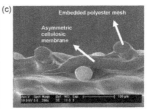

FIGURE 1.8 Crosscut images of commercial TFC desalination membranes. **(reproduced with permission from *Journal of Membrane Science* 318 (2008) 458–466).**

overall cost mitigation, both the cost of membranes per unit area and the cost of the containment vessel into which they are mounted, are equally important. Basically, the challenge is to achieve the packing of the highest possible area of membranes into the smaller possible volume to minimize the cost of the containment vessel, along with providing acceptable flow hydrodynamics in the vessel. These packages are called *membrane modules*. Plate-and-frame, tubular, spiral-wound and hollow fiber are the most popular categories of modules.

1.4.1 Tubular Modules

These modules are now generally limited to ultrafiltration applications, for which the benefit of resistance to membrane fouling outweighs the high cost. Tubular membranes contain as many as five to seven smaller tubes, each 0.5–1.0 cm in diameter, nested inside a single larger tube. In a typical tubular membrane system, a large number of tubes are manifolded in series. The permeate is removed from each tube and sent to a permeate collection header.

1.4.2 Spiral-Wound Modules

Industrial-scale modules contain several membrane envelopes, each with an area of 1–2 m^2, wrapped around the central collection pipe (Figure 1.9). Multienvelope designs minimize the pressure drop encountered by the permeate traveling toward the central pipe. The standard industrial spiral-wound module is 8 in. (20 cm) in diameter and 40 in. (~1m) long.

The module is placed inside a tubular pressure vessel. The feed solution passes across the membrane surface and a portion of the feed permeates into the membrane envelope, where it spirals toward the center and exits through the collection tube. Normally, four to six spiral-wound membrane modules are connected in series inside a single pressure vessel.

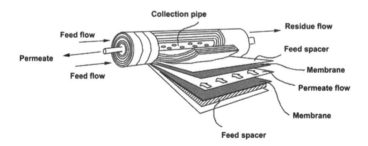

FIGURE 1.9 Spiral wound membrane module.
(reproduced with permission from Balster J. (2013) Spiral Wound Membrane Module. In: Drioli E., Giorno L. (eds) *Encyclopedia of Membranes.* **Springer, Berlin, Heidelberg).**

Introduction to Membrane Desalination

1.4.3 Hollow-Fiber Modules

Hollow-fiber modules (Figure 1.10) are characteristically 4–8 in. (10–20 cm) in diameter and 3–5 ft (1.0–1.6 m) long. Hollow-fiber units are almost always run with the feed stream on the outside of the fiber.

Water passes through the membrane into the inside or *lumen* of the fiber. A number of hollow fibers are collected together and *potted* in an epoxy resin at both ends and installed into an outer shell. Hollow-fiber membrane modules are formed in two basic geometries: (a) shell-side feed design and (b) bore-side feed design.

1.4.4 Pretreatment

The source of sea water for the desalination process is either from open intake or beach well intake. The source water composition varies due to various factors such as industrial discharge, water depth and temperature, ocean currents, and algae content. The various foulants in the sea water are chemical foulants, which are responsible for membrane scaling; particulate matter which is responsible for particle deposition on membrane surfaces; biological matter, which is responsible for biofouling of membranes; and dissolved organic matter, which interacts with membranes.

Figure 1.11 shows the conventional pretreatment process. The key contaminants to be filtered in the pretreatment stage are particulate/suspended matter, microbial contaminants, and dissolved organic matter.

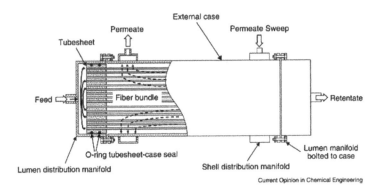

FIGURE 1.10 Hollow fiber module.
(reproduced with permission from Norfamilabinti Che Mat et.al. Current Opinion in Chemical Engineering 2014, 4:18–24).

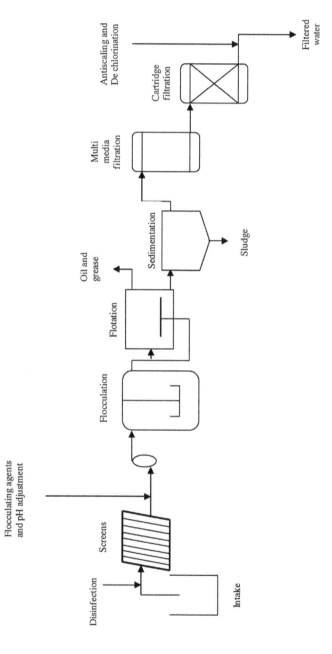

FIGURE 1.11 Conventional pretreatment system for RO plant. (reproduced with permission from J. Kavitha et al. *Journal of Water Process Engineering* 32 (2019) 100926).

REFERENCES

[1] Mekonnen MM, Hoekstra AY. (2016). Four billion people facing severe water scarcity. *Sci Adv* 2: e1500323.

[2] WWAP (UNESCO World Water Assessment Programme). (2019). *The United Nations World Water Development Report 2019: Leaving No One Behind*. Paris, UNESCO.

[3] Rodell M, Famiglietti JS, Wiese DN, Reager JT, Beaudoing HK, Landerer FW, Lo MH. (2018, May 31). Emerging trends in global freshwater availability. *Nature* 557: 651.

[4] Ali M, Nelson AR, Lopez AL, Sack D. (2015). Updated global burden of cholera in endemic countries. *PLoS Negl Trop Dis* 9(6): e0003832. doi: 10.1371/journal. pntd.0003832. www.ncbi.nlm.nih.gov/pmc/articles/PMC4455997/.

[5] WHO fact sheet. (2019). www.who.int/news-room/fact-sheets/detail/cholera (accessed 1 Oct 2019).

[6] Jones E, Qadira M, van Vlietb MTH, Smakhtina V, Kang S.-mu. (2019). The state of desalination and brine production: A global outlook. *Sci Total Environ* 657: 1343–1356.

[7] Loeb S, Sourirajan S. (1963). Sea water demineralization by means of an osmotic membrane. Saline water conversion – II, chapter 9. *Adv Chem* 38: 117–132.

[8] Breton EJ, Jr. (1957). *Water and Ion Flow through Imperfect Osmotic Membrane*. Office of Saline Water, US Department of the Interior, Research & Development Progress Report, No. 16.

[9] Altinkaya SA, Ozbas B. (2004, February 15). Modeling of asymmetric membrane formation by dry-casting method. *J Membr Sci* 230(1–2): 71–89.

[10] Jeong B-H, Hoek EMV, Yan Y, Subramanib A, Huanga X, Hurwitza G, Ghosha AK, Jawora A. (2007, May 15). Interfacial polymerization of thin film nanocomposites: A new concept for reverse osmosis membranes. *J Membr Sci* 294 (1–2): 1–7.

2 Membrane Materials Design Trends

Nanoadditives

Shadia R. Tewfik, Mohamed H. Sorour, Abdelghani M.G. Abulnour, Hayam F. Shaalan, Heba A. Hani, Marwa M.H. El-Sayed, Yomna O. Abdelrahman, Eman S. Sayed, Aya N. Mohamed, Amany A. Al-Mansoup, and Nourhan A. Shawky
National Research Centre, Dokki, Giza, Egypt

2.1 INTRODUCTION

This chapter is concerned with membrane materials design trends with emphasis on nanoadditives. Generally, dope design incorporates key polymeric substrates such as PS, PES, PAN, PVC, and cellulosic esters. Alloying polymers are generally dissolved in appropriate solvents and their blends including, but not limited to, NMP, DMF, DMAC, THF, Acetone, and DMSO. Of course, dope formulation has witnessed numerous interesting advances. Three noteworthy advances achieve remarkable progress in membrane technology.

- Functionalization of membrane surface enables chelation of desirable functionalities to achieve new targeted performance.
- Balancing solvent and nonsolvent ratios in the dope enables control of permeability of modules and AFM patterns.
- Membrane additives have been proven to introduce functional advantages and desirable morphological features. Of course, each additive is associated with its benefits and limitations, especially while consuming consortium of additives that need to be rationally adjusted in the context of intensive experimental work guided by appropriate modeling and simulation techniques. Perhaps the revolutionary role of nanotechnology has been particularly proven with the use of carbon nanotubes, graphene oxide, nano-inorganic oxides, and nanopolymers.

The main characteristics affected by the nanoadditives either positively or negatively include hydrophilicity, fouling, porosity, permeability, roughness, and certain mechanical properties.

2.1.1 Hydrophilicity

Water treatment by membranes necessitates a hydrophilic surface, which is achieved by adding nanoadditives directly or after functionalization of the nanoparticle's surface before being added to the polymer solution.

2.1.2 Fouling

Researchers have reported that the antifouling properties of membranes increase significantly by the incorporation of nanoadditives.

2.1.3 Porosity

Most of the Nanoadditives tend to increase the porosity of the membranes.

2.1.4 Permeability

It has been reported that addition of the nanoparticles at low concentration causes pore plugging with a decrease in flux. However, as the particle concentration increases, surface hydrophilicity and mean pore size become dominant, leading to maximum flux, which is then reduced by further increasing the concentration as pore-plugging assumes the main role again (Rahimpour et al., 2012).

2.1.5 Rejection

Most researchers have reported that rejection does not significantly change on adding nanoparticles.

2.1.6 Roughness

It has been generally reported that nanomaterials cause a rougher surface, but this is not detected if the quantity of additive is small.

2.1.7 Mechanical Properties

The nanoparticles interact with the polymer matrix, leading to a substantial increase in rigidity and the energy required to break the polymer chain.

In the following sections, addition of specific nanoparticles is addressed. For each category, the preparation, effect on morphology and performance are presented for specific applications.

2.2 GRAPHENE OXIDE NANOADDITIVES

Graphene and its derivatives have been the focus of a huge number of studies, owing to their outstanding mechanical, physical, thermal, and optical properties, as well as their unique structures (Potts et al., 2011). Graphene oxide (GO) is one of the most important graphene derivatives and it is prepared by various methods, such as the Brodie (Brodie, 1859) and Hummers-Offerman methods (Hummers and Offeman, 1958). GO has a layered structure with a high aspect ratio and a negatively charged hydrophilic surface, making it an attractive solution to overcome many limitations in various applications (Wang et al., 2018).

GO has found huge potential in many applications, especially in membrane separation applications in which its incorporation successfully achieved favorable results in overcoming polymeric membrane limitations, as well as in introducing much-needed enhancements to their performance (Sears et al., 2010).

There are different types of GO membranes, such as freestanding GO membranes (Cotet et al., 2017), GO-modified composite membranes (Zahid et al., 2018), and supported GO membranes (Ma et al., 2017). Freestanding GO membranes are composed of layers of GO, only as the selective layer, while GO modified composite membranes are composite membranes where GO is incorporated in the membrane casting blend during the fabrication process (Jiang et al., 2016). Finally, GO-supported membranes are membranes where GO is deposited on the surface of a substrate (Zhang et al., 2017).

GO is characterized by SEM, EDX, AFM, FTIR, Zeta potential, and many other tests to identify surface morphology, elemental composition, surface topography, and surface charge (Ma et al., 2017).

2.2.1 Preparation

The fabrication process of GO membranes is crucial to determining the efficiency and uniformity of the prepared membrane with good nanomaterial dispersion or deposition (Johnson et al., 2015). GO membranes are prepared by various techniques, depending on the membrane type, mentioned previously. The most common techniques are layer by layer assembly (LBL) (Zhang et al., 2016), casting/coating (spinning, drop casting, dip coating, spray coating) (Zahri et al., 2016), filtration-assisted (vacuum and pressure filtration), and the evaporation-assembled method (Ma et al., 2017).

2.2.2 Effect on Morphology

As shown in Table 2.1, (Huang and Feng, 2018) prepared PI/GO HF MMM via the phase inversion technique for pervaporation. They found that the morphological structure of the PI/GO membrane is similar to the pristine PI membrane, but with several differences, such as introduction of lateral pores that enhanced the membrane's performance, as well as much less denser pores and a uniform smooth surface. They confirmed the ability of GO to form porous membranes, which may be attributed to the hydrophilic nature of GO, and accordingly accelerates the rate of solvent and nonsolvent exchange during phase inversion. (Fahmi et al., 2018) prepared PES/GO flat sheet MMM for hemodialysis via phase inversion. GO incorporation once again yielded a smoother surface with the same cross-sectional morphological structure. (Alam and Ali, 2018) studied PES/GO HF membranes with different GO-loading content and found that the

TABLE 2.1

Effect of GO nanoadditives on membrane morphology and performance.

Application/ Membrane type	Mechanism	Impact				Reference
		Morphology	Performance	Mechanical strength	Others	
Pervaporation/PI/ GO MMM HF	PI/GO prepared via direct spinning (wet phase inversion)	SEM: # PI/GO have same cross-sectional structure as PI HF, but with less dense pores # Introduction of lateral pores	# At 90°C, water flux increased from 6.4 to 15.6 kg m−2 h−1, salt rejection increased (99.8%)	# Tensile strength decreased to 8.41 MPa upon GO loading while tensile modulus increased to 515 MPa	# Contact angle decreased from 92° to 59° upon GO- loading	(Huang and Feng, 2018)
Hemodialysis/PES/ GO, FS.	Phase inversion, casting	SEM: # Cross sections had symmetric structure, did not change # PES/GO had smoother surface, PES surface had small pores	# solute flux= 2.94 L m 2 h 1 # Clearance of creatinine= 78.3%	# Tensile stress increased (5.55 Mpa) # Tensile strain increased to 39% # Tensile modulus increased from 10x10-7 Pa to 16.9x10-7 Pa upon GO incorporation	# Contact angle decreased from 82° to 64.7° and hydrophilicity increased	(Fahmi et al., 2018)
UF/-PES/GO MMM HF	Dry – wet spinning, GO 0.2%, 0.5%, and 1.0 wt..%	SEM: # PES/GO: Thin top layer, a porous sublayer with larger and uniform finger-like pores and macrovoids # No morphological changes upon increasing GO loading	# Water permeability increased 36% (30 ± 1.5 Lm−2 hr−1 bar−1) at 0.5 wt..% GO loading, but decreased upon increasing GO content more than 0.5wt..%	Tensile strength increased to 1.937 MPa at 0.5 wt..% GO loading, while elongation decreased	# Best antifouling at 0.5 wt..% GO # Increased porosity of 15% at 0.5 wt..% GO # Contact angle decreased with increas-ing GO content	(Alam and Ali, 2018)

Membrane	Method	Characterization	Performance	Mechanical properties	Remarks	References
UF/PPSU/GO MMM/FS.	Phase inversion, GO content (0.2, 0.5, and 1.0 wt.%)	**AFM:** # PES/GO roughness decreased with closer valleys and ridges **SEM:** # Pores in the sub porous layer of the membrane were relatively thinner and appeared in straight finger structures with open ends # Thickness of the skin layer reduced with very fine oval-shaped pores underneath # Reduced spongy support layer thickness, and denser with interconnected pores **AFM:** Surface roughness increased upon increasing GO- loading	**0.5 wt.%:** Flux: increased from 119 $Lm^{-2}h^{-1}$ to 171 $Lm^{-2}h^{-1}$, rejection of proteins: 95%	# Tensile strength: highest (3.8 MPa) at GO 1 wt.% # Elongation decreased upon increasing GO loading reaching 9.9% at GO 1 wt.%	# MWCO increased upon increasing GO content **#Porosity:** increased upon increasing GO wt.%, highest Rm=10.6 at 0.5 wt.% **#Contact angle:** decreased upon increasing GO content, lowest 41° at 0.5 wt.% **#MWCO:** increased from 25 KDa to 40KDa # Improved fouling resistance by 58%	(Shukla et al., 2017)
HF CO_2 separation/-PSF/GO	phase inversion (dry – wet spinning technique)	# GO incorporation suppressed the development of fingerlike structures near the outer surface of the membrane	# CO_2 permanence increased by 14% #CO_2/N_2 and CO_2/CH_4 selectivity was	# tensile strength increased by 4.36% # elongation-at-break improved by 8.79%	-	(Zahri et al., 2016)

(Continued)

TABLE 2.1 (Cont.)

Application/ Membrane type	Mechanism	Impact				Reference
		Morphology	**Performance**	**Mechanical strength**	**Others**	
		# Smoother surface upon GO incorporation, fewer open pores	increased by 158% and 74%, respectively			
NF/-PAI/PEI HF coated with GO.	# PAI HF prepared via phase inversion # PAI/PEI modified HF at 30, 90 and 120 cross-linking time) # GO electrically deposited via dip coating	# rougher outer surface is observed after PEI cross-linking, while GO nanosheets are seen wrinkling on the membrane surface after GO deposition	# Water permeability increased 86% # Salt rejection increased by 15%	# Tensile strength decreased upon GO deposition while elongation-at-break increased	# Shortening of the membrane cross-linking time from 90 to 30 min # Contact angle decreased upon GO deposition	(Goh et al., 2015)
NF/-PVDF/GO MMM, FS, removal of natural organic matter (NOM)	Phase inversion induced by immersion precipitation process	SEM: # Macrovoids layer increased # Smoother surface with higher pore density AFM: # Roughness decreased from 7.48 nm to 6.7 nm at 0.5 wt..% GO	#PWP increased upon 0.5 wt..% GO incorporation, from 47L/m2 h to 97 L/m2h	-	# Contact angle decreased from 74° to 68.4° at 1 wt..% GO # Porosity increased from 59% to 80% at 0.5 wt..% GO # Pore density increased from 48.38 to 103.36 μm-2 for 0.5 wt..% GO	(Xia and Muzi, 2014)

| NF-dye removal/ PES/GO MMM FS. | Fabricated by phase inversion (immersion precipitation technique) | SEM: # PES/GO have wider finger like pores AFM: #Roughness decreased upon GO loading from 20 nm to 8 nm # As GO loading increases, roughness increases | # Water flux increased   - # Rejection of protein > 98% | #Contact angle decreased from 65° to 53°, hydrophilicity increased # 0.5 wt..% GO loading showed highest fouling resistance # 0.5 wt..% GO loading had highest mean pore radius of 4.5 nm and porosity of 83.1% # PES/GO MMM had higher dye removal capacity | (Zinadini et al., 2014) |

morphological structure changed, and closely packed fingerlike structures were formed upon adding GO, but changed little upon increasing GO content; surface roughness of the PES/GO modified membranes decreased with closer ridges and valleys, however. (Shukla et al., 2017) found that the incorporation of GO into PPSU membranes made the pores in the subporous layer of the membrane thinner, and also reduced the skin layer and rendered it reduced with denser pores. Conversely, AFM results confirmed the increase of surface roughness. (Zahri et al., 2016) investigated the effect of GO incorporation on PSf HF membranes for CO_2 separation. Results showed that the morphological structure was affected as the development of fingerlike structure near the outer surface of the membrane was suppressed, while less open pores were observed on the membrane's surface. (Goh et al., 2015) fabricated PAI/PEI modified membranes via phase inversion, and GO was electrically deposited on the surface via dip coating. SEM results revealed GO nanosheets wrinkling on the membrane surface after GO deposition, as well as a rougher outer surface after PEI cross-linking. (Xia and Muzi, 2014) fabricated PVDF/GO membranes via phase inversion for natural organic matter (NOM) removal. Results showed that upon GO incorporation, the macrovoid layer increased, and a smoother surface with higher pore density was observed, which can be attributed to the formation of larger pores, owing to the increased mass transfer between solvent and nonsolvent resulting from GO hydrophilicity. Also, surface roughness decreased from 7.48 nm to 6.7 nm at 0.5 wt. % GO, leading to better antifouling properties. (Zinadini et al., 2014) prepared PES/GO MMM via phase inversion. SEM images revealed that PES/GO have wider fingerlike pores, as well as lateral pore formation, while AFM results demonstrated a decreased surface roughness on GO loading, but as GO loading increases, roughness increases as well.

2.2.3 EFFECT ON PERFORMANCE

As shown in Table 2.1, (Huang and Feng, 2018) confirmed enhancement of PI/GO MMM desalination performance as a result of facilitated water diffusion in the membrane owing to increased hydrophilic sites in the modified membrane. (Fahmi et al., 2018) studies of PES/GO also showed noticeable improvement in the membrane's solute flux and rejection. (Alam and Ali, 2018) revealed that the water flux increased 36% on incorporating 0.5 wt. % GO into a PES membrane blend, along with increases in the MWCO and porosity, while these decreased upon increasing the GO content, which may be attributed to agglomeration of GO and bad dispersion. (Shukla et al., 2017) found that best performance was achieved at 0.5 wt. % GO loading in a PPSU membrane matrix, with highest porosity as well as MWCO, resulting in increased flux and rejection. Also, fouling resistance improved by 58%. (Zahri et al., 2016) observed an increase in CO_2 permeance by 14%, which is mainly a result of GO's high absorption properties toward CO2 gas, as well as enhancement in CO_2/N_2 and CO_2/CH_4 selectivity by 158% and 74%, respectively for Psf/GO HF MMMs. (Goh et al., 2015) found that GO deposition on PAI/PEI modified membranes increased water permeability by 86%, while salt

Membrane Materials Design Trends

permeation decreased and salt rejection increased by 15%. These results confirm the effective action of GO nanosheets on the surface as a barrier restricting water flow. (Xia and Muzi, 2014) results demonstrated an increase of pure water permeability upon 0.5 wt. % GO incorporation in PVDF membrane, from 47 L/m2.h to 97 L/m2.h, which mainly results from the increased porosity and decreased contact angle caused by the hydrophilic groups in GO. (Zinadini et al., 2014) results demonstrated the increase of Water flux of PES/GO MMM as well as protein rejection was maintained above 98% for all GO incorporation schemes. This may be explained by the decreased contact angle from 65° to 53°, increased hydrophilicity as well as 0.5 wt. % GO loading had highest mean pore radius of 4.5 nm and porosity of 83.1% . Also 0.5 wt. % GO loading showed highest fouling resistance owing to decreased surface roughness.

2.2.4 EFFECT ON MECHANICAL PROPERTIES

As shown in Table 2.1, (Huang and Feng, 2018) tested the PI/GO HF MMM tensile strength and found that it decreased upon GO incorporation, which may be attributed to the formation of lateral pores, while the tensile modulus almost doubled owing, to good GO dispersion in the PI matrix. (Fahmi et al., 2018) studied PES/GO flat sheet MMM mechanical properties and found that the tensile strength and tensile modulus were enhanced, while the tensile strain was not significantly affected. (Alam and Ali, 2018) demonstrated the enhanced tensile strength of the PES/GO modified HF membrane upon adding 0.5 wt. % GO while the elongation-at-break decreased. (Shukla et al., 2017) achieved best mechanical enhancement 1 wt. % GO loading, owing to good interaction between GO and PPSU matrices. (Zahri et al., 2016) reported the increase of tensile strength by 4.36%, as well as increased elongation-at-break by 8.79% for PSF/GO HF MMMs, while the enhancement is not of true significance because mechanical properties were not sacrificed upon modification. (Goh et al., 2015) found that tensile strength decreased upon GO nanosheet deposition, while elongation-at-break increased.

2.3 CARBON NANOTUBE (CNTS) ADDITIVES

Recently, carbon nanotubes (CNTs) have attracted attention for the synthesis of novel membranes with attractive properties for water treatment. CNTs have unique properties of high mechanical stability, high thermal stability and conductivity, good chemical stability, and high electrical conductivity (Saifuddin et al., 2012; Ihsanullah, 2019). Many researchers studied the effect of CNTs on membrane properties, either as surface modifiers or as additives.

2.3.1 PREPARATION

The most common synthesis techniques of CNTs are laser ablation, arc-discharge evaporation, and chemical vapor deposition (CVD) (Saifuddin et al., 2012). Arc discharge and laser ablation were used for preparation of carbon nanotubes at high temperature. Currently, low temperature CVD is applied (Pandey and Dahiya, 2016). By CVD, CNTs are grown on metal catalyst surfaces such as iron, nickel, and cobalt (Ahn et al., 2012).

2.3.2 Effect on Morphology

Membrane roughness increases in most cases when adding CNTs, while in some cases, membrane roughness decreases. Polyamide membrane surface modification using acidified (H_2SO_4+HNO_3) MWNTs increased membrane roughness, Also, it was observed that addition of multiwall carbon nanotubes (MWCNT) with concentrations of 0.05% and 0.1% (w/v) packs the membrane surface together (Zhang et al., 2011). It has been reported that adding interlayers of acidified (H_2SO_4+HNO_3) MWCNTs to NF membranes also increased membrane roughness (Wu et al., 2016). Typical SEM images are presented in Figure 2.1. Hydrophilic surface is more suitable for MF, UF, NF, and RO membranes, while a hydrophobic surface is required for MD membranes. Generally, CNTs have hydrophobic effects on membrane surfaces; however, it can by modified with acids to produce hydrophilic effects. Zhang studied the effect of acidified (H_2SO_4+ HNO_3) MWCNT surface modification for RO polyamide membranes and found that membranes became more hydrophilic (Zhang et al., 2011). Also, in nanofiltration membranes, when

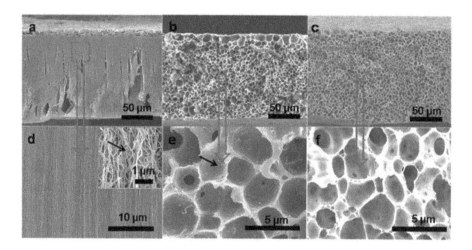

FIGURE 2.1 SEM cross-section images of CNT/ PES composite membranes with different magnifications.(a, d): vertically aligned-CNT/PES membrane; (b, e): randomly distributed CNTs/PES membrane; and (c, f): PES pure membrane (Li et al., 2014). Reprinted from (Li et al., 2014), with permission from Royal Society of Chemistry, Order License Id: 4612430890809 (FIG 1 – "Enhanced water flux in vertically aligned carbon nanotube arrays and polyethersulfone composite membranes" Li, Shaoyun, et al / *Journal of Materials Chemistry* A2 31 (2014) 12171-12176.

adding an acidified (H2SO4+HNO3) MWNTs layer between the microfiltration membrane support and interfacial skin, the polymerization membrane became more hydrophilic (Wu et al., 2016).

For DCMD PVDF-co-HFP membranes, porosity and pore size increase when adding CNTs as a dope additive, as a result of fiber structure overlapping (Tijing et al., 2016).

2.3.3 Effect on Performance

CNTs affect membrane performance (flux and rejection) in different manners, based on permeation mechanism for each type of membrane. In the case of UF and MF, membrane rejection depends mainly on pore size (Baghbanzadeh et al., 2016). In the case of membrane distillation, carbon nanotubes enhance both flux and rejection, owing to higher vapor flux and higher hydrophobicity. For RO membranes, (Zhang et al., 2011) found that acidified MWCNT enhanced membrane flux significantly, while NaCl rejection decreased dramatically. It has been reported that addition of CNTs enhances membrane flux, Also, the additive concentration of optimum CNTs enhances membrane rejection (El-Din et al., 2015). The effect of surface modification of PSF membranes using CNTs (0.12–0.32wt. %) indicated that increases in acidified MWCNTs increase membrane flux until a concentration of 0.19 wt. % is achieved; after that,

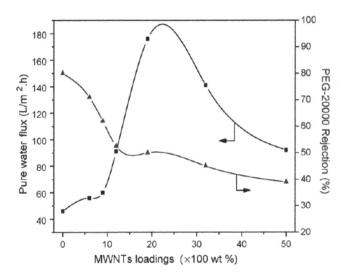

FIGURE 2.2 Effect of MWCNT content in dope solution on PWP, and rejection of ultrafiltration PSF /CNT composite membrane (Qiu et al., 2009) with permission from Elsevier Order Number: 4612470410778 (FIG 10 – "'Preparation and properties of functionalized carbon nanotube/PSF blend ultrafiltration membranes'" S. Qiu et al. / *Journal of Membrane Science* 342 (2009) 165–172).

membrane flux decreased, which indicated that there was a threshold for adding CNTs to enhance membrane flux, while membrane rejection decreased (Qiu et al., 2009). A typical performance curve is presented in Figure 2.2.

2.3.4 EFFECT ON MECHANICAL PROPERTIES

Several studies indicated that CNTs enhance membrane mechanical properties. A study reported that insertion of MWCNTs interlayer between membrane support and coating layer improved mechanical stability and allow operating at higher pressure for composite nanofiltration PES membranes (Wu et al., 2016). Also, adding CNTs as dope additives to direct contact membrane distillation (DCMD) membranes enhances mechanical stability (Tijing et al., 2016).It is worth mentioning that addition of CNTs decreases membranes' biofouling tendency because it deactivates bacteria (Baghbanzadeh et al., 2016). CNTs enhance membrane thermal stability as well (Gethard et al., 2010).

Table 2.2 summarizes the effect of CNTs on membrane morphology and performance from previous studies.

2.3.5 MECHANISM OF ACTION

Gethard et al. (2010) studied vapor permeation enhancement mechanisms in the case of CNT-enhanced MD. This study illustrated that addition of CNTs results in more hydrophobic membranes that decrease pore-wetting tendency, enhancing pure vapor transport through pores. Also, CNTs allow vapor transport via diffusion because CNTs have high adsorption and desorption capacities. Moreover, CNTs allow fast vapor diffusion along their surfaces, as well as a direct vapor transport through CNTs inner pores (Gethard et al., 2010). CNTs provide a pore network between pores inside membrane matrices, and provide additional pathways within membrane matrices that improves membrane permeability (Li et al., 2014; Baghbanzadeh et al., 2016).

2.4 ZEOLITE NANOADDITIVES

2.4.1 ZEOLITE PREPARATION

Zeolites are microporous crystalline aluminosilicate materials with uniform pore and channel size; they are used in various applications owing to their unique properties. Incorporation of zeolites into polymer matrices has attracted great attention in membrane technology, because of such excellent advantages as permeability improvement of the selective component, in addition to the enhancement of molecular sieving property, thermal resistance, and chemical stability (Maghami and Abdelrasoul, 2018).

Zeolites are commonly prepared by using hydrothermal crystallization, hydrothermal synthesis, vapor phase transport, sol – gel method, chemical growth, seeding method (embedding zeolite into a support), and microwave synthesis. Typically, hydrothermal synthesis is used to prepare zeolite on a porous

TABLE 2.2

Effect of CNTs additives on membrane morphology and performance

Application	Polymer/additive	IP % (w/v)	Characteristics	Performance	Reference
HF/NF	PS/NMP/PEG/LiCl/Triton X-100/glycerol SiO$_2$	PIP (3%) TMC (0.5%)	#CA: 41 decreased #thickness: increase #Ra: (18.9 nm) decrease #zeta potential increase negative charge	#LMH: 34.5 #R%: 19.9 for NaCl	(Abolfazli and Rahim-pour, 2017)
	PS, NMP, PEG, LiCl, Triton X-100, glycerol/TETA (4%) + SiO$_2$ (0.1%)	PIP (3%) TMC (0.5%)	#CA: 30 decreased #thickness: increase #Ra: (11.9 nm) decrease #zeta potential: increase negative charge	#LMH:31 #R% 26 for NaCl	
FS/NF	PSU, DMF/porous MCM-41 silica NPs (100 nm) (0to0.1wt.. %) MCM-41	MPD 2.0wt..% (3min) TMC 0.15wt..% (2min)	#Thickness: (300-500nm) #CA: decrease from 57 to 27.9 #Ra: increase from 135 to 159nm #zeta potential decrease from -5.71 to -9.54mV	#LMH increase from 28 to 46 #R% no change 97.5, 98.5% for (NaCl, Na$_2$SO$_4$)	(Yin et al., 2012)
	PSU, DMF/spherical silica (NPs)	—	#Thickness: (300- 500nm) #CA decrease from 57 to 30 #Ra: increase from 135 to 159 nm #zeta potential decrease from -5.71 to -9.54mV	#LMH increase from 28 to35 # R% no change 98.5% for (NaCl, Na$_2$SO$_4$)	
FS/NF	PSF/silica NPs (wt.. 15 nm) (1-3%)	PAMAM 0.5% (wt..)/SDS\SiO2 (10 min) TMC 0.3%wt..(60s)	#R increase from 8.72 to 36.5nm #zeta potential increase negativity from -16.78 to -20.56 mV	#LMH: increase to 10 # R%: 91.23(MgSO$_4$), 48.80 (MgCl$_2$), 46.01 (NaCl), 92.62 (Na$_2$SO$_4$), at1.5 % SIO2	(Jin et al., 2012)

TABLE 2.2 (Cont.)

Application	Polymer/additive	IP % (w/v)	Characteristics	Performance	Reference
FS/MD	PVDF/silica NPs	—	#Pore size increased to 140 nm #CA:94 #Porosity decreased to 50% #Ra: 27 nm	LMH: increase from 0.7 to 2.9	(Efome et al., 2015)
HF/UF	PVDF/silica NPs	—	#Viscosity: increase #CA: decrease from 82 to 53˙ #porosity: increase from 5 to 84 % #pore size: increase #Ra: increase	LMH: 301 increase	(Yu et al., 2009b)

1 dispersed in mixed acid (H2SO4/HNO3 = 3/1 by volume), then treated by ultrasonic agitation at 80°C for 6 h.
2 MWCNTs (3 g) were treated by 100 mL H2SO4/HNO3 (3/1, v/v) solution under 120°C for 2 h.
3 CA:Acetone 20:80.
4 CA:Acetone 25:75.
5 Functionalized by oxidation/ purification in a strong acidic medium to enhance their dispersion within the polymer matrix.
6 Multiwalled carbon nanotubes (MWNTs) functionalized by isocyanate and isophthaloyl chloride groups were synthesized via the reaction between carboxylated carbon nanotubes and 5-isocyanato-isophthaloyl chloride (ICIC).

Membrane Materials Design Trends

support for membrane-based separation. This approach produces a thin dense layer of zeolite membranes that enables high water permeability (Feng et al., 2015; Makhtar et al., 2020). They are generally prepared hydrothermally by mixing solutions of aluminates and silicates, often with the formation of a gel, and by maintaining the various mixing temperatures for selected periods (Barrer, 1982; Cundy and Paul, 2005; Hani et al., 2009; Dahe et al., 2012). A typical hydrothermal zeolite synthesis can be explained as follows:

- Reactants containing silica and alumina are mixed together with a cation source, usually in a basic medium.
- The aqueous reaction mixture is heated in a sealed autoclave.
- For some time after rising to synthesis temperature, an amorphous phase is formed.
- After the aforementioned *induction period*, crystalline zeolite products can be detected and gradually, all amorphous material is replaced by an approximately equal mass of zeolite crystals.
- Zeolite crystals are recovered by filtration, washing, and drying.

Zeolites are commonly categorized into three groups in terms of the Si/Al ratio in their framework: low silica ($\approx$ 1), intermediate silica (between 1 and 10), and high silica ($\geq$ 10); the higher the silica in zeolite, the higher surface hydrophobicity (Rezaei, Dasht Arzhandi et al., 2016). It was observed that zeolite inclusion into the membranes was conducted by two ways: as a coat for the membrane (by interfacial polymerization), and as an additive in the dope of the prepared membrane. In both, the most common effects of zeolite incorporation observed on the membrane performance and characteristics are illustrated in Table 2.3 and are summarized in the following four subsections.

2.4.2 EFFECT ON WATER PERMEABILITY AND HYDROPHILICITY

Fathizadeh et al. (2011) showed that A and X zeolites in sodium form [incorporated in polyamide membranes (PA)] have pore diameters of approximately 4.2 and 7.4A, respectively that is between water diameter (2.7A) and the hydrated sodium and chloride ion diameters (8–9A).Therefore, these particles provide preferential flow paths only for water and can increase the hydrophilicity of the polyamide membrane surface. This concept was validated by Lind (2009) and Yin et al. (2012). The latter added that embedding the inorganic MCM-41 nanoparticles in the PA polymer could simultaneously reduce the cross-linking in the thin-film layer and provide additional short paths for water flow through the hydrophilic ordered porous structure. An alternative explanation is that the increased negative charge in thin-film nanocomposite (TFN) membranes caused by MCM-41 NPs has contributed to salt rejection by the electrical repulsion or Donnan exclusion; accordingly, the level of salt rejection can be maintained. Zeolitic imidazolate framework-8 (ZIF-8) (with large cavity 11.16 A) revealed that the PA/ZIF-8 membrane surface has an enhanced water uptake, and a more negative charge density with increased ZIF-8 content, which increases water uptake values (Li et al., 2018).

Lind (2009) showed that the AgA-TFN membranes have higher pure water flux than NaA-TFN membranes, while maintaining similar rejections of NaCl. For perspective, when similarly tested, commercially fabricated seawater, brackish water, and high-flux RO membranes produced pure water permeabilities of 2.6, 16.8, and 27.7 μm MPa^{-1} s^{-1} and NaCl rejections of 92.0, 93.0, and 75.2%, respectively. Because the characteristic pore dimension of AgA nanocrystals (3.5 Å) is smaller than that of NaA nanocrystals, the permeability of AgA-TFN membranes was smaller.

On the other hand, Dahe et al. (2012) used Lucidot Nanozeolite in the PS dope preparation and showed that increase in zeolite-added ratio resulted in an initial increase, followed by a subsequent decrease in flux of the hollow fiber membrane (HFM). The increase in flux of HFMs at initial loading may be attributed to zeolite suspension at nanoscale, leading to enhancement in nucleation and nodule formation. Further decrease in flux could result from reduction in pore number and increase in pore size, because of agglomeration of zeolite at higher loading. It also explains the increase in pore size, which resulted in increase in NMWCO. Rezaei et al. (2016) tested ZSM5 zeolite-filled PVDF mixed matrix membranes that were used for CO_2 absorption in contactor systems. The hydrophobicity, porosity, and fingerlike macrovoids of membranes were increased. Typical performance data is presented in Figure 2.3.

2.4.3 Fouling

Lind (2009) demonstrated the strong antimicrobial reactivity of silver-exchanged LTA nanocrystals and suggested the possibility of developing more fouling-

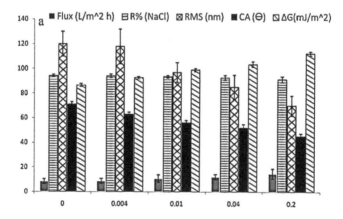

FIGURE 2.3 Permeate water flux, NaCl rejection, RMS roughness, contact angle, and interfacial free energy versus zeolite weight % in zeolite membranes (Fathizadeh et al., 2011) (License Number 4615280311102).

resistant membrane materials. In addition, Yin et al. (2012) concluded that the high hydrophilicity and negative charge of TFN membranes introduced by MCM-41 nanoparticles could improve the membranes' resistance to fouling.

2.4.4 EFFECT ON MORPHOLOGY

NaX zeolite is very hydrophilic and improves the interfacial free energy and contact angle in the TFN membrane, which decreases the surface roughness in the membrane. The variation of the RMS roughness confirmed that the surface of the membrane turns into a smooth layer after adding nano-NaX in IP polymerization. These improvements in surface properties lead to good physical and chemical stability properties (Fathizadeh et al., 2011).

Rezaei et al. (2016) tested ZSM5 zeolite-filled PVDF mixed matrix membranes and showed that the surface roughness, wettability resistance, and mechanical stability of membranes were also considerably improved.

Li et al. (2018) used zeolitic imidazolate framework-8 (ZIF-8) to prepare high-performance, thin-film-nanocomposite cation exchange membranes and showed that by raising the initial concentration of ZIF-8 to 0.08%, the surface roughness increased to 70.8 nm, with the contact angles decreasing from 78° to 71°.

Lind (2009) deduced that the addition of Nano zeolite in the polyamide thin films alter the interfacial polymerization kinetics and polyamide thin film structure. In addition, silver-exchanged LTA zeolites produced membranes with higher flux and rejection, along with smoother and more hydrophilic interfaces than those formed from as-synthesized LTA in the sodium form. Dahe et al. (2012) showed that zeolite nanoparticles participated in, and directed, the nucleation process during phase separation and the mixed matrix membrane skin consisted of nodules containing more zeolite nanoparticles. Rezaei et al. (2016) tested ZSM5 zeolite-filled PVDF mixed-matrix membranes, and fingerlike macrovoids of membranes were increased. Considering the opposing effects of the increase in thermodynamic instability (thermodynamic effect) and the increase in viscosity (kinetic effect), the enhancement of fingerlike microvoid formation by the incorporation of ZSM5 nanoparticles indicates that the thermodynamic effect dominates the formation of the cross-sectional structure.

Shi et al. (2016) proved that the use of silver-loaded zeolite Y as a carrier in dual layer PVDF HF membranes had not only effectively avoided potential aggregation of Ag nanoparticles, but had also facilitated their long term stability in the membrane outer surfaces for longer-term antifouling effects in seawater, providing additional routes to extend the service life of the current desalination units.

2.4.5 EFFECT ON MECHANICAL PROPERTIES

Dahe et al. (2012), among others, have reported the effects of nanocomposite PSF/zeolite HFMs on mechanical properties. Their work is depicted in Figure 2.4 and is included in Table 2.3.

TABLE 2.3

Effect of zeolite Nanoadditives on membrane morphology and performance.

Application	Additives	Polymer	Morphology	Mechanical	Performance	References.
			Flat sheet: zeolite added in the membrane coat			
NF	NaX zeolite (40–150 nm) (0.004–0.2 wt./vol)	PES/PVP/DMAC 15/5/80(wt..%) MPD/ TMC 2/0.1	#CA decreased from 70 to 42 #thickness decrease from 0.282 to 0.15μm	-	#LMH increase from 8.32 to 14.6 #R% decrease from 98 to 95 (NaCl)	Fathizadeh et al. (2011)
		MPD/TMC 3/0.15	#CA decreased from 65 to 45 #thickness decrease from 0.253 to 0.173μm	-	#LMH increase from 8 to 29, #R% decrease from 98 to 60 (NaCl)	
RO	silver-exchanged LTA zeolites (prepared using hydrothermal method) 0.4% (w/v)	PSF MPD/TEA/CSA/ SLS 2.3:6:0.02% (15s) TMC 0.1%(1min)	#zeta potential mV(-ve) #surface area difference % (SAD)(30–59). # Ra increase from 60 to 70	-	# LMH increased from 5.9 to 7 um/ Mpa/s #R%: no change 93–95 (NaCl)	Lind et al. (2009)
RO	(AgA) Nanocrystals with silver 0.4% (w/v): added in coating		#zeta potential (-ve) #surface area difference % (SAD) (30–19). # Ra decrease from 60 to 40	-	#LMH: increased from 5.9 to 9.5 um/Mpa/s #R%: no change 93–95 (NaCl)	

Hollow fiber: zeolite added in the membrane dope

RO	Lucidot NZL 40: 0.01–1%	PSf/TPGS/NMP/NZL: 25/1/74/0.01–0.1–1 wt..%	#MWCO: increase from 9500 to 5400 #ID: decrease (823–783) # wall thickness: increase (128–132) then decrease to 121	#Young's Modulus: decrease from 182 to 168 Mpa #Break strength: decrease from 9.44 to 8.87 Mpa #Elongation: decrease from 80.1 to 63.9%	# LMH: initial increase to 21.31, then decrease to 11.79	Dahe et al. (2012)
UF	ZSM-5: 0–5 wt..%	PVDF/NMP/LiCl/ZSM: 18/79.5/2.5/0–5 wt..%	# fully asymmetric structure with increase in of fingerlike macrovoids #Pore size: decrease from 126 to 59 nm #CA increase from 89 to 104 #Ra increase from 19 to 48	-	# CO_2 removal via membrane contactor	Rezaei, Dasht Arzhandi, et al. (2016)
UF	Ag-loaded zeolite Y	PVDF/PVP/DMAC/Z: 15/3–7/77-82/0–1	#Avoided potential aggregation of Ag nanoparticles, but also had facilitated their long term stability in the membrane outer surfaces	-		Shi et al. (2012)

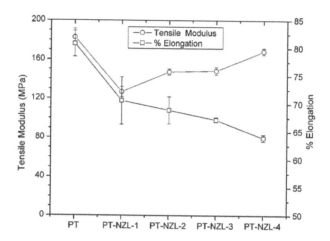

FIGURE 2.4 Mechanical properties plot of nanocomposite PSF/zeolite HFMs. (Dahe et al., 2012) (License Number4615281150855).

2.5 SILICA NANOPARTICLES

Mesoporous silica nanoparticles (MSNs) are defined as those having pore size in the range of 2–50 nm and an ordered arrangement of pores giving an ordered structure. (MSNs) have gained wide popularity over recent years. Their advantages of uniform and tunable pore size; easy, independent functionalization of the surface; internal and external pores; and pore opening gating mechanism make them distinctive and promising as carrier. (Narayan et al., 2018)

Kim et al. (2015) found some important points about membrane separation through their studies on silica nanoparticles. The mesoporous silica with two- or three-dimensional pore structure is one of the most promising types of molecular sieve materials for gas separation membranes applications. Other applications include pervaporation, microfiltration, ultrafiltration, nanofiltration, reverse osmosis, and forward osmosis, depending on the pore range.

Porous membranes with a dense layer of nanoparticles imparts useful functionality and can enhance membrane separation and antifouling properties. Asymmetric membranes facilitate control over membrane flux and selectivity, which enables the formation of stimuli-responsive, hydrogel nanocomposite membranes; they can be easily modified to introduce antifouling features. This approach forms a foundation for the formation of advanced nanocomposite membranes comprising diverse building blocks, with potential applications in water treatment, industrial separations, and as catalytic membrane reactors. (Haase et al., 2017)

2.5.1 Preparation of Functionalized Silica

There are three major techniques for the functionalization of mesoporous: condensation, impregnation, and post synthesis grafting. This approach maintains the substrate structure, and the formed amino oxides remain stable, even after several adsorption/desorption cycles. The supported mesoporous silica membranes with controlled structures, such as silica powder, are synthesized via the well-established sol – gel method, but in the presence of support materials. This technique involves hydrolysis and condensation of respective precursors to form colloidal sols, as reported by Kim et al. (2015). The nanoparticle-functionalized hollow fiber membranes are prepared by STRIPS (solvent transfer-induced phase separation) and photo polymerization have exceptionally high nanoparticle loadings (up to 50 wt..% silica nanoparticles) (Haase et al., 2017) to achieve highly permeable composite membranes (Jin et al., 2012; Yin et al., 2012; Abolfazli and Rahimpour, 2017). Porous MCM-41 and nonporous spherical silica NPs were synthesized and used as fillers to fabricate the thin-film nanocomposite (TFN) membrane. (Yin et al., 2012; Abolfazli and Rahimpour, 2017)

2.5.2 Effect on Mechanical Properties

$PVDF/SiO_2$ hollow fiber membranes prepared by SiO_2 sol – gel method showed increase in the break strength and in Young's modulus, with increase in concentration of TEOS (tetraethyl orthosilicate) (< 3) then declined with the further increase of TEOS concentration. At higher TEOS concentration, the formed SiO_2 network increased the rigidness of the membrane and confined the crystallization of PVDF, which led to the decrease of the mechanical properties, such as elongation-at-break (Yu et al., 2009b) .

2.5.3 Effect on Morphology

The morphologies, chemical composition, and surface image of membranes were investigated by using FTIR, SEM, AFM, and contact angle analyses. Abolfazli and Rahimpour (2017) studied the modification of a thin-film composite (TFC), hollow fiber polyamide membrane fabricated by interfacial polymerization, using piperazine (PIP) and trimesoyl chloride (TMC) on a porous polysulfone substrate. The effects of triethylene tetra mine (TETA) and silica nanoparticles (SiO_2) contents in the aqueous phase (as the additives) have been investigated. In the case of the SiO_2 nanoparticles, the surface was rougher than when using TETA in the modified membranes. However, the nanoparticle's size is close to the membrane pore size. As a result, the silica particles block a few pure water channels of the membranes.

Yin et al. (2012) studied the effect of silica NPs onto a thin-film, nanocomposite (TFN) membrane containing porous MCM-41. By increasing the concentration of NPs, hydrophilicity, roughness, and zeta potential increased. (Yu et al. 2009b) reported that by incorporating silica nanoparticles into the PVDF membrane, the cross-sectional morphology experienced a transition from fingerlike macrovoids to a spongelike structure.

2.5.4 Effect on Performance

Abolfazli and Rahimpour (2017) showed that the water flux increased with increasing the silica contents in the aqueous phase. The hollow fiber composite membrane depicted a salt (NaCl) rejection of about 26 %, and flux of about 31 LMH. On preparing TFC, the presence of silica NPs in the aqueous solution affected the IP reaction and the formation of the polyamide layer leading to the increase of membrane hydrophilicity. Increased hydrophilicity caused the water molecules to pass through the membrane at higher speeds (Liu et al., 2011). Yin et al. (2012) found the same effect for silica NPs, where the permeate water flux increased from 28.5±1.0 to 46.6±1.1 LMH with the incorporation of MCM-41 NPs, while maintaining high rejections of NaCl and Na_2SO_4 (97.9% and 98.5%, respectively). The internal pores of MCM-41 NPs contributed significantly to the increase of water permeability than the nonporous. Others (Jin et al., 2012) found that permeation performance for $PA-SiO_2$ membranes increased nearly 50% without loss of salt rejection rate by adding 1.0% (wt..) SiO_2 NPs in aqueous solution. The order of rejection to inorganic salts Na_2SO_4> $MgSO_4$> $MgCl_2$> NaCl revealed that both PA and $PA-SiO_2$ membranes were negatively charged. The zeta potentials testing results indicated that addition of SiO_2 increases the negative charge quantities on the surface of $PA-SiO_2$ membrane for negatively charged hydroxyl groups and silanol-covered nano-SiO_2 surfaces. The value of molecular weight cutoff (MWCO) for $PA-SiO_2$ membranes was about 1000 g/mole, and the additive of SiO_2 NPs to PA membranes enlarges the pore size slightly. The $PA-SiO_2$ membrane had a higher stable flux, and could remove nearly 50% salts when treated with oily wastewater in one-cycle filtration as a type of application.

2.5.5 Mechanism of Action

Studying the mechanism of action through the fabrication of a highly hydrophilic PVDF UF membrane as in Liang et al. (2013) via post fabrication tethering of superhydrophilic silica NPs to the membrane surface. The authors found that by plasma-induced graft copolymerization and by providing sufficient carboxyl groups as anchor sites, the binding of silica NPs were surface-tailored with amine-terminated cationic ligands. The NP binding was achieved through dip-coating, which improved the wettability of the membrane and converted the membrane surface from hydrophobic to highly hydrophilic. The irreversibly bound layer of superhydrophilic silica NPs endowed the membranes with strong antifouling performance as demonstrated by three sequential fouling filtration runs. Kim et al. (2015) studied the mechanism for the synthesis of mesoporous silica in the presence of a cationic surfactant. When dissolved in water, the cationic surfactant forms micelle structures. In this process, the cationic *heads* of the surfactant molecules are arranged to the outer side, while their hydrophobic *tails* collect in the center of each micelle. The silica source then covers the micelle surfaces. Once the surfactant is removed via calcination or extraction, the pores are activated.

Table 2.4 summarizes the effect of nanosilica on membrane characteristics and performance.

TABLE 2.4
Effect of functionalized silica NPs on membrane characteristics and performance

Application	Polymer/additive	Interfacial polymerization	Characteristics	Performance	Reference
HF/NF	PS, NMP, PEG/ LiCl, Triton X-100, glycerol, SiO_2	PIP (3 w/v %) TMC (0.5 w/v %)	#CA: 41 decreased #thickness: increase #Ra: (18.9 nm) decrease #zeta potential increase negative charge	#LMH: 34.5 #R%: 19,9 for NaCl	(Abolfazli and Rahimpour, 2017)
	PS, NMP, PEG/ LiCl, Triton X-100, glycerol, TETA (4%) + SiO_2 (0.1%)	PIP (3 w/v %) TMC (0.5 w/v %)	#CA: 30 decreased #thickness: increase #Ra: (11.9 nm) decrease #zeta potential: increase negative charge	#LMH:31 #R% 26 for NaCl	
FS/NF	PSU, DMF/ porous silica NPs and MCM-41NPs	MPD (2.0 wt.%) TMC (0.15wt.%)	#Thickness: (300 to 500nm) #CA: decrease (57 to 27.9) #Ra: increase (135 to159 nm) #zeta potential decrease (-5.71 to -9.54 mV)	#LMH increase (28 – 46) #R% no change 97.5, 98.5% for (NaCl, Na_2 SO_4)	(Yin et al., 2012)
	PSU, DMF/spherical silica (NPs)		#Thickness: (300 to 500nm) #CA decrease (57 to 30) #Ra: increase (135 to 159 nm)	#LMH increase from 28 to35 # R% no change 98.5% for (NaCl, Na_2SO_4)	

(Continued)

TABLE 2.4 (Cont.)

Application	Polymer/additive	Interfacial polymerization	Characteristics	Performance	Reference
FS/NF	PSF/ silica NPs (15 nm, 1% wt.)	PAMAM (0.5% wt.) SDS TMC (0.3% wt.)	#zeta potential decrease (-5.71 to -9.54mV) #R increase (8.72 to 36.5 nm) #zeta potential: increase negativity (-16.78 to -20.56 mV)	#LMH: increase to 10 # R%: (91.23 $MgSO_4$) (48.80 $MgCl_2$) (46.01 NaCl) (92.62 Na_2SO_4)	(Jin et al., 2012)
FS/MD	PVDF/silica NPs	/	#Pore size: increased to 140 nm #CA: 94 #Porosity: decreased to 50% #Ra: 27 nm	LMH: increase (0.7 to 2.9)	(Efome et al., 2015)
HF/UF	PVDF/silica NPs	/	#Viscosity: increase #CA: decrease (82 to 53) #porosity: increase (5 to 84 %) #pore size: increase #Ra: increase	LMH: 301 increase	(Yu et al., 2009b)

2.6 METAL OXIDES NANOPARTICLES

2.6.1 PREPARATION

2.6.1.1 Coprecipitation Method

In the coprecipitation method, the synthesis reaction is generally carried out at room temperature. The prepared NPs' properties are determined by initial reaction parameters such as pH and reaction temperature. This method is very easy and simple for the synthesis of NPs in aqueous media in presence of different surfactants (Iida et al., 2007).

2.6.1.2 Sol–Gel Method

The method involves preparation of metal oxides by hydrolysis of precursors, usually alcoxides in alcoholic solution, resulting in the corresponding oxohydroxide. Condensation of molecules by giving off water leads to the formation of metal hydroxide network. Hydroxyl species undergo polymerization by condensation and form a dense porous gel. Appropriate drying and calcinations lead to ultrafine porous oxides (Hampden-Smith and Interrante, 1998)

2.6.1.3 Thermal Decomposition

The thermal decomposition method is based on the decomposition and oxidation of several types of precursors in an organic medium by using high temperature. Reaction is usually endothermic and heat is important to break chemical bonds in the compound enduring decomposition (Navaladian et al., 2007).

2.6.1.4 Microemulsion Process

Microemulsion process is based on creating a thermodynamically stable and homogeneous dispersion of two immiscible liquids (usually water and oil solvent), using a surfactant (Vidal-Vidal et al., 2006).

2.6.2 EFFECT ON MECHANICAL PROPERTIES

Nanoparticles have the potential to improve membrane mechanical properties in UF applications as shown by tensile testing. PVDF Nanocomposite membranes incorporated with ZnO NPs exhibited increased tensile strength and elongation-at-break (Liang et al., 2012). The strength was enhanced especially at high ZnO loading, where the tensile strength was twice that of the unmodified membrane when the ZnO to PVDF ratio was at 3:15 and 4:15. Likewise, PVDF with Al_2O_3 increased tensile strength and elongation-at-break with increased Al_2O_3-particle concentration to 2% and, then declined as the Al_2O_3-particle concentration was further increased (Yan et al., 2006). PVDF/TiO_2 hollow fiber membranes prepared by either TiO_2 sol – gel or blending methods showed 30% increase in tensile strength (Yu et al., 2009a). Han et al. (2010) studied the effect of using multiple types of NPs (TiO_2 and $Al2O_3$) in PVDF hollow fiber membranes. It was noted that all NP membranes had higher tensile strength, and the best improvement was from 1.71 MPa to 3.74 MPa, with

42   Membrane Desalination

a combination of 2 wt. % TiO_2 and 1 wt.. % Al_2O_3.The improvement could be attributed to the reduced macrovoid formation observed in the NP membrane.

2.6.3 EFFECT ON PERFORMANCE AND HYDROPHILICITY

Increase in water flux has been observed with PVDF and PS/CS membranes incorporated with TiO_2 (Kumar et al., 2013), (Ong et al., 2015), (Yu et al., 2009a), (Efome et al., 2015), CuO, (Baghbanzadeh et al., 2016), and Al_2O_3 (Yan et al., 2006). This was attributed to the interplay between the hydrophilicity and viscosity of the casting solution, which resulted in the maximum pore size. Also, at high particle concentrations, sometimes pores are plugged causing a further reduction in the flux. Yan et al. (2006) studied the effect of Al_2O_3 NP concentration for PVDF UF membrane on pure water fluxes and rejection. The authors observed significant improvement in water flux by the addition of nanosized Al_2O_3 particles, which have some favorable characteristics, such as hydrophilicity and higher ratio surface areas. Although increasing the Al_2O_3-particle concentration caused a decrease of the contact angle, the porosity and rejection, were not affected.

2.6.4 EFFECT ON MORPHOLOGY

Reports on the effect of addition of NPs on porosity and mean pore size are often contradictory. In some cases, it has led to a higher porosity and mean pore size (Baghbanzadeh et al., 2016), (Kumar et al., 2013), (Ong et al., 2015), (Yu et al., 2009a), (Liang et al., 2012), (Han et al., 2010), This may be attributed to the hindrance effect of nanomaterials that reduces the interactions between solvent and polymer molecules. As a result, easier and faster diffusion of the solvent molecules from the polymer matrix to the coagulant takes place, which further increases the porosity and mean pore size (Kim et al., 2001). However, others have reported that the addition of nanomaterials led to a reduction in both porosity and mean pore size (Yu et al., 2009b), (Yan et al., 2006), which might result from the deposition of nanomaterials in the membrane pores. On reaching higher concentrations of NPs added to the membrane, owing to the irregular standing of the Nanomaterials increase in the viscosity of casting solution occurs, which results in a delayed phase separation, thus creating a dense structure in the sub-layer (Yu et al., 2009b). At the low concentrations, nanomaterials act as the nucleating agents that increase the rate of nucleation of the polymer lean phase (Wang et al., 2012) and enhance the phase separation, mostly owing to an increment in the hydrophilicity of the casting solution, which leads to larger mean pore size. The effect of TiO_2 NPs has been also investigated. It has been reported that surface images revealed neat PES membranes consisting of cellular pores, but with the addition of small amounts of TiO_2 NPs, within 1–2 wt. %, surface pores switched to a lacy structure. By further increasing the TiO_2 loading, the lacy structure returned into a cellular structure accompanied by the agglomerated NPs, resulting in a poor skin layer arrangement (Kumar et al., 2013). On

Membrane Materials Design Trends   43

using Al_2O_3 NPs, increasing the concentration caused an increase of the surface roughness of the membrane, which is attributed to the accumulation of hydrophilic Al_2O_3 particles on the membrane surface, This improves the membrane surface hydrophilicity significantly and reduces the interaction between the contaminants and the membrane surface (Yan et al., 2006). Table 2.5 summarizes the effect on metal oxide NPs on the membranes for selected applications.

2.7 METAL SALT NPS

Physicochemical properties of membrane can be enhanced by introducing metal salt NPs in the membrane structure (membrane matrix) or by coating NPs on the membrane surface. It is reported that distinct improvements on the performance of membranes, including salts separation, flux and antifouling properties are achieved.

2.7.1 SILVER NPS

Several metal salts of NPs are used for membrane modification, such as salts of silver and copper, which have exhibited antimicrobial properties. Silver NPs in the range 1nm -100 nm have good characteristics, including; good conductivity, high mechanical properties, chemical stability and antimicrobial properties. They also have the ability to adsorb oxygen on its surface and easily interact with sulfur and phosphorous groups in bacterial proteins. Thus, silver NPs have vital role in enhancing membrane antifouling properties (García-Ivars et al., 2019).

2.7.1.1 Preparation

Two methods including ex-situ and in situ reductions of silver have been reported. In the first method, silver nitrate ($AgNO_3$) is added to a solvent as a reducing agent for silver. The solution is heated under intense stirring to allow for NP formation. The formed NPs are then added to the membrane dope solution containing polymer, additives and solvent. The second method includes dissolving silver nitrate in a solvent at room temperature. The dope solution is then subjected to intense stirring under heating, and when it becomes homogeneous, the salt/solvent solution is added to initiate the reduction of silver ions to form the NPs (Taurozzi et al., 2008).

2.7.1.2 Effect of Silver NPs on Membrane Characteristics and Performance

Several studies reported the preparation of membrane NPs using silver NPs by blending or forming a coating of silver particles on the membrane surface in addition to its effect on membrane characteristics and performance. Toroghi et al. (2014) used two methods for membrane modification. The first method is coating PES membrane surface by soaking membrane in silver Nanoparticles solution where the solution is prepared by dissolving 1 g of fructose and 0.1 g of ammonium hydrogen citrate in 1 liter of distilled water and pH=9.5 and heated to 80° C, then 9.35 ml of 0.1 M silver nitrate solution was added to form silver NPs. The second method includes blending of NPs with membrane polymer solution

TABLE 2.5

Effects of metal oxide NPs on the surface and performance of composite membranes

Application	Base Material	Morphology	Performance	Mechanical	Reference
		CuO			
VMD	PVDF (15 wt..%)	#CA decrease from 92 to 85 #pore size and porosity increase 87/84%	#LMH increase from 0.5 to 1.25 @ 0.95wt. %	(-)	Baghbanzadeh et al., 2016
		TiO$_2$			
RO	PS/PA	#Ra increased	#LMH decreased from 136 to 129 $l.m^{-2}.h^{-1}$ #R% increase from 94.7% to 96%	(-)	Kim et al., 2003
Nf	PS/CS	#viscosity increased #pore size decrease from 30 to 50 nm #CA decrease from 73° to 58 at 3.0% #average pore size 60 nm at 6% and 40 and 8%	#LMH increase #R% 34% at 6% and 46% at 8% TiO$_2$NT	(-)	Kumar et al., 2013
UF	PVDF (18 wt..%)	#porosity increased from 86 to 88.6), #CA decreased from 74–68, #Pore size increased from 98 to104), #roughness increased from 10.8 to 31nm	#LMH 70.48 L/m2 h %R 99.7% for oil removal	#Tensile strength increased from 1.98–2.58 MPa	Ong et al., 2015
UF	PVDF (18 wt.%)	#viscosity increase #CA decrease #porosity decrease	#LMH increased	#Tensile strength increased	Yu et al., 2009a

		Al₂O₃			
UF	PVDF (19wt..%)	#CA decrease from 83 to 57.5 #porosity no change (54.9–55.3), #roughness increase from 64 to 114 nm	LMH increased to 100 l.m-2.h-1	# tensile strength and elongation-at-break increase more than 50%.	Yan et al., 2006
ZnO					
UF	PVDF (15 wt..%)	#CA decrease from 43 to 36 #average pore size increased from 28.1 to 31 nm	LMH increased	#tensile strength and elongation-at-break increased	Liang et al., 2012
		Mixed metal oxide Nanoparticles TiO2/Al2O3			
UF	PVDF (18 wt..%)	#viscosity increased #pore size 10nm	#LMH 179 L.m⁻².h⁻¹.bar⁻¹, # R 98.9% for bovine serum albumin	increased	Han et al., 2010
		Mixed metal oxide Nanoparticles TiO₂/Al₂O₃			
UF	PVDF (18 wt.%)	#viscosity increased #pore size 11nm	#LMH 352 L.m⁻².h⁻¹/bar, #R 89 % for bovine serum albumin	increased	Han et al., 2010

through dissolving AgNO$_3$ in dimethylformamide (DMF) and stirred at 100°C for 1 h then, from 2 to 6 wt % of silver-DMF organosol was added to the casting solution containing PES and polyvinylpyrrolidone (PVP) and dimethylacetamide (DMAc). The results showed that the amount of silver formation in the coating method was much greater than in blending method. Both methods showed good antibacterial activity against E. coli and S. aureus. The first method exhibits significant changes in morphology or permeate, while the second method showed different cross sections, in addition to decrease of flux .

Hoek et al. (2011) prepared mixed matrix membranes by dissolving silver metal powder in n-methyl-2-pyrrolidone (NMP), followed by the addition of polysulfone. Introducing silver particles led to pore size increase from 14 to 25 nm, increased roughness, and decreased, the contact angle, while tensile strength increased from 26.8±0.6 to 48.1±2.8 MPa. Dolina et al. (2018) developed a method for modification by stabilizing silver NPs on hollow-fiber polyethersulfone membranes through soaking the membrane in a solution of silver nitrate for 4 h at room temperature. The modification results showed an increase in permeability of 15% when compared to the unmodified membrane, no changes in separation properties, decrease contact angle, and no changes in the original membrane's cross-section morphology. Chou et al. (2005) studied the membrane matrix fabrication through the addition of cellulose acetate (CA) powder and 0.001–0.1 wt.. % AgNO$_3$ silver nitrate pellets to DMF solvent to form the dope solution. The modification results showed increase in permeability from 5.586 to 6.15 LMH .atm, dense structure in the sublayer and no apparent pore changes on the outer and inner surfaces. The mechanical properties showed a slight decrease in tensile strength and elongation. Li et al. (2013) prepared membrane mixed matrix through the in situ method in which 0.79 wt.. % of AgNO$_3$ was dissolved in DMF at room temperature and added to a homogenous membrane polymer solution containing PVDF, PVP and DMF and stirred at 50°C for a further 7 h. Introducing silver NPs led to increased pore size from 0.209 to 0.251 μm and to increased porosity. On the other hand, contact angle was reduced from 81° to 68°, while flux increased from 36.4 to 108.6 LMH with a decrease in rejection.

2.7.2 COPPER (CU) NPS

Copper NPs are solid particles; like other NPs, their size range is between 1nm and 200 nm. Cu NPs possess excellent antimicrobial properties, are of low cost, and are more available than other NPs (Zahid et al., 2018).

Many techniques have been used to prepare nanocomposite membranes containing Cu NPs for improving antifouling properties of the membrane surface. Xu et al. (2015) modified polyacrylonitrile (PAN) ultrafiltration membrane surfaces by immersing membranes in the solution contained polyethylenimine (PEI) and copper sulfate (CuSO$_4$) powder to obtain a poly-cation–copper(II) complex on the membrane surface, followed by cross-linking. The developed membrane antibacterial efficiency was enhanced more than 95%. Membrane micropores were more compact than in pristine membranes.

Ben-Sasson et al. (2016) developed in situ formations of Cu NPs that coated the surface of TFC-RO membranes using copper salt and reducing agent. The modification showed reduction of biofouling, minor increase in flux, and slight decrease in NaCl rejection. Zhang et al. (2017) modified PA-thin-film composite membranes by coating Cu NPs membrane surfaces with copper chloride ($CuCl_2$) aqueous solution. The modified membrane showed an enhancement in antibacterial properties, with more than 99% efficiency, but flux decreased after fouling. An overview of the effect of silver and copper metal salt NPs on membrane characteristics and performance is given in Table 2.6.

2.8 NANOCELLULOSE ADDITIVE

Nanocellulose (NC) based materials are carbon-neutral, sustainable, recyclable, and nontoxic. Cellulosic materials can be converted into cellulose nanofibers (CNFs), nanowhiskers (CNWs), and nanocrystals (CNCs) with many potential applications (Missoum et al., 2013, Mariano et al., 2014, Carpenter et al., 2015).

2.8.1 PREPARATION

NC mainly consists of chemical CNCs or mechanically extracted (nanofibrillated) NPs. First, pure cellulose products are pretreated to remove ash, wax, lignins, hemicellulose, and other noncellulosic compounds.

CNFs from precursors are prepared mechanically using ultrasonication, high-pressure homogenization, grinding/crushing, and microfluidization. CNWs, composed of short fibrils; are produced by acid hydrolysis. Cellulose Nanocrystals CNCs are high-aspect-ratio NPs produced via the acid hydrolysis of CNFs as shown in Figure 2.5 (Gopakumar et al., 2019)

2.8.2 EFFECT ON MECHANICAL PROPERTIES

A typical stress – strain curve is given in Figure 2.6. The tensile strength and modulus decreased with the addition of CNCs, which is unexpected (Nair et al. (2017). This was attributed to the aggregation of CNCs, which causes defects rather than reinforcements. In contrast, the addition of CNCs increased the strain-at-break, indicating that the addition of CNCs and introduction of porosity has a positive impact on the strain and toughness of the electrospun membranes. The tensile strength and modulus decreased with the addition of CNCs, which is also unexpected.

Conversely, Goetz et al. (2016) reported that the tensile strength and E-modulus increased from 1.43 MPa to 3.31 MPa, 0.34 GPa to 1.16 GPa, respectively by addition of ChNCs with cellulose acetate. ChNC has positively influenced the tensile which can be attributed to the stiffening effect of the ChNCs coated on individual electrospun fibers, as well as the mats However, the strain was decreased, which is attributable to the restricted slippage of the electrospun fibers past each other owing to the *tied* junction points.

TABLE 2.6

Effect of metal salt NPs on the characteristics and performance of composite membranes

			Morphology					Performance			Mechanical properties		
App.	Base polymer	Additive	Pore size	Porosity	Cross section	Roughness	Contact angle	Flux	Rejection	Antibacterial efficiency	Tensile strength	Elongation	Reference
UF	PES	AgNO$_3$ 2-4-6% (wt..)			Thick top layer and different cross section morphology	Decreased from 21.11 to 14.70, 12.64, 17.77 nm		Decreased from 33.86 to 6.86,9.61, 4.98 (kg/m^2.h)					(Toroghi et al., 2014)
UF	PS	Ag metal powder 3.7% (wt..)	Increased from 14 to 25 nm			Increased from 13 to 30 nm	Deceased from 76.2° to 74.1°				Increased from 26.8 to 48.1 MPa		(Hoek et al., 2011)
	PES	AgNO$_3$ 3.5% (wt..)			Not affected		Decreased from 91° to 86°	Increased 15%	Not affected				(Dolina et al., 2018)
UF	CA	AgNO$_3$ 0.1% (wt..).	No apparent pore was observable on the outer and inner surface		Dense structure			Increased from 5.5 to 6.15 LMH atm.			Decreased from 14.7 to 14.3 MPa	Decreased from 35 to 33.6	(Chou et al., 2005)

UF	PVDF	AgNo$_3$ 0.79% (wt..)	Increased from 0.209 to 0.251 µm	Increased from 82.7 ±0.2% to 86.7% ±0.3 %		Decreased from 81° to 68°	Increased (36.4 to 108.6 LMH).	Decreased from 91.6% to 88.1%		(Li et al., 2013)
NF	PAN	CuSO$_4$ (0–0.4gm/ 100ml PEI)						Seawater Rejection: SO^{-4} =43.5 Mg^{+2} =68.3% Na$^+$ =19.1% Cl$^-$=18.7%	More than 95%	(Xu et al., 2015)
RO	PA-TFC	CuSO$_4$				Minor increase from 2.53 to 2.97 LMH. bar		NaCl rejection slight decrease from 98.94 ± 0.23% to 98.31 ± 0.32%	decreased	(Ben-Sasson et al., 2016)
RO	PA-TFC	CuCl$_2$				decreased	Increased above 99%			(Zhang et al., 2017)

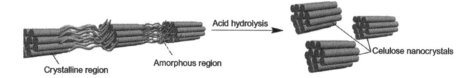

FIGURE 2.5 Acid hydrolysis of cellulose for the production of CNCs. Permission from Elsevier (Gopakumar et al., 2019).

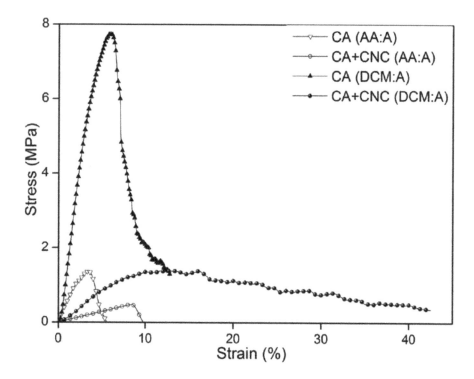

FIGURE 2.6 Stress – strain curve of electrospun membranes. Permission from Elsevier (Nair et al. (2017).

2.8.3 Effect on Morphology

On increasing the concentration of CNCs with CA dissolved in DCM: A or (AA: A) solvents, the average size of the added CNC was reduced with no gain in the surface area (Nair et al. (2017). This can be explained by the increase in the conductivity of the solvent by the addition of CNCs, which is consistent

Membrane Materials Design Trends

with the findings of Naseri et al. (2015). Smith et al. (2019) studied the addition of CNCs and TOCNs in the polyamide thin film composite membrane and produced membrane with slight decrease in the average thickness. This decrease occurred as a result of slight differences in the interfacial polymerization conditions during sample preparation. Surface roughness decreased for the membranes fabricated via the dispersion method (addition CNC), while the roughness increased for the vacuum-filtered membrane (TONC). The presence of CNCs may have affected the surface formation of the polymer, but it is more likely that these variations result from small changes in interfacial polymerization reaction conditions. The surface zeta potential was improved for CA-CNC nanocomposite fibers, compared to CA fibers (Nair et al. (2017). Goetz et al. (2016) used CA polymer-based electrospun mats to support Chitin nanocrystals (ChNC) to prepare microfiltration membranes with a hydrophilic surface. The average pore diameter decreased from 11.02 to 10.07 nm and surface area increased from 2.73 to 3.709 m2/g. Typical SEM and AFM images are given in Figures 2.7 and 2.8.

FIGURE 2.7 SEM image of fibers (a) CA(AA:A); (b)CA+CNC(AA:A); (c) CA(DCM:A); and (d) CA+CNC(DCM:A). Permission from Elsevier (Nair et al. (2017).

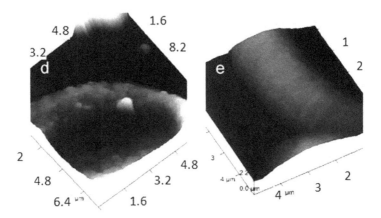

FIGURE 2.8 AFM image of porous fibers and particles d) 3D height image of CA and CA-CNC porous particle. Permission from Elsevier (Nair et al. (2017)).

2.8.4 EFFECT ON PERFORMANCE

Nanocellulose in a native form or with additional surface functionalization (tempo oxidation, enzymatic phosphorylation, cationization, etc.) have significant adsorption efficiency toward metal ions, nitrates, dyes, humic acid from industry effluents (Liu et al., 2015; Sehaqui et al., 2014).

Ma et al. (2012) studied the removal of crystal violet dyes using nanofibrous microfiltration membranes based on CNWs by adsorption. They found that the maximum adsorption capacity of the CNW-based microfiltration membrane for crystal violet dye was higher than that of GS0-22 (a commercially available membrane). This was because of the strong electrostatic attraction between protonated of crystal violet dye molecules and the highly negative carboxylate group on the cellulose Nanowhiskers' surface. These Nanofibrous microfiltration(MF) membrane system had high flux, low pressure drop, and a high retention capability against both bacteria and bacteriophages. Other researchers (Karim et al., 2014) showed that nanocrystals-based membranes prepared by freeze drying resulted in low water flux (64 LMH. MPa^{-1}), despite good functionality. This was attributed to the adoption of the freeze-drying method, which reduces H-bonding between nanocrystals and the relatively high thickness (250 μ m). These membranes had a pore diameter of 13–17 nm and had the ability to immobilize positively-charged dye molecules with reported 84%, 69% and 98% removal rates for methyl violet 2B, rhodamine 6G, and Victoria Blue 2B, respectively. Partially hydrolyzed polyacrylamide/cellulose nanocrystal (HPAM/CNC) Nanocomposites were investigated for the uptake of methylene blue (Zhou et al., 2014). The adsorption capacity of the nanocomposite increased with increasing CNC concentration to 20% and decreasing pH. A maximum capacity of 224.8 mg/g was detected for the

Membrane Materials Design Trends

nanocomposite composed of 20% CNC at pH 6.5. Recently, Gopakumar et al. (2017) removed toxic crystal violet dyes from water by using a cellulose nanofiber-based PVDF membrane with adsorption capacity 2.948 mg/g .This can be attributed to the electrostatic interaction between the protonated crystal violet and CNF-based PVDF membrane. Table 2.7 summarizes some of the reported data concerning the effect of nanocellulose materials additive on membrane characteristics and performance.

2.9 OTHER NPS

In this section, some NPs, other than those mentioned in previous sections, that are used as additives for membrane modifications are presented. These essentially include WS2 (tungsten disulfide) and PANI (polyaniline)

2.9.1 PREPARATION OF POLYANILINE (PANI)

Polyaniline (PANI) is one of the most important conductive polymers and it is prepared by various methods, namely, chemical synthesis (oxidative polymerization) and electrochemical synthesis (Stejskal et al., 1999).

2.9.1.1 Chemical Synthesis (Oxidative Polymerization)

Synthesis of PANI by chemical oxidation way involves the use of acid in the presence of ammonium persulfate as the oxidizing agent in the aqueous medium Chemical synthesis requires three reactants: aniline, an acidic medium (aqueous or organic) and an oxidant. The more common acids are essentially hydrochloric acid (HCl) and sulfuric acid (H_2SO_4). Ammonium persulfate ((NH_4)$_2S_2O_8$), potassium dichromate ($K_2Cr_2O_7$), cerium sulfate ($Ce(SO_4)_2$), sodium vanadate ($NaVO_3$), potassium ferricyanide ($K_3(Fe(CN)_6)$), potassium iodate (KIO_3), hydrogen peroxide (H_2O_2) are recommended as oxidants (Malinauskas, 2001)

2.9.1.2 Electrochemical Synthesis

The electrochemical synthesis of conducting polymers is similar to the electrodeposition of metals from an electrolyte bath; the polymer is deposited on the electrode surface and also in the in situ doped form. Three electrochemical methods can be used to PANI synthesis Nalwa (1997)

- Galvanostatic method when applied a constant current,
- Potentiostatic method with a constant potential, and
- Potentiodynamic method where current and potential varies with time.

2.9.2 EFFECT ON MORPHOLOGY

Lin et al. (2013), showed that the EDX spectrum indicates that oxygen element existing in the nanoparticles samples, accounted for 5.53±0.98%., the sulfur 18.21±1. 66%, less than the theoretical weight (25.8%) in pure nanoWS$_2$,

TABLE 2.7

Effect of nanocellulose additives on membrane characteristics and performance

Application	base material	morphology	Mechanical	flux LMH	%R/Qe	References
					performance	
		cellulose nanocrystals (CNCs)				
MF/ads	*CS **Cs/GA	#thickness increase 250–270 µm # Average pore size increase 13-17nm, # surface area increase 2.9–3.1 m2/g	#tensile stress increase .98–1.1 Mpa #Strain slightly decrease from 0.28 to 0.23 % #Modulus elasticity increase from 128 to 318 MPa	64 Lm-2h-1	98%, 84%, 70%, for positively dye	Karim et al., 2014
RO	PES/PA	#Thickness decrease from 360 to 311 nm, # roughness increase 180 to 158 nm	(-)	# LMH increase from 9.4 to 11.4 Lm−2 h−1	# % R no change 98.8 for NaCl	Smith et al., 2019
MF	PVDF	#surface area increase from 4.6 to 5.2 m2/g #contact angle decrease 42-126	(-)	(-)	# increase adsorption capacity of dye	(Gopakumar et al., 2017)
		Cellulose Nanowhisker				
MF	PAN nanofibrous scaffold/ poly(ethylene terephthalate) (PET)	#maximum pore size decrease from 0.7 to 0.4 µm #mean pore size decrease from 0.38 to 0.22 µm # CA: decrease from 50.6° to 16.9°	#tensile stress increase from 8.5 to 1.3 MPa #Young's Modulus increase from 226 to 375 MPa	#LMH decrease from 83 to 59 Lm −2 h−1 kPa−1	# increase adsorption capacity for dye,	(Ma et al., 2012)
		cellulose Nanocrystals (CNCs)/TOCNs				
RO	PES/PA	#Thickness increase from 360-365nm, #roughness decrease from 180 to 158nm in CNC, 212 in TOCN	(-)	#LMH decrease from 9.4 to 5.5 Lm−2 h−1	#% R decrease from 98.8 to 97.8 % @ 2000ppm NaCl)	Smith et al., 2019

Adsorption	CA dissolved in AA:A	#Diameter decrease from 1 to 0.2 μm #surface area decrease from 5.7 to 5.6 m2/g #average pore diameter decrease 7.5 to 6.6nm #Zeta potential increase -14 to -23mV	#tensile stress decrease from 1.4 to 0.7 MPa # elongation at break increase from 5.4 to 9.7 % #Young's Modulus decrease from 0.07 to 0.01 GPa	(-)	# R increase from 17.5 to 67.9 %	Nair et al., 2017
Adsorption	CA dissolved in DCM:A	#Diameter decrease 10.8 to 7.8 μm #surface area decrease from 3.5 to 1.9 m2/g #average pore diameter decrease from 10.6 to 7.3nm #Zeta potential increase -27 to -65mV	#tensile stress decrease from 6.8 to 1.3 MPa #elongation at break increase from 12.9 to 40 %: #Youngs Modulus decrease from 0.18 to 0.02 GPa	(-)	# R increase from 25.6 to 98 %	Nair et al., 2017
Chitin Nanocrystals (ChNC)						
MF	CA	#Zeta – potential decrease from -30 to -4.7 mV #CA decrease from 136 to 0.0 ° #Average Pore diameter. decrease from 11 nm to 10 nm . #surface area increase from 2.73 to 3.7 m2/g #Porosity decrease 88-85%	#tensile strength: increase from 1.4 to 3.3Mpa # strain decrease from 6.2 to 3.4% #Youngs Modulus increase from 0.3 to 1.16 GPa	#LMH increase from 13,400 to 14,000 $lm^{-2}h^{-1}$		Goetz et al., 2016
CNF/SIO$_2$						
UF	cellulose Nanofibers/ PAE	#Thickness 20 μm	(-)	110 at 2bar	molecular weight cut off 300 kDa	Varanasi et al., 2105
UF	Calcium cross linking 2,3-dicarboxylic acid cellulose Nanofibers on filter paper	#Thickness 0.85 μm	(-)	105 at 2bar	R 80 % PEG (150,000)	Visanko et al., 2014

* uncrosslinked
** crosslinked

because of the presence of WO$_3$. In general, the membrane morphology changes as the nanofiller content increased.

Fan et al. (2008), found that the peaks of PANI aqueous dispersion were centered on 340, 440, and 800 nm. The spectra showed that PANI nanofibers were in the emeraldine oxidation state. The emeraldine oxidation state of PANI contains half imine and half amine nitrogen and could be represented by the formula [–N C6H4 N–C6H4–NH–C6H4–NH–C6H4–]n. The author used SEM images of PS membrane surface, PANI/PS Nanocomposite membrane surface and cross section. It can be seen that PANI nanofibers are assembled evenly on the PS membrane surface and porous structure is formed. The nanofiber layer cracks might result from the distortion during the preparation of SEM samples (Figure 2.9).

Kajekar et al. (2015), found that PANI-nanofibers have high surface energy and hydrophilicity, thus when the hollow fiber membrane was immersed in a water bath, the PANI-nanofibers may migrate from the polymer matrix towards the water bath to reduce the interfacial energy between the two phases. The migration would leave cavities in the polymer matrix, which in turn would increase the membrane porosity and have interconnection between the finger like pores and macrovoids.

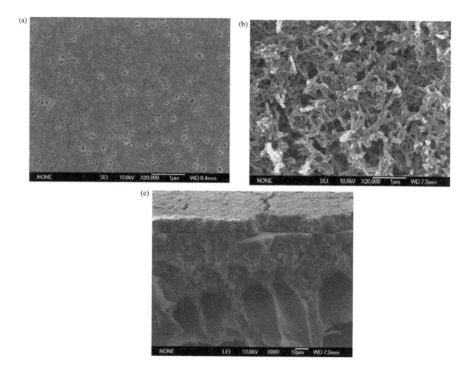

FIGURE 2.9 Fan et al. (2008), depicts SEM images of the membranes: (a) PS membrane surface; (b) PANI/PS nanocomposite membrane surface; and (c) PANI/PS nanocomposite membrane cross-section.

Membrane Materials Design Trends

2.9.3 CHARACTERISTICS AND PERFORMANCE

Lin et al. (2013), reported that at nanofiller contents above 0.1% the permeation of membrane decreased although the hydrophilicity still increased. The increase of membrane permeability by 36% at a lower amount of WS_2 was attributed to the improvement of the membrane hydrophilicity and change in membrane structure.

Fan et al. (2008), found that the reduction of contact angle to 37∘, was mainly caused by the hydrophilicity of PANI nanofibers and the surface roughness of the nanofiber layer. According to the surface AFM image of nanocomposite membrane, it was estimated that about 12.5∘ decrease of the contact angle was attributed to the surface roughness difference between the PANI/PS composite membrane and the PS membrane. Thus, about 24.5∘ decrease of the contact angle resulted from the hydrophilicity of PANI nanofibers.

Kajekar et al. (2015), found that the contact angle reduces with increasing concentration of PANI-nanofiber indicating an increase in the hydrophilicity of the of PSF/PVP/PANI-nanofiber hollow fiber membrane cross-sections. This may be caused by PANI nanofibers, which migrate towards the polymer–water interface and aggregate at the membrane surface. The PANI-nanofibers being highly hydrophilic in nature may be responsible for reducing the water contact angle on the membrane surface.

Table 2.8 summarizes some of the reported data concerning the effects of WS2 and nano- PANI additive on membranes characteristics and performance.

TABLE 2.8

Effect of WS2 and nano – PANI additives on membranes characteristics and performance

Application	Polymer	Additive	Morphology	Performance	Reference
UF	PES/ NMP	WS2 (0.025– 0.25%) WS/ PES	# Nanoparticle size is around 80–140 nm, # CA: increase 36–54 WS/PES ratio%: 0–0.25 #Porosity max at 0.1% of nanofillers	# Permeability increase by 36% at lower WS2	Lin et al., 2013
UF	PS	PANI	# The average diameter and length of the fibers were 43 and 259 nm # contact angle decrease to 37∘		Fan et al., 2008
NF	PS/ NMP	PANI (0-1%)/ PVP(2%)	# asymmetric pores and nonuniform distribution of pores # CA decrease from 93 to 88		Kajekar et al. (2015)

ACKNOWLEDGEMENT

This work was undertaken by the National Research Centre, Dokki, Giza, Egypt through a project entitled "Development of a solar powered, zero liquid discharge Integrated desalination membrane system to address the needs for water of the Mediterranean region" within the scope of ERANETMED, an EU FP7 initiative dealing with the Mediterranean region. The work is funded by the local agency Science and Technology development Fund, STDF, Ministry of Scientific Research, Egypt; Project No. 30280.

REFERENCES

Abolfazli, Z. and A. Rahimpour. 2017. Fabrication and modification of thin-film composite hollow fiber NF membranes. *Journal of Membrane Science and Research* 3 (1):42–49.

Ahn, C.H., Y. Baek, C. Lee, S.O. Kim, S. Kim, S. Lee, S.-H. Kim, S.S. Bae, J. Park, and J. Yoon. 2012. Carbon nanotube-based membranes: Fabrication and application to desalination. *Journal of Industrial and Engineering Chemistry* 18(5):1551–1559.

Alam, J. and A. Ali. 2018. Graphene oxide, an effective nano-additive for a development of hollow fiber nanocomposite membrane with antifouling properties. *Advances in Polymer Technology* 37:1–12.

Baghbanzadeh, M., R. Dipak, C.Q. Lan, and T. Matsuura. 2016. Effects of inorganic Nanoadditives on properties and performance of polymeric membranes in water treatment. *Separation & Purification Reviews* 45(2):141–167.

Barrer, R.M. 1982. *Hydrothermal Chemistry of Zeolites*, Academic Press, London.

Ben-Sasson, M., X. Lu, S. Nejati, H. Jaramillo, and M. Elimelech. 2016. In situ surface functionalization of reverse osmosis membranes with biocidal copper Nanoparticles. *Desalination* 388:1–8.

Brodie, B.C. 1859. On the atomic weight of graphite. *Philosophical Transactions of the Royal Society of London* 14:249–259.

Carpenter, A.W., C.F. De-lannoy, and M.R. Wiswener. 2015. Cellulose nanomaterials in water treatment technologies. *Environmental Science & Technology* 49:5277–5287.

Chou, W., Y. Da-Guang, and M. Yang. 2005. The preparation and characterization of silver-loading cellulose acetate hollow fiber membrane for water treatment. *Polymers for Advanced Technologies* 16(8):600–607.

Cotet, L.C., K. Magyari, M. Todea, M.C. Dudescu, V. Danciu, and L. Baia. 2017. Versatile self-assembled graphene oxide membranes obtained under ambient conditions by using a water-ethanol suspension. *Journal of Materials Chemistry* 5:2132–2142.

Cundy, C.S. and A.C. Paul. 2005. The hydrothermal synthesis of zeolites: Precursors, intermediates and reaction mechanism. *Microporous and Mesoporous Materials* 82:1–78.

Dahe, G.J., R.S. Teotia, and J.R. Bellare. 2012. The role of zeolite nanoparticles additive on morphology, mechanical properties and performance of polysulfone hollow fiber membranes. *Chemical Engineering Journal* 197:398–406.

Dolina, J., Z. Gončuková, M. Bobák, and L. Dvořák. 2018. Modification of a hollow-fibre polyethersulfone membrane using silver nanoparticles formed in situ for biofouling prevention. *RSC Advances* 8(26):14552–14560.

Efome, J.E., M. Baghbanzadeh, D. Rana, T. Matsuura, and C.Q. Lan. 2015. Effects of superhydrophobic SiO2 nanoparticles on the performance of PVDF flat sheet membranes for vacuum membrane distillation. *Desalination* 373:47–57.

El Badawi, N., A.R. Ramadan, A.M.K. Esawi, and M. El-Morsi. 2014. Novel carbon nanotube–cellulose acetate nanocomposite membranes for water filtration applications. *Desalination* 344:79–85.

El-Din, L.A.N., A. El-Gendi, N. Ismail, K.A. Abed, and A.I. Ahmed. 2015. Evaluation of cellulose acetate membrane with carbon nanotubes additives. *Journal of Industrial and Engineering Chemistry* 26:259–264.

Fahmi, M.Z., M. Wathoniyyah, M. Khasanah, Y. Rahardjo, and S. Wa. 2018. Incorporation of graphene oxide in polyethersulfone mixed matrix membranes to enhance hemodialysis membrane performance. *RSC Advances* 8(2):931–937.

Fan, Z., Z. Wang, M. Duan, J. Wang, and S. Wang. 2008. Preparation and characterization of polyaniline/polysulfone nanocomposite ultrafiltration membrane. *Journal of Membrane Science* 310:402–408.

Fathizadeh, M., A. Aroujaliana, and A. Raisi. 2011. Effect of added NaX nano-zeolite into polyamide as a top thin layer of membraneon water flux and salt rejection in a reverse osmosis process. *Journal of Membrane Science* 375:88–95.

Feng, C., K.C. Khulbe, T. Matsuura, R. Farnood, and A.F. Ismail. 2015. Review paper recent progress in zeolite/zeotype membranes. *Journal of Membrane Science and Research* 1:49–72.

García-Ivars, J., M.-J. Corbatón-Báguena, and I.-C. María-Isabel 2019. Development of mixed matrix membranes: Incorporation of metal nanoparticles in polymeric membranes.

Gethard, K., O. Sae-Khow, and S. Mitra. 2010. Water desalination using carbon-nanotube-enhanced membrane distillation. *ACS Applied Materials & Interfaces* 3(2):110–114.

Goetz, L.A., B. Jalvo, R. Rosal, and A.P. Mathew. 2016. Superhydrophilic anti-fouling electrospun cellulose acetate membranes coated with chitin nanocrystals for water filtration. *Journal of Membrane Science* 510:238–248.

Goh, K., L. Setiawan, L. Wei, S. Rongmei, A.G. Fane, R. Wang, and Y. Chen. 2015. Graphene oxide as effective selective barriers on a hollow fiber membrane for water treatment process. *Journal of Membrane Science* 474:244–253.

Gopakumar, D.A., V. Arumughan, D. Pasquini, S.-Y.B. Leu, A. Khalil H.P.S., and S. Thomas. 2019. Nanocellulose-Based Membranes for Water Purification. In *Micro and Nano Technologies*, Elsevier, pp. 59–85.

Gopakumar, D.A., D. Pasquini, M.A. Henrique, L.C. Morais, Y. Grohens, and S. Thomas. 2017. Meldrum's acid modified cellulose nanofiber-based polyvinylidene fluoride microfiltration membrane for dye water treatment and nanoparticle removal. *ACS Sustainable Chemistry & Engineering* 5(2):2026–2033.

Haase, M.F., H. Jeon, N. Hough, J.H. Kim, K.J. Stebe, and D. Lee. 2017. Multifunctional nanocomposite hollow fiber membranes by solvent transfer induced phase separation. *Nature Communications* 8(1):1234.

Hampden-Smith, M.J. and L.V. Interrante. 1998. *Chemistry of Advanced Materials: An Overview*, Wiley, New York.

Han, L.F., Z.-L. Xu, L.-Y. Yu, Y.-M. Wei, and Y. Cao. 2010. Performance of PVDF/multi-nanoparticles composite hollow fibre ultrafiltration membranes. *Iranian Polymer Journal* 19:553–565.

Hani, H.A., S.R. Tewfik, M.H. Sorour, and N.A. Monem. 2009. Prediction and verification of the conditions governing the synthesis of tailored zeolite a for heavy metals removal. *Eurasian Chemico-Technological Journal* 11:84–91.

Hoek, E.M.V., A.K. Ghosh, X. Huang, M. Liong, and J.I. Zink. 2011. Physical–chemical properties, separation performance, and fouling resistance of mixed-matrix ultrafiltration membranes. *Desalination* 283:89–99.

Huang, A. and B. Feng. 2018. Synthesis of novel graphene oxide-polyimide hollow fi ber membranes for seawater desalination. *Journal of Membrane Science* 548:59–65.

Hummers, W.S. and R.E. Offeman. 1958. Preparation of graphitic oxide. *Journal of the American Chemical Society* 80:1339.

Ihsanullah. 2019. Carbon nanotube membranes for water purification: Developments, challenges, and prospects for the future. *Separation and Purification Technology* 209:307–337.

Iida, H., K. Takayanagi, T. Nakanishi, and T. Osaka. 2007. Synthesis of Fe3O4 nanoparticles with various sizes and magnetic properties by controlled hydrolysis. *Journal of Colloid and Interface Science* 314(1):274–280.

Jiang, Y., P. Biswas, and J.D. Fortner. 2016. A review of recent developments in graphene-enabled membranes for water treatment. *Environmental Science: Water Research & Technology* 2(6):915–922.

Jin, L., S. Yu, W. Shi, X. Yi, N. Sun, Y. Ge, and C. Ma. 2012. Synthesis of a novel composite nanofiltration membrane incorporated SiO2 nanoparticles for oily waste-water desalination. *Polymer* 53(23):5295–5303.

Johnson, D.W., B.P. Dobson, and K.S. Coleman. 2015. A manufacturing perspective on graphene dispersions. *Current Opinion in Colloid and Interface Science* 20 (5–6):367–382.

Kajekar, A.J., B.M. Dodamani, A.M. Isloor, Z.A. Karim, and N.B. Cheer. 2015. Preparation and characterization of novel PSf/PVP/PANI-nanofiber nanocomposite hollow fiber ultrafiltration membranes and their possible applications for hazardous dye rejection. *Desalination* 365:117–125.

Karim, Z., A.P. Mathew, M. Grahn, J. Mouzon, and K. Oksman. 2014. Nanoporous membranes containing cellulose nanocrystals as functional additive in chitosan: Removal of positively charged dyes from water. *Carbohydrate Polymers* 112:668–676.

Kim, H.-J., H.-C. Yang, D.-Y. Chung, I.-H. Yang, Y.J. Choi, and J.-K. Moon. 2015. Functionalized mesoporous silica membranes for CO2 separation applications. *Journal of Chemistry* 2015: 1–9.

Kim, I., K. Lee, and T. Tak. 2001. Preparation and characterization of integrally skinned uncharged polyetherimide asymmetric nanofiltration membrane. *Journal of Membrane Science* 183(2):235–247.

Kim, S., S. Kwak, B. Sohn, and T. Park. 2003. Design of TiO2 nanoparticle self-assembled aromatic polyamide thin-film-composite (TFC) membrane as an approach to solve biofouling problem. *Journal of Membrane Science* 211(1):157–165.

Kumar, R., A.M. Isloor, A.F. Ismail, S.A. Rashid, and A. Al Ahmed. 2013. Permeation, antifouling and desalination performance of TiO2 nanotube incorporated PSf/CS blend membranes. *Desalination* 316:76–84.

Li, J., Z. Zhao, S. Yuan, J. Zhu, and B. Van der Bruggen. 2018. High-performance thin-film-nanocomposite cation exchange membranes containing hydrophobic zeolitic imidazolate framework for monovalent selectivity. *Applied Sciences* 8:759–773.

Li, S., G. Liao, Z. Liu, Y. Pan, Q. Wu, Y. Weng, X. Zhang, Z. Yang, and O.K.C. Tsui. 2014. Enhanced water flux in vertically aligned carbon nanotube arrays and polyethersulfone composite membranes. *Journal of Materials Chemistry A* 2 (31):12171–12176.

Li, X., R. Pang, L. Jiansheng, X. Sun, J. Shen, W. Han, and L. Wang. 2013. In situ formation of Ag nanoparticles in PVDF ultrafiltration membrane to mitigate organic and bacterial fouling. *Desalination* 324:48–56.

Liang, S., X. Kang, M. Yinghui, and H. Xia. 2012. A novel ZnO nanoparticle blended polyvinylidene fluoride membrane for anti-irreversible fouling. *Journal of Membrane Science* 394:184–192.

Liang, S., Y. Kang, A. Tiraferri, E.P. Giannelis, X. Huang, and M. Elimelech. 2013. Highly hydrophilic polyvinylidene fluoride (PVDF) ultrafiltration membranes via postfabrication grafting of surface-tailored silica nanoparticles. *ACS Applied Materials & Interfaces* 5(14):6694–6703.

Lin, J., R. Zhanga, W. Yea, N. Julloka, A. Sottoa, and B.V. Bruggen. 2013. Nano-WS2 embedded PES membrane with improved fouling and permselectivity. *Journal of Colloid and Interface Science* 396:120–128.

Lind, M.L., B. Jeong, A. Subramani, X. Huang, and E.M.V. Hoek. 2009. Effect of mobile cation on zeolite-polyamide thin film nanocomposite membranes. *Journal of Materials Research* 24:1624–1631.

Liu, Y., B. He, J. Li, R.D. Sanderson, L. Li, and S. Zhang. 2011. Formation and structural evolution of biphenyl polyamide thin film on hollow fiber membrane during interfacial polymerization. *Journal of Membrane Science* 373(1–2): 98–106.

Liu, P., P.F. Borrell, M. Božić, V. Kokol, K. Oksman, and A.P. Mathew. 2015. Nanocelluloses and their phosphorylated derivatives for selective adsorption of Ag+, Cu2+ and Fe3+ from industrial effluents. *Journal of Hazardous Materials*, 294: 177–185.

Ma, H., C. Burger, B.S. Hsiao, and B. Chu. 2012. Nanofibrous microfiltration membrane based on cellulose nanowhiskers. *Biomacromolecules* 13:180–186.

Ma, J., D. Ping, and X. Dong. 2017. Recent developments of graphene oxide-based membranes: A review. *Membranes* 7:3.

Maghami, M. and A. Abdelrasoul. 2018. Zeolite Mixed Matrix Membranes (Zeolite-mmms) for Sustainable Engineering. In *Zeolites and Their Applications*, IntechOpen, pp. 115–134.

Makhtar, S.N.N.M., M.Z.M. Pauzi, N.M. Mahpoz, et al. 2020. Preparation, characterization and performance evaluation of supported zeolite on porous glass hollow fiber for desalination application. *Arabian Journal of Chemistry* 13: 3429–3439.

Malinauskas, A. 2001. Chemical deposition of conducting polymers. *Polymer* 42:3957–3972.

Mariano, M., N.E. Kissi, and A. Dufresne. 2014. Cellulose nanocrystals and related nanocomposites: Review of some properties and challenges. *Journal of Polymer Science Part B: Polymer Physics* 52:791–806.

Missoum, K., M.N. Belgacem, and J. Bras. 2013. Nanofibrillated cellulose surface modification: A review. *Materials* 6:1745–1766.

Nair, S.S. and A.P. Mathew. 2017. Porous composite membranes based on cellulose acetate and cellulose nanocrystals via electrospinning and electrospraying. *Carbohydrate Polymers* 175:149–157.

Nalwa, H.S. 1997. *Handbook of Organic Conductive Molecules and Polymers*, Vols I–IV, John Wiley and Sons Ltd. ISBN-13: 978-0125139205.

Naseri, N., A.P. Mathew, L. Girandon, M. Fröhlich, and K. Oksman. 2015. Porous electrospun nanocomposite mats based on chitosan–cellulose nanocrystals for wound dressing: Effect of surface characteristics of nanocrystals. *Cellulose*, 22: 521.

Narayan, R., U. Nayak, A. Raichur, and S. Garg. 2018. Mesoporous silica nanoparticles: A comprehensive review on synthesis and recent advances. *Pharmaceutics* 10(3):118.

Navaladian, S., B. Viswanathan, R.P. Viswanath, and T.K. Varadarajan. 2007. Thermal decomposition as route for silver nanoparticles. *Nanoscale Research Letters* 2(1):44–48.

Ong, C.S., W.J. Lau, P.S. Goh, B.C. Ng, and A.F. Ismail. 2015. Preparation and characterization of PVDF–PVP–TiO2 composite hollow fiber membranes for oily wastewater treatment using submerged membrane system. *Desalination and Water Treatment* 53 (5):1213–1223.

Pandey, P. and M. Dahiya. 2016. Carbon nanotubes: Types, methods of preparation and applications. *International Journal of Pharmaceutical Science and Research* 1(4): 15–21.

Potts, J.R., D.R. Dreyer, C.W. Bielawski, and R.S. Ruoff. 2011. Graphene-based polymer nanocomposites. *Polymer* 52(1):5–25.

Qiu, S., W. Liguang, X. Pan, L. Zhang, H. Chen, and C. Gao. 2009. Preparation and properties of functionalized carbon nanotube/PSF blend ultrafiltration membranes. *Journal of Membrane Science* 342(1–2):165–172.

Rahimpour, A., M. Jahanshahi, S. Khalili, A. Mollahosseini, A. Zirepour, and B. Rajaeian. 2012. Novel functionalized carbon nanotubes for improving the surface properties and performance of polyethersulfone (PES) membrane. *Desalination* 286:99–107.

Rezaei, M., A. Dasht Arzhandi, F. Ismail, et al. 2016. Zeolite ZSM5-filled PVDF Hollow fiber mixed matrix membranes for efficient carbon dioxide removal via membrane contactor. *Industrial & Engineering Chemistry Research* 5549:12632–12643.

Saifuddin, N., A.Z. Raziah, and A.R. Junizah. 2012. Carbon nanotubes: A review on structure and their interaction with proteins. *Journal of Chemistry* 2013: 1–18.

Sears, K., L. Dumée, J. Schütz, M. She, C. Huynh, S. Hawkins, M. Duke, and S. Gray. 2010. Recent developments in carbon nanotube membranes for water purification and gas separation. *Materials* 3(1):127–149.

Sehaqui, H., U.P. de Larraya, P. Liu, N. Pfenninger, A.P. Mathew, T. Zimmermann, P. Tingaut. 2014. Enhancing adsorption of heavy metal ions onto biobased nanofibers from waste pulp residues for application in wastewater treatment. *Cellulose.* 21: 2831–2844.

Shi, H., L. Xue, A. Gao, and Q. Zhou. 2016. Dual layer hollow fiber PVDF ultra-filtration membranes containing Ag nano-particle loaded zeolite with longer term anti-bacterial capacity in salt water. *Water Science & Technology* 73:2159–2167.

Shukla, A.K., J. Alam, M. Alhoshan, L.A. Dass, and M.R. Muthumareeswaran. 2017. Development of a nanocomposite ultrafiltration membrane based on polyphenyl-sulfone blended with graphene oxide. *Nature Publishing Group, Scientific Reports* 7:41976.

Smith, E.D., K.D. Hendren, J. Haag, E.J. Foster, and S.M. Martin. 2019. Functionalized cellulose nanocrystal nanocomposite membranes with controlled interfacial transport for improved reverse osmosis performance. *Nanomaterials* 9(125):1–25.

Stejskal, J., I. Sapurina, J. Prokeš, and J. Zemek. 1999. In-situ polymerized polyaniline films. *Synthetic Metals* 105:195–202.

Taurozzi, J.S., H. Arul, V.Z. Bosak, et al. 2008. Effect of filler incorporation route on the properties of polysulfone–silver nanocomposite membranes of different porosities. *Journal of Membrane Science* 325(1):58–68.

Tijing, L.D., Y.C. Woo, W.-G. Shim, H. Tao, J.-S. Choi, S.-H. Kim, and H.K. Shon. 2016. Superhydrophobic nanofiber membrane containing carbon nanotubes for high-performance direct contact membrane distillation. *Journal of Membrane Science* 502:158–170.

Toroghi, M., A. Raisi, and A. Aroujalian. 2014. Preparation and characterization of polyethersulfone/silver nanocomposite ultrafiltration membrane for antibacterial applications. *Polymers for Advanced Technologies* 25(7):711–722.

Vidal-Vidal, J., J. Rivas, and M.A. López-Quintela. 2006. Synthesis of monodisperse maghemite nanoparticles by the microemulsion method. *Colloids and Surfaces A: Physicochemical and Engineering Aspects* 288(1–3):44–51.

Visanko, M., H. Liimatainen, J.A. Sirviö, A. Haapala, R. Sliz, J. Niinimäki, and O. Hormi. 2014. Porous thin film barrier layers from 2,3-dicarboxylic acid cellulose nanofibrils for membrane structures. *Carbohydrate Polymers* 102:584–589.

Wang, X., Y. Zhao, E. Tian, L. Jing, and Y. Ren. 2018. Graphene oxide-based polymeric membranes for water treatment. *Advanced Materials Interfaces* 5(15):1701427, 1–20.

Wang, Z., Y. Hairong, X. Jianfei, Z. Feifei, L. Feng, X. Yanzhi, and L. Yanhui. 2012. Novel GO-blended PVDF ultrafiltration membranes. *Desalination* 299:50–54.

Wu, M.-B., Y. Lv, H.-C. Yang, L.-F. Liu, X. Zhang, and Z.-K. Xu. 2016. Thin film composite membranes combining carbon nanotube intermediate layer and microfiltration support for high nanofiltration performances. *Journal of Membrane Science* 515:238–244.

Xia, S. and N. Muzi. 2014. Preparation of poly(vinylidene fluoride) membranes with graphene oxide addition for natural organic matter removal. *Journal of Membrane Science* 473:54–62.

Xu, J., L. Zhang, X. Gao, H. Bie, F. Yunpeng, and C. Gao. 2015. Constructing antimicrobial membrane surfaces with polycation–copper (II) complex assembly for efficient seawater softening treatment. *Journal of Membrane Science* 491:28–36.

Yan, L., S.L. Yu, B.X. Chai, and X. Shun. 2006. Effect of nano-sized Al2O3-particle addition on PVDF ultrafiltration membrane performance. *Journal of Membrane Science* 276(1–2):162–167.

Yin, J., E.-S. Kim, J. Yang, and B. Deng. 2012. Fabrication of a novel thin-film nanocomposite (TFN) membrane containing MCM-41 silica nanoparticles (NPs) for water purification. *Journal of Membrane Science* 423:238–246.

Yu, L.Y., H.M. Shen, and Z.L. Xu. 2009a. PVDF–TiO2 composite hollow fiber ultrafiltration membranes prepared by TiO2 sol–gel method and blending method. *Journal of Applied Polymer Science* 113(3):1763–1772.

Yu, L.-Y., Z.-L. Xu, H.-M. Shen, and H. Yang. 2009b. Preparation and characterization of PVDF–SiO2 composite hollow fiber UF membrane by sol–gel method. *Journal of Membrane Science* 337(1–2):257–265.

Zahid, M., A. Rashid, S. Akram, Z.A. Rehan, and W. Razzaq. 2018. A comprehensive review on polymeric nano-composite membranes for water treatment. *Journal of Membrane Science & Technology* 8(1): 1–20.

Zahri, K., K.C. Wong, P.S. Goh, and A.F. Ismail. 2016. Graphene oxide/polysulfone hollow fiber mixed matrix membranes for gas separation. *RSC Advances* 6(92): 89130–89139.

Zhang, A., Y. Zhang, G. Pan, X. Jian, H. Yan, and Y. Liu. 2017. In situ formation of copper nanoparticles in carboxylated chitosan layer: Preparation and characterization of surface modified TFC membrane with protein fouling resistance and long-lasting antibacterial properties. *Separation and Purification Technology* 176:164–172.

Zhang, L., G.-Z. Shi, S. Qiu, L.-H. Cheng, and H.-L. Chen. 2011. Preparation of high-flux thin film nanocomposite reverse osmosis membranes by incorporating functionalized multi-walled carbon nanotubes. *Desalination and Water Treatment* 34(1–3):19–24.

Zhang, Y., S. Zhang, J. Gao, and T.S. Chung. 2016. Layer-by-layer construction of graphene oxide (GO) framework composite membranes for highly efficient heavy metal removal. *Journal of Membrane Science* 515:230–237.

Zhou, C., Q. Wu, T. Lei, and I.I. Negulescu. 2014. Adsorption kinetic and equilibrium studies for methylene blue dye by partially hydrolyzed polyacrylamide/cellulose nanocrystal nanocomposite hydrogels. *Chemical Engineering Journal* 251:17–24.

Zinadini, S., A.A. Zinatizadeh, M. Rahimi, V. Vatanpour, and H. Zangeneh. 2014. Preparation of a novel antifouling mixed matrix PES membrane by embedding graphene oxide nanoplates. *Journal of Membrane Science* 453:292–301.

APPENDIX

Nomenclature

AA:A	Binary solvent system of Acetic acid (AA) and Acetone
Ag A	Silver Exchanged zeolite A
Al_2O_3	Aluminum Oxide
CA	Contact angle
CHNC	Chitin nano-crystals
CNTs	Carbon Nanotubes
CSA	(+)-10-champhor sulfonic acid
CuO	Cupric oxide
CVD	Chemical vapor deposition
DCM	A binary highly volatile solvent system of Dichloromethane (DCM) and Acetone (A)
DCMD	Direct contact membrane distillation
DMAC	dimethylacetamide
DMF	N, N-dimethyl formamide
FS	Flat Sheet
GO	Graphene oxide
HF	Hollow Fiber
HFM	hollow fiber membranes
LiCl	Lithium chloride
LMH	water flux unit (L m-2 h-1)
MCM-41	Mobil Crystalline Materials or Mobil Composition of Matter (Silica Nanoparticle type)
MD	membrane distillation
$MgCl_2$	Magnesium Chloride
$MgSO_4$	Magnesium Sulfate
MMM	mixed matrix membrane
MPD	m-phenyl diamine
MWCNT	Multi-walled carbon Nanotubes
MWCO	molecular weight cut-off
Na2SO4	Sodium Sulfate
NaCl	Sodium Chloride
$NaIO_4$	Sodium periodate
NF	Nano Filtration
NIPS	non-solvent induced phase separation
NMP	N-Methyl-2-Pyrrolidone
NPs	Nanoparticles
PA	Polyamide
PAE	Polyamide-amine- epichlorohydrin
PAI	poly(amideimide)
PAMAM	poly(amidoamine)
PANI	Polyaniline

(Continued)

Membrane Materials Design Trends

(Cont.)

Nomenclature

PEEK	poly ether ketone
PEG	Polyethylene glycols
PEI	polyethyleneimine
PEO	Poly(ethylene oxide)
PES	polyethersulfone
PET	Poly(ethylene terephthalate)
PI	Polyimide
PIP	Piperazine
PPSU	Polyphenyl sulfone
PS, PSU, PSF	Polysulfone
PVDF	Polyvinylidene fluoride
PVP	polyvinylpyrrolidone
R%	Salt rejection
Ra	Roughness
SDS	Sodium Dodecyl Sulfate
SiO_2	silica Nanoparticles
SiO_2	Silica Oxide
SLS	sodium lauryl sulfate
TEMPO	2,2,6,6-Tetramethylpiperidine-1-oxyl
TEA	Triethyl amine
TEOS	(tetraethyl orhosilicate)
TETA	Triethylene tetramine
TiO_2	Titanium oxide
TOCNs	Oxidized cellulose Nanocrystals
TMC	1,3,5-benzene tricarbonyl trichloride, Trimesoyl chloride
TPGS	D-a-tocopheryl polyethylene glycol 1000 succinate
WS2	Tungsten disulfide
ZnO	Zinc oxide

3 Functionalization of Carbon-Based Additives

Myrsini Kyriaki Antoniou, Andreas Sapalidis, and Zili Sideratou
Institute of Nanoscience and Nanotechnology
National Centre for Scientific Research Demokritos
Agia Paraskevi Attikis, Greece

3.1 INTRODUCTION

Nanocomposite materials have driven considerable attention on the part of many researchers for the development of membranes with enhanced antifouling and antibacterial properties for water treatment. In particular, polymeric nanocomposite membranes are widely used in desalination and can be fabricated by complete dispersion of nanoadditives, such as nanoparticles, nanosheets, nanotubes or nanofibers, into the polymer matrix through various synthetic techniques (e.g., phase inversion, Figure 3.1, Guillen et al. 2011, Esfahani et al. 2019). The ability to disperse the nanoadditives homogeneously throughout the matrix, and the compatibility between them and the matrix, are two import factors that can determine the performance of the final nanocomposite membrane.

The incorporation of these nanoadditives into polymers can tune the structure and the physicochemical properties of the membranes, such as porosity, hydrophilicity and chemical stability (Kim and Van der Bruggen 2010, Pendergast and Hoek 2011, Baghbanzadeh et al. 2015). Moreover, mechanical properties can be improved as a result of strong interfacial interactions between the nanoadditives and the polymer matrix (Rong et al. 2001). Nanoadditives that alter the porosity and pore size of the membranes can influence the water permeability and solute rejection of the nanocomposite membrane (Esfahani et al. 2019). Various nanoadditives, such as zeolites, ZnO, TiO_2, SiO_2 and metal – organic frameworks, have been used for the development of the polymer nanocomposite membranes (Baghbanzadeh et al. 2015). Nevertheless, carbon nanotube – based (Sianipar et al. 2017) and graphene-based (Hegab et al. 2015) nanomaterials have drawn particular attention for desalination and water purification applications due to their unique properties. Adding carbonaceous additives to the polymeric matrix not only improves the mechanical, chemical, and thermal properties of the membranes but also increases water purification performance/permeability through an enhanced antifouling mechanism (Esfahani et al. 2019). These carbon additives can be incorporated in the polymeric matrix with or without surface

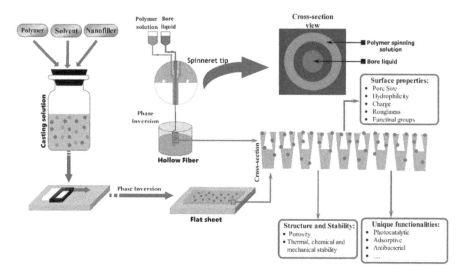

FIGURE 3.1 Fabrication procedures of the flat sheet and hollow fiber Mixed Matrix Nanocomposite Membranes (MMNMs) through the phase inversion process.
(Reproduced with permission of Esfahani et al. 2019, Elsevier Ltd.)

functionalization. Nonfunctionalized carbon nanomaterials have a strong tendency to aggregation (Punetha et al. 2017) due to strong intermolecular interactions, such as van der Waals forces, which can negatively affect their mechanical properties and render them incompatible with polymer matrices. Furthermore, permeation and antifouling performance of the nanocomposite membranes may be limited by this aggregation. For this reason, the functionalization of carbon additives has been considered one of the best approaches to achieving outstanding mechanical strength, and increased permeation of the resultant membrane through dispersions, when blended with polymer matrices. Furthermore, it is proved that the functional groups (such as HO, COOH και C-O-C) on the surface of the carbon additives increase the hydrophilicity of the nanocomposite membranes, resulting in higher fouling resistance because proteins, such as Bovine Serum Albumin (BSA), are hydrophobic in nature. (Hebbar et al. 2017, Punetha et al. 2017).

3.2 GRAPHENE-BASED ADDITIVES

Yu et al. (2013) reported the fabrication of nanocomposite ultrafiltration membranes with antifouling and antibacterial properties. In particular, Graphene Oxide (GO) nanosheets were first functionalized by hyperbranched polyethylenimine (HPEI) in order to improve the compatibility with the polymer, and were then blended into polyethersulfone (PES) solution through the phase inversion method (Figure 3.2).

Carbon-Based Additives

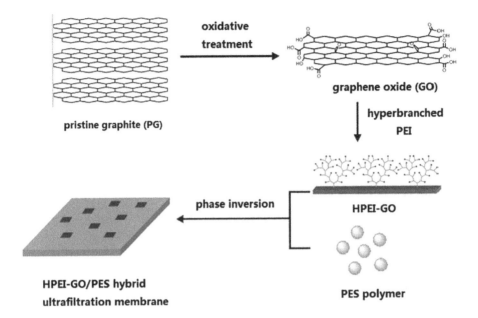

FIGURE 3.2 Preparation process of HPEI-GO/PES hybrid membrane: Step1, the preparation and modification of GO; Step 2, the preparation of HPEI-GO/PES hybrid membrane. **(Reproduced with permission of Yu et al. 2013, Elsevier Ltd.)**

Mechanical tests proved that HPEI-GO/PES membranes exhibited better tensile strength and Young's modulus, compared to pure PES membranes. Water contact angle measurements showed enhanced hydrophilicity of the membranes after the incorporation of HPEI-GO nanoparticles with a decrease from 85.5°C to 63.1°C, while a smoother surface was observed for the hybrid membranes by the SEM images. Furthermore, HPEI-GO/PES membranes presented lower protein adsorption amounts, effective antibacterial performance (Figure 3.3) against *Escherichia coli* (with a high bacteriostasis rate of 74.88%), and improved antifouling properties compared to pure PES membranes.

Xu et al. (2014) incorporated different ratios of GO and functionalized GO (with 3-aminopropyltriethoxysilane (APTS)), into polyvinylidene fluoride (PVDF) matrix via phase inversion induced by the immersion precipitation technique (Figure 3.4).

The produced hybrid ultrafiltration membranes, f-GO/PVDF, presented higher hydrophilicity, water flux, and rejection rates than pure PVDF membranes and GO/PVDF membranes. This significant improvement in membrane performance is attributed to the better dispersion of the functionalized GO in the PVDF matrix and to the strong interfacial interaction between them. The addition of GO in the polymeric matrix resulted in a 6.58% decrease in tensile strength compared with pure PVDF membranes, while tensile strength was

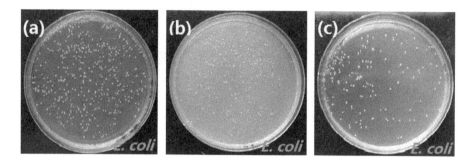

FIGURE 3.3 Measurement of antibacterial property of membranes by the bacteriostasis rate: (a) control membrane; (b) PES membrane with 3% of HPEI; and (c) PES membrane with 3% of HPEI-GO nanosheets.
(Reproduced with permission of Yu et al. 2013, Elsevier Ltd.)

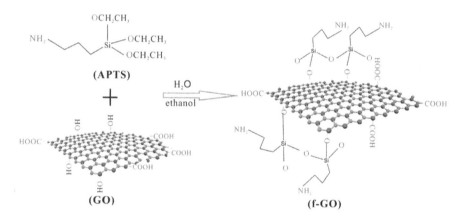

FIGURE 3.4 Illustration of the reaction between GO and APTS.
(Reproduced with permission of Xu et al. 2014, Elsevier Ltd.)

increased by 57.89% for the f-GO/PVDF membranes. These results indicated that the functionalization of GO with APTS played an important role in enhancing the mechanical strength of the ultrafiltration membranes. Antifouling measurements showed that the f-GO/PVDF membranes exhibit improved antifouling capability than GO/PVDF membranes, indicating an overall superior performance of the functionalized hybrid ultrafiltration membranes (Figure 3.5).

Zhao et al. (2013) studied the performance of hybrid ultrafiltration membranes after the insertion of isocyanate-treated GO (iGO) in polysulfone (PSF) matrix via the phase inversion method. The iGO exhibited good dispersion in organic solvent and excellent compatibility with the polymeric components. An increase

Carbon-Based Additives 71

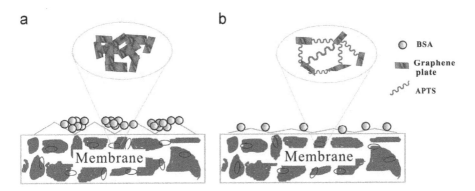

FIGURE 3.5 Schematic illustration of states of additives in (a) P/GO; and (b) P/f GO membranes and the relationship between surface morphology and fouling behavior.
(**Reproduced with permission of Xu et al. 2014, Elsevier Ltd.**)

in hydrophilicity, along with an improved surface smoothness (Figure 3.6) was observed after the addition of iGO to the PSF membrane, while the zeta potential became more negative with an increasing iGO content (Figure 3.7).

These results are connected with the improved antifouling properties that the iGO/PSF membranes exhibited, compared to the poor fouling resistance of the PSF membrane.

PVDF ultrafiltration membranes with incorporated GO and sulfonated GO (SGO) nanoparticles via a phase inversion process were synthesized by Ayyaru and Ahn (2017). Both nanocomposite membranes exhibited improved pore structure and higher surface roughness than the pure PVDF membrane. In particular, the sulfonic acid groups led to greater roughness (Figure 3.8).

Compared to the PVDF membranes, SGO/PVDF membranes showed 53.3% increase in the water permeation flux. The functionalization of GO resulted in a highly hydrophilic nanocomposite additive compared to GO, due to higher water uptake capacity of the sulfonic acid groups than the carboxyl groups of GO. This increase in hydrophilicity enhanced the antifouling performance of the SGO/PVDF membrane (Figure 3.9) with a low irreversible fouling of 11.2%, compared to 24% of the GO/PVDF membrane.

Partially reduced GO (rGO)/TiO$_2$ nanocomposites with different molar ratios were fabricated by Safarpour et al. (2015) by a hydrothermal method. The nanocomposites were then blended with PVDF matrix via phase inversion method to improve the hydrophilicity and antifouling performance of the hybrid ultrafiltration membranes. Because of the carbon based structure of GO, the rGO/TiO$_2$ additive is well dispersed in the PVDF matrix and as the EDX measurements proved, no agglomeration in the membranes was observed. The rGO/TiO$_2$/PVDF membranes showed increased hydrophilicity due to the various oxygen groups on the rGO/TiO$_2$, while a higher pure water

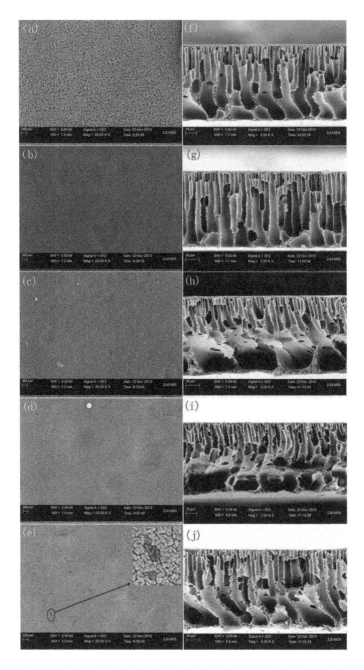

FIGURE 3.6 SEM photographs of plane (a–e) and cross-sections (f–j) of membranes with different iGO contents of 0.0, 0.025, 0.05, 0.10, and 0.15%.

Carbon-Based Additives

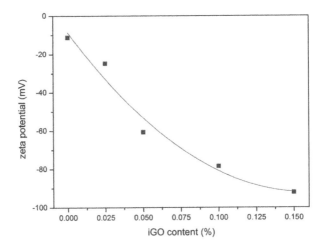

FIGURE 3.7 Zeta potential of prepared membrane with different iGO contents. **(Reproduced with permission of Zhao et al. 2013, RSC.)**

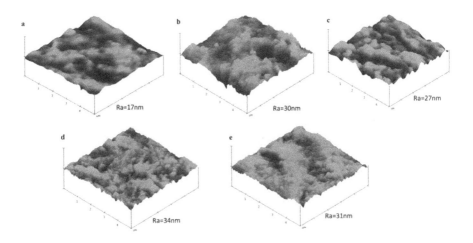

FIGURE 3.8 AFM three-dimensional surface images and surface roughness of the membranes: (a) PVDF; (b) P-GO 0.8; (c) P-SGO 0.4; (d) P-SGO 0.8; and (e) P-SGO 1.2. **(Reproduced with permission of Ayyaru and Ahn (2017), Elsevier Ltd.)**

flux and flux recovery ratio was observed compared to the pure PVDF membranes (Figure 3.10). Particularly, the hybrid membrane with GO to TiO_2 ratio of 70:30 showed improved permeability and antifouling performance (Figure 3.11).

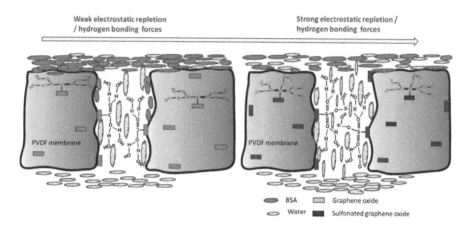

FIGURE 3.9 Schematic illustration of the antifouling mechanisms of P-GO membranes (left) and P-SGO membranes (right).
(Reproduced with permission of Ayyaru and Ahn (2017), Elsevier Ltd.)

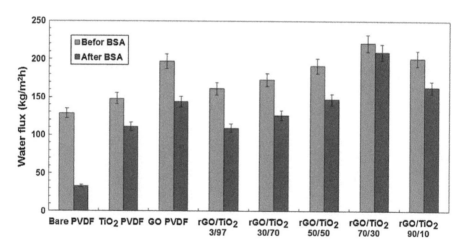

FIGURE 3.10 Water flux of the prepared PVDF membranes before and after BSA filtration (after 90 min at 0.3 MPa, additive concentration = 0.05 wt.%).
(Reproduced with permission of Safarpour et al. 2015, Elsevier Ltd.)

The same research group (Safarpour et al. 2016) has observed similar results for rGO/TiO$_2$/PES membranes. rGO/TiO$_2$ was uniformly dispersed in the polymer matrix while the nanocomposite membranes displayed improved water flux (45.0 kg/m^2 h) and antifouling properties, compared to the bare PES (Figure 3.12).

Carbon-Based Additives

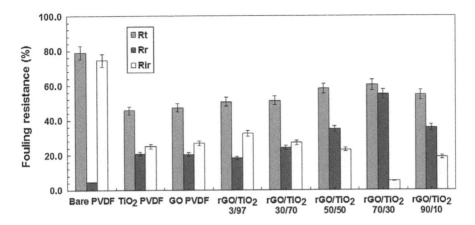

FIGURE 3.11 Fouling resistance ratio of the prepared PVDF membranes. [total fouling ratio (Rt), reversible fouling ratio (Rr) and irreversible fouling ratio (Rir)].
(Reproduced with permission of Safarpour et al. 2015, Elsevier Ltd.)

Finally, a great performance of 99.4% dye removal was achieved using 0.1 wt.% rGO/TiO2/PES membrane, while for the same dye, the bare PES membrane showed 93.2 % rejection performance (Figure 3.13).

Zambare et al. (2017) developed mixed matrix membranes by blending amine functionalized graphene oxide (fGO) in polysulfone. Three polyamines, ethylenediamine (EDA), diethylenetriamine (DETA), and triethylenetetramine (TETA), with different chain lengths were used in order to functionalize GO, and it was shown that by increasing the interlayer space of GO nanosheets with polyamines, the dispersion in polysulfone–NMP solution is improved. The presence of polyamines in the membranes resulted in the increase of hydrophilicity and the formation of finely porous structures, which enhanced the membrane's permeability. The antifouling properties of the membranes were tested by using BSA as a foulant. BSA rejection was greater than 90% for all fGO based polysulfone mixed matrix membranes, while a remarkably high pure water flux of 170.5 LMH/bar was displayed (Figure 3.14).

Nanocomposite membranes were prepared by Wu et al. (2014) by doping SiO_2-GO nanohybrids with different wt% in polysulfone (PSf). Because of the silica nanoparticles, which increased the interlayer space of GO, the SiO_2-GO nanohybrid showed a good dispersion and compatibility with the PSf matrix. The presence of SiO_2-GO positively influenced the hydrophilicity (Figure 3.15), the water permeation rate, the protein rejection, and the antifouling ability of the SiO_2–GO/PSf membranes.

Particularly when the content of SiO_2–GO in the membrane was 0.3 wt%, the water flux reached ~375 $L/m^2 h$, which is nearly twice of that of the pure PSf membrane (Figure 3.16).

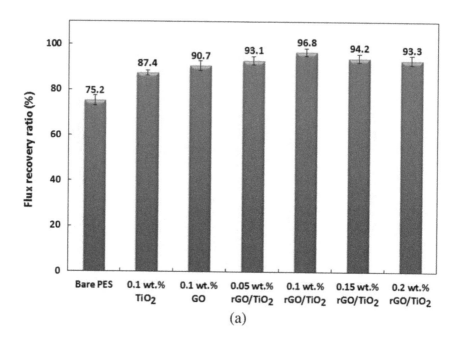

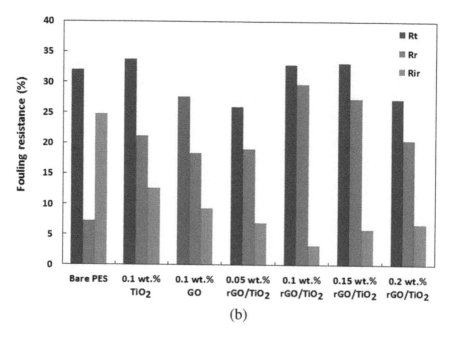

FIGURE 3.12 (a) Flux recovery ratios (average of four replications); and (b) fouling resistance ratio of the prepared membranes.

(Reproduced with permission of Safarpour et al. 2016, Elsevier Ltd.)

Carbon-Based Additives

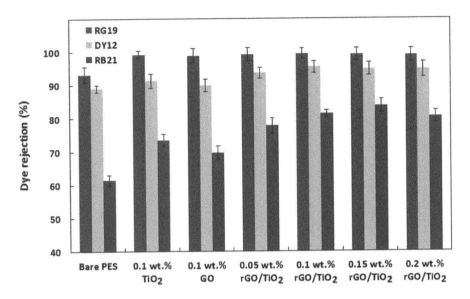

FIGURE 3.13 Dye rejection performance of the prepared nanofiltration membranes (after 60 min at 0.5 MPa, pH= 7.0 ± 0.1, and 100 mg/L dye).
(Reproduced with permission of Safarpour et al. 2016, Elsevier Ltd.)

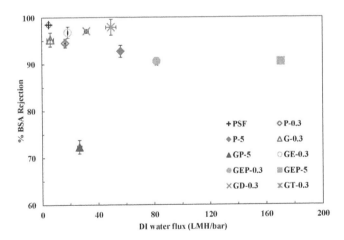

FIGURE 3.14 Plot of % BSA rejection versus water permeability of the membranes.
(Reproduced with permission of Zambare et al. 2017, Elsevier Ltd.)

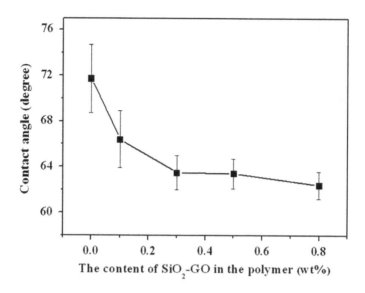

FIGURE 3.15 Contact angles of SiO$_2$–GO/PSf hybrid membranes against water as a function of SiO$_2$–GO content in the polymer.
(Reproduced with permission of Wu et al. 2014, Elsevier Ltd.)

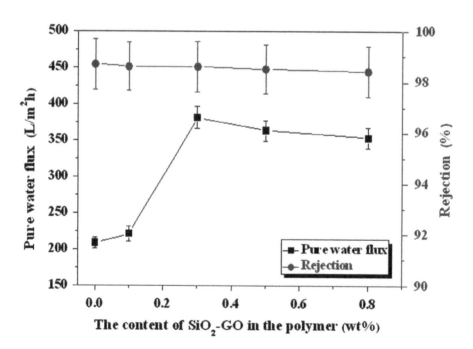

FIGURE 3.16 Effect of SiO$_2$–GO content on the pure water flux and rejection to egg albumin of SiO$_2$–GO/PSf hybrid membrane at 0.2MPa operation pressure.
(Reproduced with permission of Wu et al. 2014, Elsevier Ltd.)

3.3 CARBON NANOTUBE – BASED ADDITIVES

Rahimpour et al. (2012) introduced amine functional groups on Multi Walled Carbon NanoTubes (MWCNTs) in order to fabricate nanocomposite PES membranes. The MWCTNs were treated by strong acids and 1,3- phenylenediamine (mPDA) amine, and the functional groups were formed on the surface of MWCNTs (Figure 3.17).

Functionalized MWCNTs with various concentrations were blended in the polymer matrix to form the ultrafiltration nanocomposite membranes via phase inversion induced by immersion precipitation. The hydrophilicity of the PES/F-MWCNT membranes enhanced significantly with addition of F-MWCNTs in the polymer matrix while the porosity, pore size, surface roughness, and pure water flux showed an increment by addition of F-MWCNTs up to 1 wt.%. Higher concentrations of F-MWCNTs resulted in a less porous membrane and lower performance. Finally, the BSA rejection of the PES/F-MWCNT was improved by increasing the concentration of F-MWCNTs, providing good antifouling properties to the nanocomposite membranes (Figure 3.18).

Majeed et al. (2012) prepared ultrafiltration membranes by blending three different concentrations of hydroxyl functionalised MWCNTs (0.5, 1 and 2 wt%) with polyacrylonitrile (PAN) via phase inversion technique. A good dispersion of the functionalized MWCNTs and good interaction with PAN was observed. The mechanical stability of the nanocomposite membranes was enhanced due to the

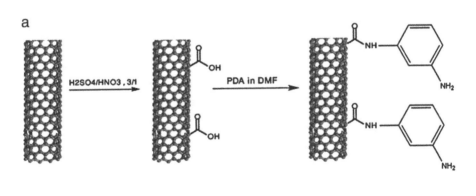

FIGURE 3.17 Procedure for (a) carboxylic and amine functionalization of MWCNTs; and (b) mechanism of amine functionalization of MWCNTs.
(Reproduced with permission of Rahimpour et al. 2012, Elsevier Ltd.)

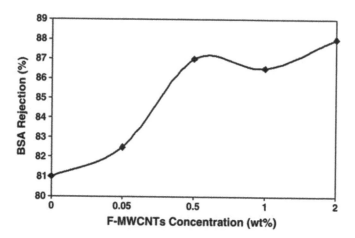

FIGURE 3.18 BSA protein rejection of different membranes.
(Reproduced with permission of Rahimpour et al. 2012, Elsevier Ltd.)

reinforcement properties of the MWCNTs, in particular, their tensile strength at 2 wt% functionalized MWCNTs loading increased over 97%, compared to the pure PAN membranes (Figure 3.19).

An increment of the hydrophilicity was noted after the addition of the functionalized MWCNTs, and this influenced the water flux of the membranes, which increased by 63% at 0.5 wt% loading of functionalized MWCNTs, compared to

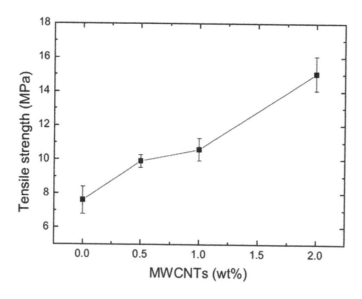

FIGURE 3.19 Tensile strength at break as a function of MWCNTs loading (wt%).
(Reproduced with permission of Majeed et al. 2012, Elsevier Ltd.)

Carbon-Based Additives

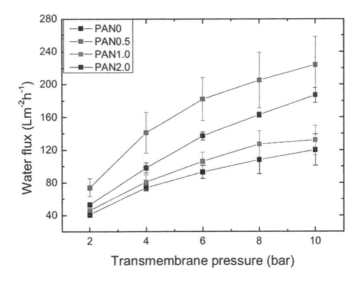

FIGURE 3.20 Water flux of PAN nanocomposite membranes as function of transmembrane pressure _P.
(Reproduced with permission of Majeed et al. 2012, Elsevier Ltd.)

the PAN membranes. Further increase in the concentration decreased the water flux, but it still remained higher, compared to the PAN membranes (Figure 3.20).

The dispersibility of different concentrations of hyperbranched poly (amine-ester) functionalized multiwalled carbon nanotubes (MWNTHPAE) in PVDF matrix was studied by Zhao et al. (2012). Furthermore the group investigated the hydrophilicity, permeability and antifouling performance of the PVDF/MWNTHPAE nanocomposite membranes. MWNTHPAE showed good and stable dispersion in the polymer matrix and no agglomeration was observed (Figure 3.21).

The hydrophilicity of the membranes increased due to the presence of the hydrophilic hyperbranched poly (amine-ester) (HPAE) groups, while high water transport was obtained by the dual effect of hydrophilic MWNTHPAE and the pore structure of the membranes. Furthermore, less protein adsorption was achieved by increasing MWNTHPAE content (Figure 3.22), thanks to the hydrogen bonding interactions between the hydrophilic groups and the water molecules.

Vatanpour et al. (2014) functionalized MWCNTs with amino groups to increase their dispersion and stability in PES matrix. Different additive contents were used for the preparation of NH_2-MWCNTs/PES nanocomposite nanofiltration membranes to improve their performance. With increase of the NH_2-MWCNTs content, the hydrophilicity and the pure water flux (from 13.6 to 23.7 L/m^2 h) of the nanocomposite membranes increased as well. In addition, greater salt rejection was achieved with the increase of NH_2-MWCNTs content at pH=7 (Figure 3.23), while an increase in the pH of the solution improved the rejection of Na_2SO_4 even more (Figure 3.24).

FIGURE 3.21 The dispersion of MWNTs and MWNTHPAE in DMF and casting solution: (a) The dispersion of MWNTs in DMF; (b) The dispersion of MWNTHPAE in DMF; (c, e) The dispersion of MWNTHPAE in casting solution; and (d, f) The dispersion of MWNTs in casting solution. All samples were prepared by gentle sonication for 30 min and then kept for several weeks before photographing.

(Reproduced with permission of Zhao et al. 2012, of Elsevier Ltd.)

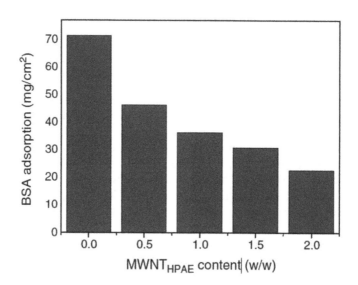

FIGURE 3.22 The amount of BSA adsorbed on PVDF/MWNTHPAE nanocomposite membranes.

(Reproduced with permission of Zhao et al. 2012, Elsevier Ltd.)

Carbon-Based Additives

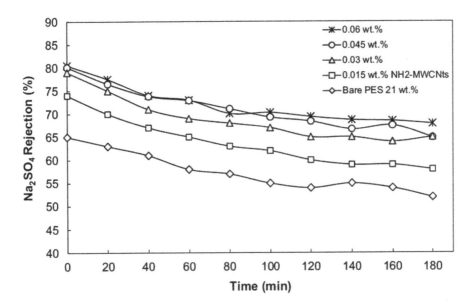

FIGURE 3.23 Na$_2$SO$_4$ retention of amine functionalized MWCNT – embedded membranes versus time (4 bar, pH¼7.070.1, 200 ppm Na$_2$SO$_4$).
(Reproduced with permission of Vatanpour et al. 2014, Elsevier Ltd.)

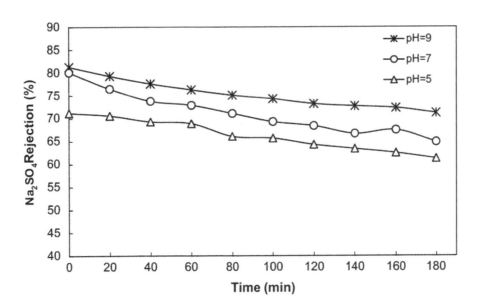

FIGURE 3.24 Effect of pH on the Na$_2$SO$_4$ rejection of 0.045 wt% NH$_2$-MWCNTs – embedded PES nanofiltration membrane.
(Reproduced with permission of Vatanpour et al. 2014, Elsevier Ltd.)

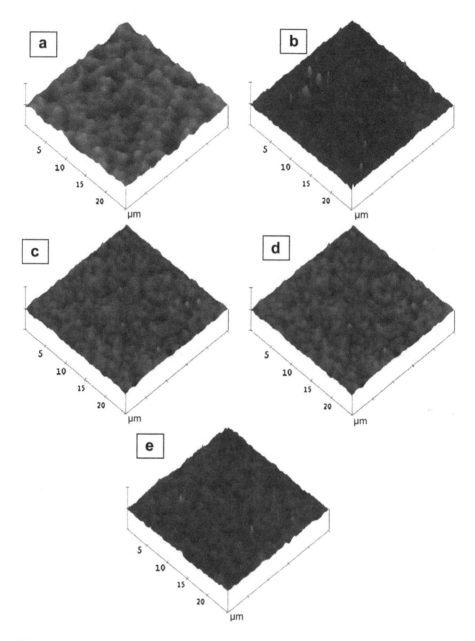

FIGURE 3.25 AFM images of blend membranes with different N-CNT loadings: (a) pristine PES; (b) 0.02%; (c) 0.04%; (d) 0.08%; and (e) 0.5%.

Carbon-Based Additives

Finally, the antifouling measurements showed that antifouling properties of the nanocomposite membranes are associated with the hydrophilicity, surface charge, and surface roughness, and that the fouling recovery ratio improved by the addition of amine-functionalized MWCTNs.

Phao et al. (2013) treated the surface of CNTs with HNO_3, obtaining nitrogen-doped CNTs (N-CNTs), which were then blended to a PES matrix for the development of nanocomposite membranes for water treatment. Through characterization, a good dispersion and superior compatibility of the N-CNTs with the PES matrix was observed compared to undoped CNTs, due to hydrogen bonding interactions. Low roughness values were recorded with an increase in N-CNT loading (Figure 3.25), resulting in more hydrophilic membrane surface.

After the addition of N-CNTs, the mechanical properties of the N-CNT/PES membranes improved while the water flux of the nanocomposite membrane increased up to 70%. The N-CNT/PES membranes showed an increase in rejecting polyethyleneglycol (PEG) from water, from 65% (PES membranes) to 77% (Figure 3.26).

Different contents (0.05 to 1.0% w/w) of quaternized hyperbranched polyethyleneimine (PEI) functionalized multiwalled carbon nanotubes (oxCNTs@QPEI) were dispersed in PVA matrix by Sapalidis et al. (2018), in order to obtain nanocomposite films with antibacterial properties via a solvent casting technique. The presence of functionalized hyperbranched dendritic polymer increased the stability (Figure 3.27) and hydrophilicity of the CNTs and consequently their compatibility with the PVA matrix.

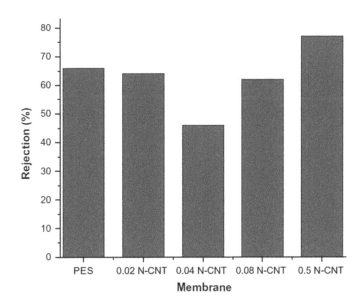

FIGURE 3.26 Rejection of a 500 ppm PEG in aqueous solution using the N-CNT/PES blend membranes.
(Reproduced with permission of Phao et al. 2013, Elsevier Ltd.)

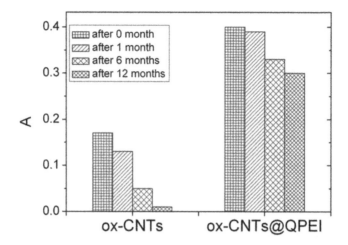

FIGURE 3.27 Sedimentation behavior of oxidized carbon nanotubes (oxCNTs) and oxCNTs@QPEI at different aging time.
(Reproduced with permission of Sapalidis et al. 2018, of Frontiers in Materials.)

Depending on the ox-CNTs@QPEI loading, the nanocomposite membranes exhibited improved mechanical properties and high antibacterial behavior against Gram-negative bacteria (Figure 3.28).

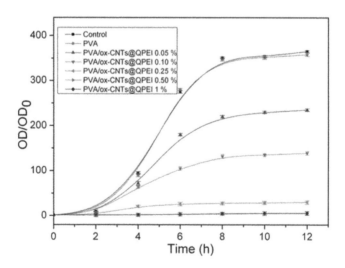

FIGURE 3.28 Growth curves of Gram negative *E. coli* bacteria incubated in the presence of PVA/oxCNTs@QPEI nanocomposites. The control was bacteria treated with distilled water.
(Reproduced with permission of Sapalidis et al. 2018, Frontiers in Materials.)

Carbon-Based Additives

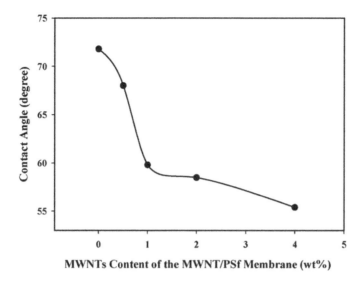

FIGURE 3.29 Contact angles of the surface of the MWNTs/PSf blend membranes against water as a function of the contents of MWNTs used.
(Reproduced with permission of Choi et al. 2006, Elsevier Ltd.)

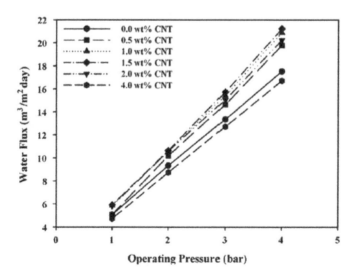

FIGURE 3.30 Pure water flux of the MWNTs/PSf blend membranes as a function of operating pressure.
(Reproduced with permission of Choi et al. 2006, Elsevier Ltd.)

Choi et al. (2006) treated MWCNTs with sulfuric and nitric acids, dispersed them in NMP solvent, and finally blended them in various concentrations with a PSf matrix. The functionalization with strong acids made the MWCNTs easily dispersed and stable in polar organic solvents such as NMP, so they can be compatible and well blended with the polymer matrix. The functional groups developed on the surface of MWCNTs increased the hydrophilicity (Figure 3.29) and conductivity of the MWCNTs/PSf membranes.

The MWCNTs loading in the nanocomposite membranes affected their properties and performance, such as the pore size, flux, and solute rejection, compared to the pure PSf membrane (Figure 3.30).

REFERENCES

Sivasankaran Ayyaru, Young-Ho Ahn. (2017). Application of sulfonic acid group functionalized graphene oxide to improve hydrophilicity, permeability, and antifouling of PVDF nanocomposite ultrafiltration membranes. *Journal of Membrane Science* 525:210–219.

Mohammadali Baghbanzadeh, Dipak Rana, Christopher Q. Lan, Takeshi Matsuura. (2015). *Effects of inorganic nano-additives on properties and performance of polymeric membranes in water treatment.* Separation and Purification Reviews, Taylor & Francis, Abingdon, 141–167.

Jae-Hyun Choi, Jonggeon Jegal, Woo-Nyon Kim. (2006). Fabrication and characterization of multiwalled carbon nanotubes/polymer blend membranes. *Journal of Membrane Science* 284:406–415.

Milad Rabbani Esfahani, Sadegh Aghapour Aktij, Zoheir Dabaghian, Mostafa Dadashi Firouzjaei, Ahmad Rahimpour, Joyner Eke, Isabel C. Escobar, Mojtaba Abolhassani, Lauren F. Greenlee, Amirsalar R. Esfahani, Anwar Sadmani, Negin Koutahzadeh. (2019). Nanocomposite membranes for water separation and purification: Fabrication, modification, and applications. *Separation and Purification Technology* 213:465–499.

Gregory R. Guillen, Yinjin Pan, Minghua Li, Eric M. V. Hoek. (2011). Preparation and characterization of membranes formed by nonsolvent induced phase separation: A review. *Industrial & Engineering Chemistry Research* 50:3798–3817.

Raghavendra S. Hebbar, Arun M. Isloor, Inamuddin, Abdullah M. Asiri. (2017). Carbon nanotube- and graphene-based advanced membrane materials for desalination. *Environmental Chemistry Letters* 15:643–671.

Hanaa M. Hegab, Linda Zou. (2015). Graphene oxide-assisted membranes: Fabrication and potential applications in desalination and water purification. *Journal of Membrane Science* 484:95–106.

Jeonghwan Kim, Bart Van der Bruggen. (2010). The use of nanoparticles in polymeric and ceramic membrane structures: Review of manufacturing procedures and performance improvement for water treatment. *Environmental Pollution* 158:2335–2349.

Shahid Majeed, Daniel Fierro, Kristian Buhr, Jan Wind, Du Bing, Adriana Boschetti-de-fierro, Volker Abetz. (2012). Multiwalled carbon nanotubes (MWCNTs) mixed polyacrylonitrile (PAN) ultrafiltration membranes. *Journal of Membrane Science* 403–404:101–109.

MaryTheresa M. Pendergast, Eric M.V. Hoek. (2011). A review of water treatment membrane nanotechnologies. *Energy & Environmental Science* 4:1946–1971.

Neo Phao, Edward N. Nxumalo, Bhekie B. Mamba, Sabelo D. Mhlanga. (2013). A nitrogen-doped carbon nanotube enhanced polyethersulfone membrane system for water treatment. *Physics and Chemistry of the Earth* 66:148–156.

Vinay Deep Punetha, Sravendra Rana, Hye Jin Yoo, Alok Chaurasia, James T. McLeskey, Jr., Madeshwaran Sekkarapatti Ramasamy, Nanda Gopal Sahoo, Jae Whan Cho. (2017). Functionalization of carbon nanomaterials for advanced polymer nanocomposites: A comparison study between CNT and graphene. *Progress in Polymer Science* 67:1–47.

Ahmad Rahimpour, Mohsen Jahanshahi, Soodabeh Khalili, Arash Mollahosseini, Alireza Zirepour, Babak Rajaeian. (2012). Novel functionalized carbon nanotubes for improving the surface properties and performance of polyethersulfone (PES) membrane. *Desalination* 286:99–107.

Min Zhi Rong, Ming Qiu Zhang, Yong Xiang Zheng, Han Min Zeng, R. Walter, K. Friedrich. (2001). Structure–property relationships of irradiation grafted nano-inorganic particle filled polypropylene composites. *Polymer* 42:167–183.

Mahdie Safarpour, Alireza Khataee, Vahid Vatanpour. (2015). Effect of reduced graphene oxide/TiO_2 nanocomposite with different molar ratios on the performance of PVDF ultrafiltration membranes. *Separation and Purification Technology* 140:32–42.

Mahdie Safarpour, Alireza Khataee, Vahid Vatanpour. (2016). Preparation and characterization of graphene oxide/TiO_2 blended PES nanofiltration membrane with improved antifouling and separation performance. *Desalination* 393:65–78.

Andreas Sapalidis, Zili Sideratou, Katerina N. Panagiotaki, Elias Sakellis, Evangelos P. Kouvelos, Sergios Papageorgiou, Fotios Katsaros. (2018). Fabrication of antibacterial poly(vinyl alcohol) nanocomposite films containing dendritic polymer functionalized multiwalled carbon nanotubes. *Frontiers of Materials* 5:1–10.

Merry Sianipar, Seung Hyun Kim, Khoiruddin, Ferry Iskandar, I Gede Wenten. (2017). Functionalized carbon nanotube (CNT) membrane: Progress and challenges. *RSC Advances* 7:51175–51198.

Vahid Vatanpour, Majid Esmaeili, Mohammad Hossein Davood Abadi Farahani. (2014). Fouling reduction and retention increment of polyethersulfone nanofiltration membranes embedded by amine-functionalized multi-walled carbon nanotubes. *Journal of Membrane Science* 466:70–81.

Huiqing Wu, Beibei Tang, Peiyi Wu. (2014). Development of novel SiO_2–GO nanohybrid/polysulfone membrane with enhanced performance. *Journal of Membrane Science* 451:94–102.

Zhiwei Xu, Jiguo Zhang, Mingjing Shan, Yinglin Li, Baodong Li, Jiarong Niu, Baoming Zhou, Xiaoming Qian. (2014). Organosilane-functionalized graphene oxide for enhanced antifouling and mechanical properties of polyvinylidene fluoride ultrafiltration membranes. *Journal of Membrane Science* 458:1–13.

Liang Yu, Yatao Zhang, Bing Zhang, Jindun Liu, Haoqin Zhang, Chunhua Song. (2013). Preparation and characterization of HPEI-GO/PES ultrafiltration membrane with antifouling and antibacterial properties. *Journal of Membrane Science* 447:452–462.

Rahul S. Zambare, Kiran B. Dhopte, Anand V. Patwardhan, Parag R. Nemade. (2017). Polyamine functionalized graphene oxide polysulfone mixed matrix membranes with improved hydrophilicity and antifouling properties. *Desalination* 403:24–35.

Haiyang Zhao, Wu Liguang, Zhijun Zhou, Lin Zhang, Huanlin Chen. (2013). Improving the antifouling property of polysulfone ultrafiltration membrane by incorporation of isocyanate-treated graphene oxide. *Physical Chemistry Chemical Physics* 15:9084–9092.

Xiaoyu Zhao, Jun Ma, Zhenghui Wang, Gang Wen, Jin Jiang, Fengmei Shi, Lianxi Sheng. (2012). Hyperbranched-polymer functionalized multi-walled carbon nanotubes for poly (vinylidene fluoride) membranes: From dispersion to blended fouling-control membrane. *Desalination* 303:29–38.

4 Nanohybrid Graphene-Based Materials for Advanced Wastewater Treatment
Adsorption and Membrane Technology

George Z. Kyzas and Athanasios C. Mitropoulos
International Hellenic University
Department of Chemistry
Kavala, Greece

Evangelos P. Favvas
National Centre for Scientific Research Demokritos
Agia Paraskevi Attikis, Greece

4.1 INTRODUCTION

Graphene is considered to be one of the most *hot* materials of recent years because of its numerous unique properties (mechanical, optical, environmental, etc). In 1986, Boehm et al. [1] described in detail the structure of graphite having a single atomic sheet [1]. The major turn in graphene history occurred from 2000–2010, when it was proved that 2-D crystals (such as graphene) had no thermodynamic stability, suggesting their nonexistence at room temperature [2]. In particular, graphene sheet is thermodynamically unstable if its size is less than about 20 nm (graphene is the least stable structure if lower than about 6,000 atoms); it becomes the most stable fullerene (as within graphite) only for molecules larger than 24,000 atoms [3].

The leader of graphene science is Konstantin Novoselov who successfully isolated and characterized an exfoliated graphene monolayer with various techniques [4]; A.K. Geim and K.S. Novoselov were awarded the Nobel Prize in 2010 for their impact on graphene science. But what is graphene? The reply was clear and was given by IUPAC: graphene is a carbon layer (single) of graphite, having a structure/nature similar or analogous to an aromatic hydrocarbon (polycyclic) of quasi-infinite size [5]. This means that graphene is a flat monolayer of hybridized sp^2 atoms of carbon, which are densely packed together into an ordered two-dimensional honeycomb network [6]. An hexagonal unit cell of graphene comprises two equivalent sublattices of carbon atoms, joined together by sigma (σ) bonds with a carbon – carbon bond length of 0.142 nm [7]. Each carbon atom in the lattice has a π-orbital that contributes to a delocalized network of electrons, making graphene sufficiently stable, compared to other nanosystems [8]. The applicability of graphene is based on an advantageous network provided by this material: combination of high three-dimensional aspect ratio and large specific surface area, superior mechanical stiffness and flexibility, remarkable optical transmittance, exceptionally high electronic and thermal conductivities, and impermeability to gases, as well as many other supreme properties. Due to all of these characteristics, Novoselov characterized graphene as a miracle material [9].

In some of the most important applications of graphene in wastewater treatment, the oxidized form of graphene (graphene oxide) is used. The large scale production of functionalized graphene at low cost should result in good adsorbents for water purification [10], due to the two-dimensional layer structure, large surface area, pore volume, and presence of surface functional groups in these materials; the inorganic nanoparticles also prevent the adsorbent aggregation. Water, as it is known (i.e., from several good handbooks), can be treated and purified by multiple techniques such as desalination, filtration, membranes, flotation, adsorption, disinfection, and sedimentation. Certainly, adsorption holds advantages over other methods (various methods are discussed in the following sections), such as ease of operation and comparatively low cost. Adsorption is the surface phenomenon in which pollutants are adsorbed on the surface of a material (adsorbent) via physical and/or chemical forces. It depends on many factors such as temperature, solution pH, concentration of pollutants, contact time, particle size, temperature, and nature of the adsorbate. In addition, apart from its use in adsorption, graphene oxide can be used as a supplement in membrane technology, especially in nanofiltration. Therefore, in this chapter the use of graphene oxide is analyzed for wastewater treatment and many examples for membranes (antifouling properties) and adsorbents are given.

4.2 SYNTHESIS PROCEDURES

Before further analysis, it is mandatory to give a definition about graphene composites. Graphene composites are considered to include all graphene-based materials that have been modified (e.g., grafting with reactive groups, functionalizations with polymers, complexes with other sources).

Graphene oxide (GO), which is considered to be the most known graphene composite material, results from the chemical exfoliation of graphite. It is a highly oxidized form of graphene, consisting of numerous and varying types of oxygen functionalities. Many theories have been developed in the past for the determination of the exact chemical structure of GO [11, 12]. This is mainly because of the complexity of the material (including sample-to-sample variability), and, of course, its amorphous, berthollide character i.e. nonstoichiometric atomic composition [13]. The Lerf-Klinowski model describes a theory according to which, the carbon plane in GO is decorated with hydroxyl and epoxy(1,2-ether) functional groups [14]. The consideration for the existence of some carbonyl groups is correct, most likely as carboxylic acids along the sheet edges, but also as organic carbonyl defects within the sheet [15, 16]. The synthesis of GO is based on three preparation methods: (i) Brodie's method [17], (ii) Staudenmaier's method [18], or Hummers' method [19].The major part of all methods is the chemical exfoliation of graphite using oxidizing agent in the presence of mineral acid. Two methods (Brodie's method and Staudenmaier's method) apply a combination of $KClO_4$ with HNO_3 to oxidize graphite. Hummer's method adds graphite to potassium permanganate and H_2SO_4. The oxidation of graphite breaks up the π-conjugation of the stacked graphene sheets into nanoscale graphitic sp2 domains surrounded by highly disordered oxidized domains (sp^3 C\C) as well as defects of carbon vacancies [20]. The GO sheets produced consist of phenol, hydroxyl and epoxy groups, mainly at the basal plane, and carboxylic acid groups at the edges [21]; they can thus readily exfoliate to form a stable, light- brown – colored, single-layer suspension in water [20].

Next, some major synthesis procedures are analyzed in detail to describe the nanohybrid graphene oxide preparation.

4.2.1 Synthesis of GO

A very crucial step in preparing an efficient graphene-based *material*, either for modification of membranes or for simple use as adsorbent, is to select the correct synthesis route. Some slightly different ways of modified graphenes are described next.

4.2.1.1 Nanohybrid GO

A classic synthesis way was described by Suresh Kumar et al. [22]. This team used graphite for the synthesis of GO. Specifically, 3 g of graphite flakes were dissolved in a mixture of H_2SO_4/H_3PO_4 (360/40 mL), and then 18 g of $KMnO_4$ were added slowly under continuous stirring for 12 h at 50°C. After this process, in order to exfoliate GO into single layers, the mixture was cooled down to room temperature and it was mixed with a solution ~400 mL ice water with concentration 3 mL of 30% hydrogen peroxide followed by sonication for 0.5 h. In addition, the diluted mixture was centrifuged at 10,000 rpm for 15 min. Multiple washings (30% HCl) were made with water to remove the residues from the solid particles. Then, the experimental process was followed by a vacuum drying process at room temperature for 12 h [22].

The next step was the preparation of Mn Fe_2O_4 nanoparticles. Briefly, 100 mL of deionized water were used to dilute 0.845 g of $MnSO_4$ H_2O and 2.7 g of $FeCl_3·6H_2O$ (the molar ratio of Mn:Fe in the mixture was 1:2). The mixture was then continuously stirred, and heated at temperature 80°C.

To slightly modify the pH of the mixture to 10.5, drops of 8 M NaOH were inserted slowly to the same temperature for 5 min and then the mixture was cooled down to room temperature. To separate the blackish precipitates, a magnetic process was used. The obtained blackish precipitates were then washed with excess of water to remove the unreacted quantity, and were then washed with propanol. Finally, the obtained blackish precipitates were dried at room temperature for 24 h [22].

The final step was the synthesis of Go-MnFe$_2$O$_4$ nanohybrids. Therefore, in the final experimental process, 0.5 g of GO were inserted to 400 mL of water and were then dispersed by ultrasonication for 5 min. 0.845 g of MnSO$_4$ H$_2$O and 2.7 g FeCl$_3$·6H$_2$O were added to the colloidal GO mixture and after this, this solution was stirred for 30 min (until the increase of the mixture temperature to 80°C). Similarly, to increase the pH of the solution, 8 M NaOH was added dropwise and heated up to the same temperature with the mixture. The experimental reaction process was then continued for 5 min and then the yielded mixture cooled down to room temperature. A magnetic process was then used to separate the nanohybrid particles. The nanohybrid particles were washed with excess of H$_2$O and propanol and dried for 24 h at room temperature [22].

4.2.1.2 Preparation of GO and Reduced GO (rGO) Mixture

Lee et al. [23] synthesized neat GO from natural graphite according to modified Hummers method. Potassium permanganate, nitric acid, sulfuric acid, and hydrogen peroxide were used as reagents to oxidize graphite to GO; 0.12 g of graphite flakes were added to a mixture of concentrated H$_2$SO$_4$/HNO$_3$ (6 mL:0.132 mL) and 0.72 g potassium permanganate was then inserted gradually to the mixture under stirring for 2 h to temperature 35°C–45°C. After this step, the mixture was heated up to 100°C and stirred for 30 min, and 42 mL of water and 1.2 mL hydrogen peroxide were added. The mixture was cooled down to room temperature to remove the acidic supernatant from the mixture, and was then centrifuged at 13,000 rpm for 15 min. After the removal of the acidic supernatant by centrifugation process, distilled water was inserted to the mixture to dilute the acidic remnant from GO. To redisperse the pellets, the mixture of GO pellets and distilled water was vortex-stirred for 1 min. This process was done to obtain nearly neutral aqueous mixture, and was continuously repeated with centrifugation and vortex, alternatively. The graphite oxide pellets were then added to N-methyl pyrrolidone (NMR), followed by sonication. A tip sonicator, was used for the exfoliation, of graphite oxide to GO. The ultrasonication also included an ice water bath for 1 h. Subsequently, the preparation of reduced graphene oxide (rGO), was achieved by reducing dispersed GO in the resultant homogeneous GO mixture using hydrazine (N$_2$H$_4$, Sigma-Aldrich) in the ratio of 0.7 mg/mg of GO. Then, the mixture stirred for 10 h at 80°C [23].

To prepare the respective membranes, N-methyl pyrrolidone (NMR) was used for the exfoliation of graphene oxide with the sonication process. To obtain various GO concentrations (0.02, 0.05, 0.14, 0.20 and 0.39 wt%) an NMR mixture of polysulfone (PSf) (15 wt% PSf and 85 wt% of NMR) was used to disperse the

Nanohybrid Graphene-Based Materials 95

nanoplatelets of GO. The solution was then stirred to 60°C and kept at room temperature overnight. The mixture was sonicated for 1 h so that the bubbles in the solution would disappear. Then, an Elcometer 3570 (Micrometric Film Applicator) was used to cast the polymer mixture on a nonwoven polyester fabric. A 24-h water bath was used for the immersion of the produced membranes to achieve a complete liquid – liquid demixing. The yielded PSf/GO membranes had GO:PS ratios of 0.16, 0.32, 0.92, 1.30, and 2.60 wt% [23].

4.2.1.3 Synthesis of Fe_3O_4/rGO

In another study, Wang et al. [24] used 120 mg of graphite oxide under the ultrasonication process for 2 h to be dispersed in deionized water. The yielded exfoliated GO nanosheets were stuck to Fe_3O_4 nanoparticles with the application of the postoxidation method. A solution of 1 M $FeCl_3 \cdot 6H_2O$ was added to GO and extrasonicated for 1 h. Then, with the use of a dropping funnel, a KBH_4 solution (1.65 M) was injected to the dispersion solution of GO with vigorous stirring. The reaction was finished after 4 h and the mixture was cooled down to room temperature. The synthesized black solution was filtered and washed several times with water and ethanol to remove residual acid and dissociative Fe(II), To obtain Fe_3O_4/rGO, the yielded solid was soaked for 60 min with the addition of anhydrous ethanol. Finally, the yielded solid residues were dried at 60°C in vacuum atmosphere [24].

The synthesis of Ppy-decorated Fe_3O_4/rGO was done by using ammonium persulfate (APS) as the oxidant Ppy was embedded on the Fe_3O_4/rGO composite surface. The process conducted by in situ polymerization under nonacidic conditions. In order to form dispersion solution, the magnetic Fe_3O_4/rGO nanoparticles were first sonicated for 3 h into 100 mL hexadecyltrimethylammonium bromide (CTAB) solution. A 1 mL pyrrole monomer was inserted and dissolved into this solution. Then, 20 mL of APS solution was dropped slowly, after cooling to 273 K, and was stirred at this temperature for 8 h. The yielded product was then washed with water and anhydrous ethanol and was finally dried at 60°C in a vacuum oven. The modification process of the Ppy-Fe_3O_4/rGO composite is schematically illustrated in Figure 4.1 [24].

4.2.1.4 Preparation of GO with One-step Ultrasonication

Zhibin Wu et al. [25] also followed the modified Hummers method for the preparation of GO from graphite powder (particle size≤30 μm). Briefly, a beaker 250 mL was used to place 1 g of sodium nitrate and 2 g of graphite. During the stirring process in ice bath, 46 mL of H_2SO_4 (98%) was added; 1 g of sodium nitrate and 6 g potassium permanganate were slowly dropped under vigorous stirring (283 K) to the suspension. After vigorous stirring for 2 h in ice bath, the mixture was stirred for 30 min to 303 K. As the reaction process continued, the color of the solution gradually transformed to brownish paste. The new yielded paste was dissolved into 92 mL of ultrapure water under vigorous agitation for 30 min at 368 K. This had as a consequence, to change the color of suspension to bright yellow. 10 mL of hydrogen peroxide (30 wt%) was added to the solution to terminate the reaction process, and was then stirred at room

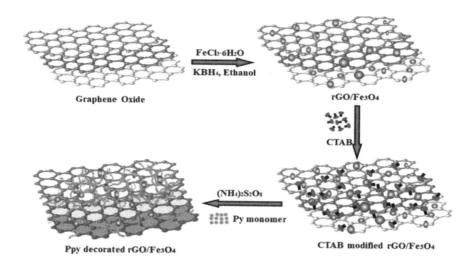

FIGURE 4.1 Schematic illustration of the ternary composites preparation. Reprinted with permission from Hou Wang et al. [24]. Copyright (2015) Elsevier.

temperature for 2 h. This process was achieved, when the suspension temperature was 333 K. The precipitate after the centrifugation process was washed repeatedly with 5% hydrochloric acid to remove residual metal ions, and was then washed with deionized water to remove the sulfate ions. Finally, the yielded precipitate-GO was sonicated and dried at 338 K under vacuum [25].

As presented in Figure 4.2, the synthesized RL-GO hybrid composite was prepared by a one-step ultrasonication process. 200 mg of GO were used and dissolved into 100 mL of dimethylformamide (DMF), followed by 1 h sonication. Then, 600 mg of rhamnolipid were added to the GO suspension and sonicated under vigorous stirring until complete dissolution. After that, 1 g of N-(3-dimethylaminopropyl-N-ethylcarbodiimide) hydrochloride and 200 mL of 4-(dimethylamino) pyridine were added to the suspension. The reaction process under stirring and ultrasonication was allowed to progress for over 3 h. Then, the procedure continued under vigorous stirring by adding methanol to the precipitation of the suspension. The black solid precipitate yielded with the centrifugation process. The process contained washing of black solid precipitate five times with anhydrous ethanol and then twice with ultrapure water. Finally, freeze-drying under vacuum was used to the yielded rhamnolipid-functionalized GO composite. The obtained GO composite was suitable for adsorption applications [25].

4.2.2 Synthesis of Magnetic Chitosan Functionalized with GO

A very promising approach regarding the use of nanohybrid GO in wastewater treatment is the combination with magnetic chitosan detailed by Fan et al. [26].

Nanohybrid Graphene-Based Materials

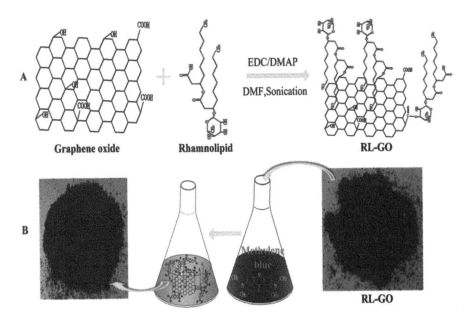

FIGURE 4.2 Schematic depiction of the formation of RL – GO and application for removal of MB. Reprinted with permission from Zhibin Wu et al. [25]. Copyright (2014) Elsevier.

25 mL of double distilled water, 1.7312 g of $FeCl_3$ 6 H_2O and 0.6268 g of Fe Cl_2 4 H_2O were inserted to ammonia solution, which was purged with N and stirred in a bath of water for 3 h at 90°C. Magnetic separation yielded the magnetic particles that were used in the chitosan coating. In order to give a final content of 1.5 % (w/v), 0.3 g chitosan was diluted in 30 mL 3% of acetic solution. A four-neck rounded bottom flask was used to add the chitosan mixture and 0.1 g magnetic particles. After that, 2.0 mL neat glutaraldehyde was inserted into a reaction flask to blend with the mixture and was stirred for 2 h at 60°C. The obtained precipitate was purged with petroleum ether, ethanol, and distilled water until pH was about 7, and was then dried at 50°C in a vacuum oven. The resulting product was magnetic chitosan [26].

On the other hand, GO was synthesized from purged natural graphite according to the modified Hummers method. Briefly, potassium permanganate and natural graphite were agitated for 12 h at 60°C with the mixed acid (HNO_3:H_2SO_4=1:9), and were then added to the mixed liquor hydrogen peroxide and agitated for 1 h, to yield a bright yellow material. This obtained bright yellow material was pureed with 2 M HCl to remove bisulfate ions and was then additionally washed with abundant amount of water to become the solution neutral. In addition, the GO was obtained by centrifuging, and was then dried in a vacuum desiccator [26].

To prepare the functionalized final material, a special procedure was followed. Ultrapure water was used to sonicate GO for 3 h to obtain a GO dispersion. Then, to activate the carboxyl groups of GO, a mixture of 0.05 M NHS and 0.05 M EDC was inserted into the GO dispersion under continuous stirring for 2 h. Then, in order to maintain the pH of the obtained solution at 7.0 was used dissolved sodium hydroxide. After that, the activated GO mixture and 0.1 g of magnetic chitosan (MC) were inserted in a flask and dispersed by ultrasonic dispersion in distilled water for 10 min, and the blended solutions were agitated for 2 h at 60°C. The yielded precipitate in turn until pH was about 7.0, was washed with 2 % (w/v) NaOH and distilled water. Furthermore, the yielded material was collected with the use of a magnet, and in order to obtain the final MCGO product dried at 50°C in a vacuum oven. Figures 4.3 and 4.4 present the synthesis of magnetic chitosan and GO and their application, respectively [26].

4.3 CHARACTERIZATIONS

Some AFM and SEM results from a prepared GO hybrid material indicate that the average flake size was about 2 μm. The average thickness results of the GO flake was ~1 nm as presented in Figure 4.5a. The average size was

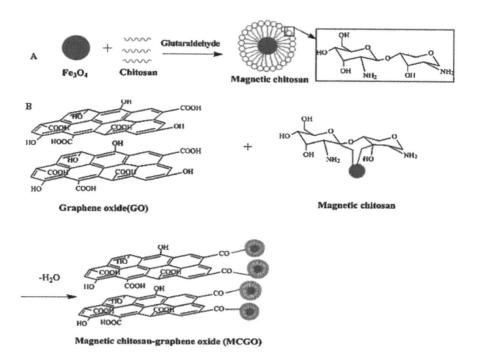

FIGURE 4.3 Schematic depiction of the formation of: (a) magnetic chitosan; and (b) MCGO. Reprinted with permission from Lulu Fan et al. [26]. Copyright (2012) Elsevier.

Nanohybrid Graphene-Based Materials

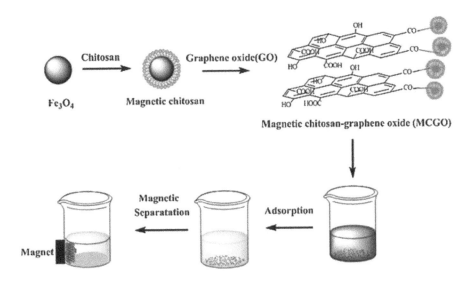

FIGURE 4.4 Synthesis of MCGO and their application for adsorption of methylene blue (MB) with the use of an external magnetic field. Reprinted with permission from Lulu Fan et.al [26]. Copyright (2012) Elsevier.

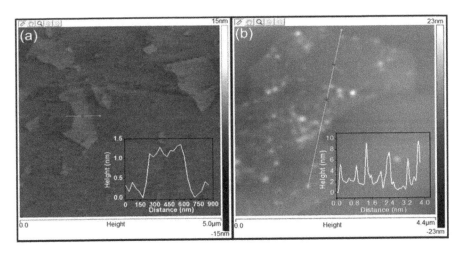

FIGURE 4.5 AFM image of (a) GO; and (b) GONH Reprinted with permission from Surech Kumar et al. [22]. Copyright (2014) American Chemical Society.

~6 nm as presented in Figure 4.5b, and refers to the NP grown on the surface of GO as measured by AFM. Furthermore, XRD patterns of GO, were confirmed using Cu Ka radiation (λ=1,542 Å). Figure 4.6a presents the GO-XRD pattern with a diffraction peak at scattering angle 2θ=9.4°, which was attributed

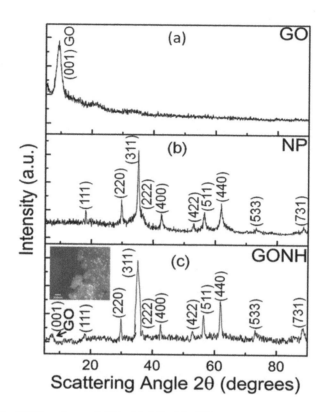

FIGURE 4.6 XRD pattern of GO: (a) NP; (b) GONH; and (c) and typical FESEM image of GONH (inset of (c)). Reprinted with permission from Surech Kumar et al. [22], Copyright (2014) American Chemical Society.

to (001) plane with an interlayer separation ~9.5 Å. Figure 4.6b presents the XRD pattern from the NP [22].

The Debye-Scherrer equation was used to calculate the average particle size, which seems to be ~11 nm (corresponding to (311) line). Figure 4.6c displays the XRD pattern of the GONH, revealing both the NPs and the GO flakes' diffraction peaks. It shows that during the experimental preparation of the nanohybrids, because of the partial reduction of the graphene oxide, the consequent decrease in the peak due to (001) reflection plane. The diffraction peaks corresponded to – Fe_2O_3 and – MnO_2, on the surface of the NPs were absent in the XRD pattern, which confirms the formation of NPs and GONH. XRD analysis pictures displayed that the average size of the nanoparticles was ~7.5 nm.

The peak of GO, in the XRD pattern of the nanohybrids, was very much reduced due to the fact, that the NP grown on the epiphany of graphene oxide averts its restacking. Furthermore, the fact that the size of the NPs was decreased could be attributed to the reason that on the side of the NPs, growth was blocked in the case of in situ growth onto the graphene surface. Figure 4.6c presents the typical SEM surface micrograph of the synthesized nanohybrid. The coverage of the NPs on graphene was confirmed to be uniform for different samples [22].

Figure 4.7a presents the FTIR spectrum of GO, where the characteristic absorption peaks were observed at 1236, 1046, 1415, 1620 and 1729 cm^{-1}, which can be attributed to the epoxy C-O stretching vibrations, alkoxy C-O stretching, O-H deformation, and due to adsorbed water molecules, C=C in-plane stretching vibrations or C=O stretching, respectively.

All of these measurements, clearly confirm the formation of GO and the appearance into the graphene skeleton of oxygenated functionalities, which have been used to the experimental process for the growth of magnetic NPs. Furthermore, the pH of the mixture affects the charge to the – OH and – COOH groups. Figure 4.7b displays the absorption peaks of NP at 577 and 490 cm^{-1}, due to manganese ferrite (metal–O stretching vibrations).

FTIR measurements confirm the formation of MnFe$_2$O$_4$ NPs. Figure 4.7c presents the FTIR spectrum of the GONH and presents the characteristic peaks of both NP and GO. These results confirm the successful preparation of the nanohybrids [22].

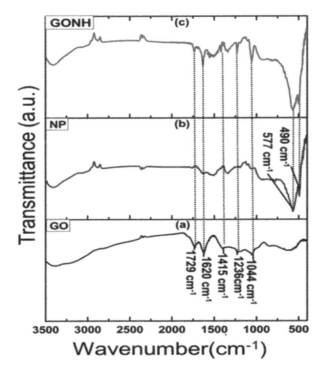

FIGURE 4.7 FTIR spectra of: (a) GO; (b) NP; (c) GONH. Reprinted with permission from Surech Kumar et al. [22]. Copyright (2014) American Chemical Society.

Some interesting microscopic images can also be a tool for the examination of the surface of nanohybrid graphenes. Figure 4.8 depicts the low and high magnification images of GO, Fe$_3$O$_4$/rGO, Ppy/rGO and Ppy-Fe$_3$O$_4$/rGO. As presented in Figure 4.8a, the prepared GO was sheetlike in shape morphology, with a slick surface and single layer structure with wrinkled edges. The formation of loosely packed Fe$_3$O$_4$ nanoparticles onto rGO sheets, as displayed in Figure 4.8b, was achieved while a big amount of granular particles attach on the rGO surface after the addition of Fe^{3+}. The diameter of the spherical Fe$_3$O$_4$ nanoparticles was 50–80 nm. Figure 4.8c presents the rough surface morphology of the Ppy decorated in the rGO sheets. In Figure 4.8d the ternary hybrids are observed [24].

4.4 PREPARATION METHODS OF GO MEMBRANES

Based on stable aqueous dispersity, as well on as the high aspect ratio structure of GO, GO membranes can be easily fabricated via different methods such as the filtration-assisted method, casting/coating-assembly method, and layer-by-layer (LbL) assembly method. Additionally, the evaporation-assisted method, templating method, shear-induced alignment method, and hybrid method are also applied to prepare GO membranes. The different preparation methods for GO membranes are detailed in Table 4.1

FIGURE 4.8 SEM images of (a) GO; (b) Fe$_3$O$_4$/rGO; (c) rGO/Ppy; and (d) Ppy-Fe$_3$O$_4$/rGO composites. The inset is the corresponding lower magnification images. Reprinted with permission from Hou Wang et al. [24]. Copyright (2015) Elsevier.

TABLE 4.1
Methods for the preparation of GO membranes

Method	Description	Note
Filtration-assisted	Vacuum filtration Pressure filtration	Good nanoscale control over the membrane thickness; laminar structure of GO membranes is dictated by the filtration force; highly scalable
Casting/coating-based	Spinning-casting/coating Drop-casting Dip-coating Spray-coating Doctor blade-casting	Nonuniform deposition of GO nanosheets; poor control over the membrane thickness; producing highly continuous GO membranes; highly scalable
LbL assembly	Layer-by-layer assembly	Easily control of the GO layer number, packing, and thickness
Others	Hybrid approach	Easily control of the GO assembly, industrial-scalability, rapid throughput.
	Evaporation-assembled method	Scale-up, easily control of the membrane thickness and size
	Templating method	–
	Langmuir-Blodgett (LB) assembly	Producing highly uniform, close-packed monolayered GO membrane
	Shear-alignment method	Scale-up, industrial-scalability, producing large-area GO membrane, rapid throughput

4.4.1 FILTRATION-ASSISTED METHOD

The filtration-assisted method, including vacuum filtration and pressure-assisted filtration, is a widely used approach to prepare GO membranes at present, especially for the free-standing GO membranes. Dikin et al. [27] fabricated a free-standing GO membrane by vacuum filtration, in which GO nanosheets were bonded together in a near-parallel manner. They reported that the physicochemical property of GO nanosheets did not change during the preparation process. Tsou et al. [28] investigated the influence of GO membrane structure prepared via three distinct self-assembly methods (pressure-, vacuum-, evaporation-assisted technique) on membrane separation performance. Results showed that the GO membrane obtained via pressure-assisted technique exhibited exceptional PV performance and superior operating stability at a high temperature (70°C) due to its dense packing and highly ordered laminate structure. In another study, a highly ordered GO/mPAN (modified polyacrylonitrile) composite membrane was prepared via pressure-assisted self-assembly (PASA) technique [29]. The resultant GO/mPAN composite membrane exhibited excellent PV performance for an isopropyl alcohol (IPA)/water mixture. They reported that the membrane thickness could be readily adjusted by changing the concentration and volume of GO solutions. From the aforementioned discussion, we can conclude that filtration-assisted method allows reasonable and easy control over the membrane thickness and microstructure, and is a potential route for large-scale preparation of GO membrane.

4.4.2 Casting/Coating-Assisted Method

At present, many GO membranes have been developed based on casting/coating-assembly method, which includes drop-casting, dip-coating, spraying-coating /casting, and spin-coating approaches. Park et al. [25] fabricated several layered GO membranes via spin-coating method on a polyethersulfone (PES) substrate and studied their gas separation performance. They reported that high gas separation selectivity could be achieved by controlling gas flow channels through adjusting stacking manner of GO nanosheets. Robinson et al. [30] proposed that large-area and ultrathin GO membranes with excellent mechanical properties could be obtained by a modified spin-coating method. In this procedure, dry nitrogen was utilized to accelerate GO solution evaporation, which correspondingly obtained continuous GO membranes with strong interfacial adhesion force between GO nanosheets and substrate surfaces. Meanwhile, membrane thickness could be controlled on nanometer scales through varying GO concentration in solution or volume of GO suspension. Individual GO nanosheets within GO membranes fabricated by casting/coating-assembly method are strongly held together with hydrogen bonding and van der Waals force.

4.4.3 Layer-by-Layer Assembly Method

Recently, LbL assembly approach has been attracting great attention for the preparation of GO membranes. An interlayer stabilizing force can be conveniently introduced into laminate GO membranes by electrostatic interaction or covalent bonding through this method. Hu et al. [31] have developed a new type of water purification membrane through this approach. The negatively charged GO nanosheets were interconnected with positively charged poly (allylamine hydrochloride) (PAH) via electrostatic interaction and then assembled onto a porous PAN support. Results showed that the resultant GO membrane reserved a compact structure in solutions of low ionic strength and showed excellent separation performance. Typically, the membrane thickness can be easily adjusted by changing the number of LbL deposition cycles [31].

4.4.4 Other Methods

Apart from the aforementioned methods for the preparation of GO membrane, some novel preparation methods such as the evaporation-assisted method, templating-assisted method [32], Langmuir-Blodgett assembly method [33, 34], hybrid method [55], and shear-induced alignment method [56] have also been utilized to fabricate GO membranes. Recently, facile engineering of GO membranes

Nanohybrid Graphene-Based Materials

was realized via a hybrid approach by Guan et al. [35], in which spray-coating and solvent evaporation-induced assembly techniques were included. The team reported that the membrane structure could be finely and conveniently manipulated by adjusting the spraying times and evaporation rate. The resultant GO membranes with ordered and compact structure presented excellent gas separation performance, which exceeded the upper bound of most polymeric membranes. Specifically, this process was less time consuming and more productive compared with the filtration method. This study provided a rather facile and productive approach for large-scale preparation of defect-free GO membranes.

Chen et al. [36] fabricated large-area free-standing GO membranes via an evaporation-driven self-assembly method. They reported that the thickness and area of the membrane could be readily adjusted by controlling the evaporating time and the liquid/air interface area. This is a facile and scale-up approach for preparation of GO membranes. Akbari et al. [56] provided a rapid, scalable, and industrially adaptable method, shear-induced alignment method, to produce large-area GO-based membranes by taking advantage of the flow properties of a discotic nematic GO fluid. The resultant membranes had a large in-plane stacking order of GO sheets and showed remarkable enhancement in water permeability with comparable or better retention of small organic molecules and ions by molecular sieving and electrostatic repulsion. Meanwhile, the obtained membranes showed good stability in aqueous environments and excellent fouling resistance due to the hydrophilic groups on GO membranes. This shear-alignment processing method is conducive to bridging laboratory curiosity to industrial productivity for GO membranes.

From the previous description, it can be concluded that various methods have been developed and utilized to fabricate GO-based membranes. Specifically, it should be noted that the structure and separation performance of the resultant GO membranes significantly depend on the fabrication method and corresponding fabrication conditions. Hence, in a specific practical application, a desired GO membrane can be obtained by appropriate preparation methods and optimized fabrication conditions.

4.5 ADSORPTION APPLICATION OF NANOHYBRID GO

4.5.1 ADSORPTION ISOTHERMS

It is important to shape the most proper adsorption equilibrium connection in the endeavor to find new materials to access a perfect adsorption framework [37], which is key for steady prediction of adsorption factors and correlation (mostly quantitative) of material's behavior for different adsorbent systems (or for different experimental parameters) [29, 38].

Explaining the phenomenon through which the preservation (or release) or mobility of a substance from the aqueous porous media or aquatic environments to a solid-phase at a persistent temperature and pH takes place, in broad spectrum, an adsorption isotherm is an invaluable curve [39, 40]. The mathematical association that establishes a significant role towards the modeling analysis, operational design, and applicable practice of the adsorption systems is normally represented by plotting a graph between the solid-phase and its residual concentration [41].

When the concentration of the solute remains unchanged as a result of zero net transfer of solute adsorbed and desorbed from sorbent surface, a condition of equilibrium is achieved. These associations between the equilibrium concentration of the adsorbate in the solid and liquid phase at persistent temperature are defined by the equilibrium sorption isotherms. Linear, favorable, strongly favorable, irreversible, and unfavorable are some of the isotherm shapes that may form. Understanding of the mechanism of adsorption, surface properties, along with the extent of affinity of the adsorbents are delivered by the physicochemical parameters accompanied by the fundamental thermodynamic suppositions [42].

In the past, some basic adsorption equations were described in detail, revealing their diversity (Langmuir (L), Freundlich (F), Langmuir-Freundlich (L-F), Brunauer-Emmett-Teller, Redlich-Peterson (R-P), Dubinin-Radushkevich, Temkin, Toth, Koble-Corrigan, Sips, Khan, Hill, Flory-Huggins and Radke-Prausnitz isotherm) [43]. Although thermodynamics are very important in adsorption theory explanations, the first successful approach was the adsorption dynamics (kinetics); thermodynamics were then explained in the second stage. Thermodynamics correlate the stage of dynamic equilibrium (having both adsorption and desorption movements/rates) [44, 45]. The differentiation in the physical explanations of adsorption parameters for each model separately shows that the single approach was not adequate [46]. Nowadays, the most widely used models are Langmuir, Freundlich and Langmuir-Freundlich (Table 4.2)

The uptake of each pollutant in equilibrium (Q_e in mg/g) is estimated using the mass balance equation [49]:

$$Q_e = \frac{(C_0 - C_e)V}{m} \tag{4.1}$$

TABLE 4.2

Widely used isotherm equations in adsorption works

Isotherm	Equation	Ref
Langmuir	$Q_e = \frac{Q_m K_L C_e}{1 + K_L C_e}$	[47]
Freundlich	$Q_e = K_F (C_e)^{1/n_f}$	[48]
Langmuir-Freundlich	$Q_e = \frac{Q_m K_{LF} C_e^{1/n}}{1 + K_{LF} C_e^{1/n}}$	[49]

Q_m (mg/g) is the maximum amount of adsorption; K_L (L/mg) is the Langmuir adsorption equilibrium constant; (K_F (mg$^{1-1/n}$ L$^{1/n}$/g) is the Freundlich constant representing the adsorption capacity; n_f (dimensionless) is the Freundlich constant depicting the adsorption intensity; K_{LF} (L/mg)$^{1/n}$) is the L-F constant; n (dimensionless) is the L-F heterogeneity constant.

Nanohybrid Graphene-Based Materials   107

where C_0 and C_e in mg/L express the initial and equilibrium concentration of pollutants; V in L is the adsorbate volume; m in g expresses the adsorbent's mass.

Although the Langmuir and Freundlich isotherms were first introduced about 90 years ago, they still remain the two most commonly used adsorption isotherm equations. Their success undoubtedly reflects their ability to fit a wide variety of adsorption data quite well. The Langmuir model represents chemisorption on a set of well-defined localized adsorption sites having the same adsorption energies independent of surface coverage and no interaction between adsorbed molecules. The Langmuir isotherm assumes monolayer coverage of adsorbate onto adsorbent. The Freundlich isotherm gives an expression encompassing the surface heterogeneity and the exponential distribution of active sites and their energies. This isotherm does not predict any saturation of the adsorbent surface; thus, infinite surface coverage is predicted, indicating physisorption on the surface. One of the most promising extensions to the Langmuir and Freundlich isotherms is the Langmuir-Freundlich (L-F) equation, which is a general purpose isotherm for heterogeneous surfaces. The L-F isotherm is essentially the Freundlich isotherm with a suitable asymptotic property in the sense that it approaches a maximum at high concentrations. Moreover, L-F trends to the Langmuir isotherm when the heterogeneity parameter (b) is set to unity.

4.5.2 EXAMPLES

A recent example of nanohybrid GO for adsorption applications was given by Kumar et al. [22]. Aqueous mixtures of metal ions Pb(II), As(III), and As(V) were prepared with different concentrations, and were treated with NPs and GONH. The degree of surface charge ionization and speciation of the adsorbate is affected by the pH of the mixture, which is a very important factor in the end. In the case of Pb(II), As(V), and As(III), the adsorption data of NPs and GONH with concentration of 100 mg/L are revealed in Figure 4.9a. The pH values showed that the maximum adsorption was for As(III) at pH=6.5, As(V) at pH=4, and Pb(II) at pH=5. At the initial and final pH values there was a small change [22].

After finding the optimum pH, the adsorption on different quantities of metal ions was investigated. In order to find the maximum adsorption Q_m, the equilibrium data were fitted to the Langmuir adsorption isotherm.

The pH of the metal solution has a strong influence on its chemistry (i.e., redox, hydrolysis, reactions, polymerization, and coordination), as well as on the ionic state of the functional groups on the adsorbent surface. Figure 4.10 displays the Cr(VI) adsorption for the initial pH in aqueous solution. The removal capacity of $Ppy-Fe_3O_4/rGO$ composite, which has lower surface area, is higher, compared with only Fe_3O_4/rGO nanoparticles. Thus, it is determined that during the adsorption process, the surface area is not the crucial factor. The Cr(VI) adsorption happens mainly though electrostatic attraction for Fe_3O_4/rGO composites. Furthermore, the Cr (VI) removal by the $Ppy-Fe_3O_4/rGO$ composite during the

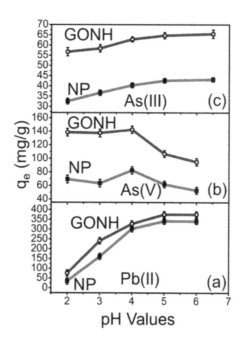

FIGURE 4.9 (a) Effect of pH at different pH values with initial concentration of 100 mg/L for (a) Pb(II), (b) As(V), and (c) As(III); (b) Langmuir adsorption isotherm for (a) Pb(II), (b) As(V), and (c) As(III) with varied initial heavy metal ion concentration ranging from 0 to 400 mg/L pH values kept at 5, 4, and 6.5 for Pb(II), As(V) and As(III), respectively. Reprinted with permission from Surech Kumar et al. [22]. Copyright (2014) American Chemical Society.

adsorption process may still be included in ion exchange and chemical reduction processes, in addition to electrostatic interaction [24].

One very interesting finding is the effect of adsorbent's dosage. So, Figure 4.11 depicts the effects of the RL-GO (or GO) dosage on the removal of MB. It was observed that when increased, the RL-GO (or GO) dosage increased, along with the removal percentage of MB. Moreover, it was observed that when the adsorbent dose was increased, the removal percentage increased slightly. It also could be observed that, when the adsorbent dose increased, the removal capacity of MB decreased. This happened because a higher adsorbent dose provides a large excess in the active sites, which, as a result have a lower utility of the sites at a specific concentration of MB solution. Meanwhile, it was observed under the experimental conditions, that the obtained RL-GO have superior properties, compared to GO in MB adsorption capacity. This can be confirmed by the measurements from XRD and XPS. The results indicate that RL-GO have had more layer spacing and contained more oxygen in the functional groups, compared to GO for MB adsorption [25].

Nanohybrid Graphene-Based Materials 109

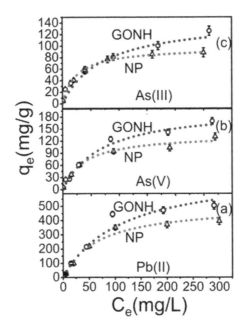

FIGURE 4.9B presents the measurements for As(V). The adsorbent results were Q_m =136 mg/g for NP and 207 mg/g for GONH. As(III) showed Q_m = 97 mg/g for NP and 146 mg/g for GONH [22]. The maximum adsorption capacities for Pb(II) by adding NP and GONH adsorbents were 488 mg/g and 673 mg/g, respectively.

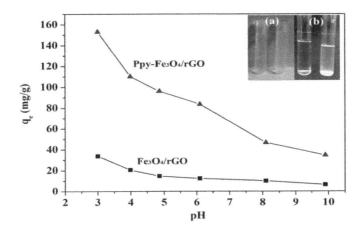

FIGURE 4.10 Cr (VI) adsorption and the pH effect for Fe_3O_4/rGO and Ppy-Fe_3O_4/rGO. C (Cr(VI)) initial=48.4 mg/L, m/V=0.25 g/L, T=303 K. The inserted photograph presents: (a) the Cr(VI) removal ability of Ppy-Fe_3O_4/rGO; and (b) the chemical experiment for SO_4^{-2} and after Ba^{2+} injection. Reprinted with permission from Hou Wang et al. [24]. Copyright (2015) Elsevier.

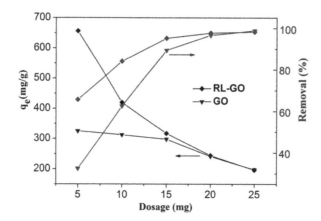

FIGURE 4.11 Effect of adsorbent dosage on adsorption capacity of MB. Reprinted with permission from Zhibin Wu et al. [25]. Copyright (2014) Elsevier.

In order to analyze the MB adsorption kinetics onto RL-GO, the present study used the intraparticle diffusion, Boyd's film-diffusion, the pseudo-first-order, and pseudo-second- order models to examine three different initial MB concentrations. During the adsorption process, the pseudo-first-order model was widely used in its linearized integral form. Table 4.3 presents the adsorption kinetic parameters of MB onto RL-GO [25]

The plots for MB adsorption are presented in Figure 4.12.

4.5.3 Applications of Nanohybrid GO Membranes

Wu et al. [50] studied the fouling property of SiO_2-GO/PSf hybrid membranes; in order to examine the long-term performance of hybrid membranes, the antifouling experiment was performed and three circles were operated. The fouling properties of membranes were determined by calculation of FRR, Rt,

TABLE 4.3
Adsorption kinetic parameters of MB onto RL-GO (pH value, 7.0; RL-GO dose, 400 mg/L, temperature 298 K)

	Pseudo-first order kinetic				Pseudo-second order kinetic			
C_0 (mg/L)	q_e,exp (mg/g)	k_1 (1/h)	q_e,cal (mg/g)	R^2 (-)	q_e,exp (mg/g)	k_2 (g/mg.h)	q_e,cal (mg/g)	R^2 (-)
100	242	0.19	55.31	0.688	242	0.0208	241.55	0.999
150	305	0.28	152.56	0.985	305	0.0068	309.60	0.999
200	365	0.25	127.13	0.926	365	0.0093	364.96	0.999

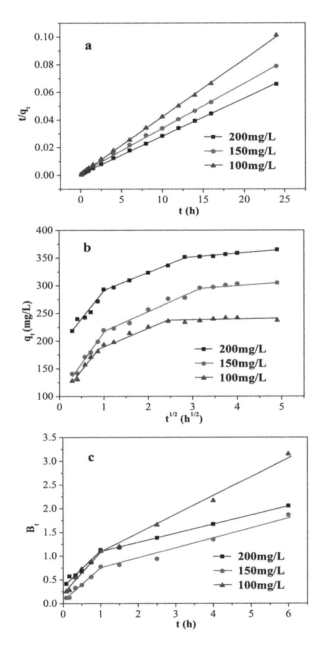

FIGURE 4.12 (a) Pseudo-second-order plots for MB adsorption; (b) Intra-particle diffusion plots for MB adsorption; and (c) Boyd plots for MB adsorption. Reprinted with permission from Zhibin Wu et al. [25]. Copyright (2014) Elsevier.

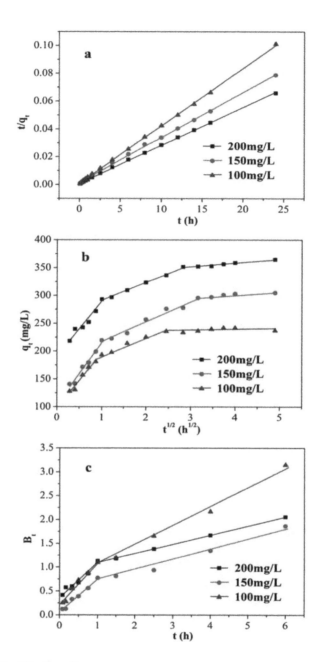

FIGURE 4.12 (Continued)

Nanohybrid Graphene-Based Materials

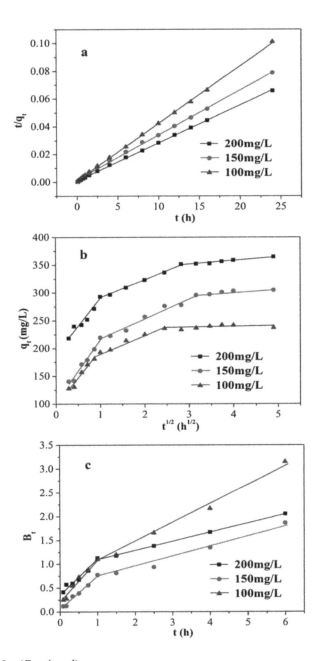

FIGURE 4.12 (Continued)

Rr, and Rir values. The first circle of membrane fouling and washing was employed to estimate the values of fouling properties. Figure 4.13a presents the value of FRR for a pure PSf membrane. Figure 4.13b shows that the value of Rir is as high as one-half of total fouling. FRR value augments to 72% when doping the PSf membrane with SiO$_2$-GO, and the value of Rir in maximum fouling simultaneously reduced sharply to one- third. Finally, the SiO$_2$-GO/PSf hybrid membrane presents a good performance stability through long-time operation and better fouling ability, compared to the other hybrid membranes from GO and SiO$_2$ [50].

In another study, Zhu et al. [27] investigated the adsorption and desorption ability of PVDF/GO/LiCl nanohybrid membranes. In order to make the solution with dye for testing, 10 mg/L Rhodamine B particles were diluted in distilled water.

The PVDF/GO/LiCl membrane surface functional groups are mainly hydroxyl and carboxyl groups, because of GO and LiCl synergistic effects. Figure 4.14

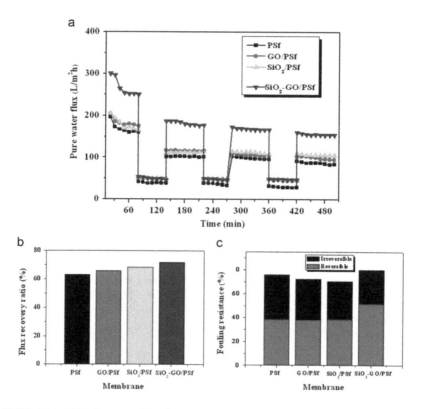

FIGURE 4.13 Membranes with different inorganic materials and the time-dependent fluxes during fouling experiment with BSA filtration [1 mg/mL, pH=7.4 at 0.2 MP a, b) FRR of the prepared membranes and (c) Fouling resistance of the prepared membranes. Reprinted with permission from Huiqing Wu et al. [50]. Copyright (2014) Elsevier.

Nanohybrid Graphene-Based Materials 115

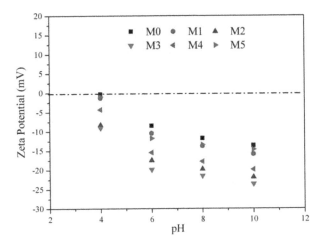

FIGURE 4.14 Zeta potential of PVDF/GO/LiCl nanohybrid membranes. Reprinted with permission from Zhenya Zhu et al. [27]. Copyright (2017) Elsevier.

displays the nanohybrid membranes of PVDF/GO/LiCl that, at pH 8, changed the surface zeta potential from −13.47 to −21.49 mV.

After the rhodamine B-ethanol solutions became stable, their absorbance values were confirmed. Calculations confirmed that the rates of decolorization (DR, %) of all types of nanohybrid membranes exceeded 85%. DRs of M1–M5 nanohybrid membranes still exceeded 80% after 20 cycles. These results confirm that Rhodamine B adsorbed and desorbed better with nanohybrid membranes of PVDF/GO/LiCl, and could be used for dye removal. Figure 4.15 displays images of PVDF/GO/LiCl nanohybrid membranes [27].

Figure 4.16 and 4.17 display the flux decline curves at 100 kPa and flux recovery ratio of the M0–M5 PVDF/GO/LiCl nanohybrid membranes, respectively [27].

Peng et al. [28] prepared a reduced graphene oxide composite nanohybrid membrane. In order to obtain an acceptable rejection ratio at a high permeation flux, the composite films were fabricated with various mass ratios of SiO2 to GO, and the rejection ratios and permeation fluxes were measured while removing SDS/diesel oil/H_2O emulsion and methylene blue (MB) solution.

As presented in Figure 4.18, the M1 PDA (GO/SiO_2 (mass ratio) – 2 mg/ 0.67 mg immersed into PDA for 24 h) membrane reveals a sharp decrease to 133.2 L m^{-2} h^{-1} at pure water flux, compared with pure PVDF membrane at pure water flux (1389.1 L m-2 h-1). Occasionally, the M1 PDA membrane revealed ultrahigh removal ratios for diesel and MB, which were 99.2% and 99.8%, respectively. It is worth noting that no visibly change of rejection ratio with more SiO_2 content was observed, while the rejection ratio of MB remained 98% with mass ratio of SiO2:GO 4:3.

As presented in Figure 4.19, the retention capacity of MB dye was limited, with a feed solution at a low pH value, but at a respectively high pH value, the

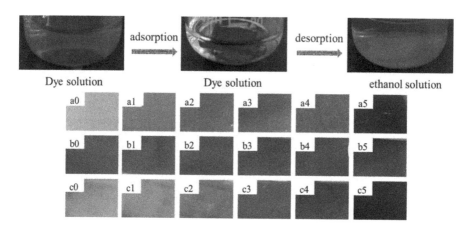

FIGURE 4.15 The images of PVDF/GO/LiCl nanohybrid membranes (a0–a5: pristine membranes; b0–b5: after immersing the Rhodamine B solution for 36 h; c0–c5: after decoloration by ethanol alcohol). Reprinted with permission from Zhenya Zhu et al. [27]. Copyright (2017) Elsevier.

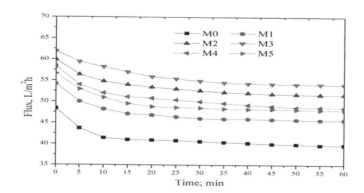

FIGURE 4.16 M0–M5 PVDF/GO/LiCl nanohybrid membranes and the flux decline curves at 100 kPa. Reprinted with permission from Zhenya Zhu et al. [27]. Copyright (2017) Elsevier.

composite membrane was revealed to have better ability to remove MB. This could be explained by the function groups of Si – OH, – COOH, – OH in the membrane functioning as receptors for hydrogen ions, with the composite membrane having a negative charge at higher pH values. It is worth noting that, the repulsion between the positive charge of the membrane and positive charge of MB lead to the low removal ratio of MB [28].

Moreover, Wang et al. [51] achieved dye removal using graphene oxide and polyelectrolyte complexes (GO/PECs). To enhance the quality of the dye

Nanohybrid Graphene-Based Materials

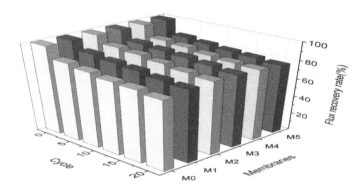

FIGURE 4.17 Flux recovery ratio of PVDF/GO/LiCl nanohybrid membranes. Reprinted with permission from Zhenya Zhu et al. [27]. Copyright (2016) Elsevier.

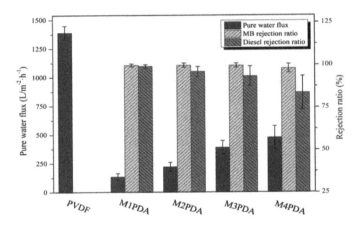

FIGURE 4.18 Pure water flux and removal ratio under 0.09 MPa [membrane M1 PDA has (GO/SiO$_2$) (mass ratio) – 2 mg/0.67 mg and was immersed into PDA for 24 h; membrane M2 PDA has (GO/SiO$_2$) (mass ratio) – 2 mg/1.34 mg and was immersed into PDA for 24 h; membrane M3 PDA has (GO/SiO$_2$) (mass ratio) – 2 mg/2 mg and was immersed into PDA for 24 h; membrane M4 PDA has (GO/SiO$_2$) (mass ratio) – 2 mg/2.67 mg and was immersed into PDA for 24 h]. Reprinted with permission from Yixin Peng et al. [28]. Copyright (2018) Elsevier.

product, a well-known vital process involves dye desalting and purification. Furthermore, when the dye molecules have impurities, one of the most promising technologies to clean them from impurities is Nanofiltration (NF). Table 4.4 presents the retention of various dyes with the GO/PECs membranes. For instance, after incorporating GO, the retention of Congo red increased from 91.5% to 99.5%; this result is beneficial for the improvement of membrane selectivity [51].

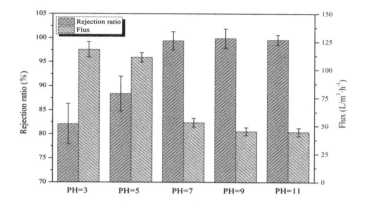

FIGURE 4.19 MB removal ratio and flux by M1 PDA at different pH values. Reprinted with permission from Yixin Peng et al. [28]. Copyright (2018) Elsevier.

TABLE 4.4
Dye retention with different membranes

Membrane	Dye Molecule	Retention (%)	Permeance (kg/m² h MPa)	Pressure (MPa)	Ref.
(NaSS-AC)/PS	Acid red	96	50-58	0.4	[52]
CMCNa/PP	Congo red	99.9	77.5	0.08	[30]
CMCNa/PP	Sunset yellow	82.2	86.3	0.08	[30]
PES-TA	Methyl green	98	20	0.5	[31]
Sulfonated PES	Reactive orange 16	92.1	32	2	[35]
PEI/PAA/PVA/GA[a]	Congo red	91.5±0.4	8.1±0.3	0.5	[51]
PEI/PAA/PVA/GA[a]	Methyl blue	87.3±0.6	8.5±0.2	0.5	[51]
PEI/PAA/PVA/GA[a]	Methyl orange	83.4±0.7	8.9±0.2	0.5	[51]
(PEI-modified GO)/PAA/PVA/GA[b]	Congo red	99.5±0.1	8.4±0.3	0.5	[51]
(PEI-modified GO)/PAA/PVA/GA[b]	Methyl blue	99.3±0.1	8.7±0.2	0.5	[51]
(PEI-modified GO)/PAA/PVA/GA[b]	Methyl orange	87.6±0.3	9.6±0.5	0.5	[51]

[a] Hydrolysis conditions for PAN support membrane: hydrolysis temperature 65°C, hydrolysis time 30 min; preparative conditions: 30 min filtration time, 0.25 wt.% PEI aqueous solution, 0.05 wt.% PAA aqueous solution, 0.5 wt.% PVA aqueous solution, 3 wt.% glutaraldehyde aqueous solution.

[b] Hydrolysis conditions for PAN support membrane: hydrolysis temperature 65°C, hydrolysis time 30 min; preparative conditions: 30 min filtration time, 0.25wt.% PEI aqueous solution with 0.1 g/L GO, 0.05 wt.% PAA aqueous solution, 0.5 wt.% PVA aqueous solution, 3 wt.% glutaraldehyde aqueous solution.

Nanohybrid Graphene-Based Materials

Table 4.5 shows that when increasing the processing temperature, the permeate flux increased and the water content in the permeate stream decreased. The diffusion coefficients and the increases in the water vapor pressure on the feed side caused the increase in flux. Because the apparent activation energy of water (E_p=13.12 kJ mol^{-1}) was much lower than that of ethanol (Ep=28.13 kJmol^{-1}) a decrease in selectivity was revealed. The Arrhenius law was used to calculate the apparent activation energy [51]

Recently, Athanasekou et al. [33] studied a large series of self-standing and supported GO membranes that were prepared via a facile synthetic approach involving the filtration of GO suspensions through polymeric and ceramic macroporous filters. The main aim of this work was to develop a membrane that would be almost impermeable to helium and hydrogen, exhibiting in parallel very high water vapor and moderate alcohol vapor permeability, properties that constitute this type of membranes as very promising for pervaporation and gas separation applications. Several of the derived self-standing membranes, especially those developed using aqueous GO suspensions of low concentration in GO, have met the aforementioned requirements. It was obvious that self-standing GO membranes present enhanced probability of being devoid of mesoporous and macroporous defects, compared to the supported membranes, and that an important pore structural characteristic to fine tune during membrane preparation is the size of the open apertures on

TABLE 4.5
Pervaporation performance of different membranes

Membrane	Feed solution	Water content in permeate (wt.%)	Flux (g/(m² h))	T (°C)	Ref.
Chitosan/TiO$_2$	90% Ethanol/water	96.8	278	80	[53]
Zeolite-filled chitosan	90% Ethanol/water	99.7	16	30	[53]
ZSM-5 incorporated in polyimide	90% Isopropanol/water	99.55	936	60	[54]
PVA-zeolite 4A	76.43% Ethanol/water	99.55	936	60	[36]
PSS-ZrO$_2$/PDDA-ZrO$_2$	95% Ethanol/water	99.9	340	50	[32]
PEI/PAA/PVA/GA	95% Ethanol/water	91.78±0.3	229±17	50	[51]
((PEI-modified GO)/PAA)$_1$/PVA/GA^a	95% Ethanol/water	94.4±0.13	368±15	80	[51]
((PEI-modified GO)/PAA)$_1$/PVA/GA^a	95% Ethanol/water	95.4±0.1	268±10	50	[51]
((PEI-modified GO)/PAA)$_1$/PVA/GA^a	95% Ethanol/water	96.9±0.1	196±13	40	[51]
((PEI-modified GO)/PAA)$_1$/PVA/GA^a	95% Ethanol/water	98.1±0.1	156±4	50	[51]

[a] Hydrolysis conditions for PAN support membrane: hydrolysis temperature 65°C, hydrolysis time 30 min; preparative conditions: 30 min filtration time, 0.25 wt.% PEI aqueous solution, 0.05 wt.% PAA aqueous solution, 0.5 wt.% PVA aqueous solution, 3 wt.% glutaraldehyde aqueous solution.

the outermost layer of the GO membrane. These apertures act as gates that control the accessibility of the gas molecules to the bulk of the membrane and should be of appropriate size to allow the unimpeded passage of water molecules, especially when the target is a future application in desalination.

SAXS analysis confirms that larger repeated units (stacks) were formed of uniformly arranged GO flakes, and therefore, more diluted GO suspensions have final stacking configurations that are closer to the ideal configuration presented in Figure 4.20 (left). The permeabilities of several gases and vapors show that the gas permeability of the membranes does not correlate with the d-distance between the GO flakes.

From a comparison of the gas and vapor permeability properties of two studied membranes that were developed using the same volume and GO concentration, but different GO production methods, it is concluded that the vapor permeation properties are greatly influenced by the type and population of the oxygenated groups attached on the surfaces and edges of the graphene flakes. More specifically, the two membranes were prepared from GO precursors produced with the Hummers and Brodie methods, respectively. The Brodie method is known to introduce smaller amount of oxygen containing groups as compared to the Hummers method and favors the formation of conjugated epoxy and hydroxyl groups [34].

As previously mentioned, it is also important to note that there is no need to develop a membrane with molecular sieving characteristics order to achieve separation of water from alcohols. Indeed, one of the produced membranes exhibiting Knudsen type of He/N_2 separation, and at the same time, was also very efficient in performing the separation of water from alcohol vapors ($Pe_{H_2O}/Pe_{MeOH} = 40$) while exhibiting high water flux that was twice as high as that of membranes with molecular sieving (He/N_2) characteristics.

4.6 CONCLUSIONS

The most interesting findings are as follows: In the case of GONH, it was observed that for concentration 100 mg/L of As(V), Pb(II) and As(III), the greatest treatment was achieved at optimum adsorbent pH of 4, 5, and 6.5,

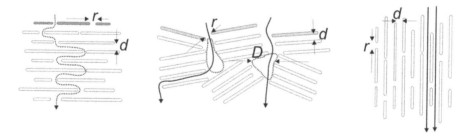

FIGURE 4.20 Possible stacking configurations of GO sheets for membranes developed with the vacuum filtration technique. Copyright (2017) Elsevier.

Nanohybrid Graphene-Based Materials

respectively. The results for As(V) was Q_m=136 mg/g for NP and 207 mg/g for GONH, while for As(III) presented Q_m=97 mg/g for NP and 146 mg/g for GONH. Finally, for GONH the maximum adsorption capacity was observed at 488 mg/g for NP and 673 mg/g for GONH. However, the Ppy-Fe$_3$O$_4$/rGO composite revealed better Cr(VI) adsorption with q_e=155 mg/g, compared to Fe$_3$O$_4$/rGO composite with q_e=36 mg/g. The better adsorption capacity of Ppy-Fe$_3$O$_4$/rGO composite could be attributed to electrostatic interaction, ion exchange, and chemical reduction. The aforementioned (and not only these that are explained in this chapter) are indicative examples of how big the impact of nanohybrids in decontamination technology is.

REFERENCES

[1] Boehm, H.P., R. Setton, and E. Stumpp, Nomenclature and terminology of graphite intercalation compounds. *Carbon*, 1986. 24(2): p. 241–245.

[2] Nemes-Incze, P. et al., Anomalies in thickness measurements of graphene and few layer graphite crystals by tapping mode atomic force microscopy. *Carbon*, 2008. 46(11): p. 1435–1442.

[3] Shenderova, O.A., V.V. Zhirnov, and D.W. Brenner, Carbon nanostructures. *Critical Reviews in Solid State and Materials Sciences*, 2002. 27(3–4): p. 227–356.

[4] Novoselov, K.S. et al., Electric field in atomically thin carbon films. *Science*, 2004. 306(5696): p. 666–669.

[5] Iupac, Recommended terminology for the description of carbon as a solid (IUPAC recommendations 1995). *Pure and Applied Chemistry*, 1995. 67: p. 491.

[6] Ivanovskii, A.L., Graphene-based and graphene-like materials. *Russian Chemical Reviews*, 2012. 81(7): p. 571–605.

[7] Avouris, P. and C. Dimitrakopoulos, Graphene: Synthesis and applications. *Materials Today*, 2012. 15(3): p. 86–97.

[8] Zhu, Y. et al., Graphene and graphene oxide: Synthesis, properties, and applications. *Advanced Materials (Weinheim, Germany)*, 2010. 22(35): p. 3906–3924.

[9] Novoselov, K.S. et al., A roadmap for graphene. *Nature*, 2012. 490(7419): p. 192–200.

[10] Kemp, K.C. et al., Environmental applications using graphene composites: Water remediation and gas adsorption. *Nanoscale*, 2013. 5(8): p. 3149–3171.

[11] Compton, O.C. et al., Crumpled graphene nanosheets as highly effective barrier property enhancers. *Advanced Materials (Weinheim, Germany)*, 2010. 22(42): p. 4759–4763.

[12] Compton, O.C. and S.T. Nguyen, Graphene oxide, highly reduced graphene oxide, and graphene: Versatile building blocks for carbon-based materials. *Small*, 2010. 6(6): p. 711–723.

[13] Dreyer, D.R. et al., The chemistry of graphene oxide. *Chemical Society Reviews*, 2010. 39(1): p. 228–240.

[14] Lerf, A. et al., Structure of graphite oxide revisited. *Journal of Physical Chemistry B*, 1998. 102(23): p. 4477–4482.

[15] Pei, S. and H.M. Cheng, The reduction of graphene oxide. *Carbon*, 2012. 50(9): p. 3210–3228.

[16] Pei, Z. et al., Adsorption characteristics of 1,2,4-trichlorobenzene, 2,4,6- trichlorophenol, 2-naphthol and naphthalene on graphene and graphene oxide. *Carbon*, 2013. 51(1): p. 156–163.

[17] Brodie, B.C., On the atomic weight of graphite. *Philosophical Transactions of the Royal Society of London*, 1859. 149: p. 249–259.

[18] Staudenmaier, L., Verfahren zur darstellung der graphitsaure. *Berichte der Deutschen Chemischen Gesellschaft*, 1898. 31: p. 1481–1487.

[19] Hummers, W.S., Jr. and R.E. Offeman, Preparation of graphitic oxide. *Journal of the American Chemical Society*, 1958. 80(6): p. 1339.

[20] Krishnan, D. et al., Energetic graphene oxide: Challenges and opportunities. *Nano Today*, 2012. 7(2): p. 137–152.

[21] Singh, V. et al., Graphene based materials: Past, present and future. *Progress in Materials Science*, 2011. 56(8): p. 1178–1271.

[22] Kumar, S. et al., Graphene oxide-MnFe2O4 magnetic nanohybrids for efficient removal of lead and arsenic from water. *ACS Applied Materials and Interfaces*, 2014. 6(20): p. 17426–17436.

[23] Lee, J. et al., Graphene oxide nanoplatelets composite membrane with hydrophilic and antifouling properties for wastewater treatment. *Journal of Membrane Science*, 2013. 448: p. 223–230.

[24] Wang, H. et al., Facile synthesis of polypyrrole decorated reduced graphene oxide-Fe3O4 magnetic composites and its application for the Cr(VI) removal. *Chemical Engineering Journal*, 2015. 262: p. 597–606.

[25] Wu, Z. et al., Adsorptive removal of methylene blue by rhamnolipid-functionalized graphene oxide from wastewater. *Water Research*, 2014. 67: p. 330–344.

[26] Fan, L. et al., Fabrication of novel magnetic chitosan grafted with graphene oxide to enhance adsorption properties for methyl blue. *Journal of Hazardous Materials*, 2012. 215–216: p. 272–279.

[27] Zhu, Z. et al., Preparation and characteristics of graphene oxide-blending PVDF nanohybrid membranes and their applications for hazardous dye adsorption and rejection. *Journal of Colloid and Interface Science*, 2017. 504: p. 429–439.

[28] Peng, Y. et al., A novel reduced graphene oxide-based composite membrane prepared via a facile deposition method for multifunctional applications: Oil/water separation and cationic dyes removal. *Separation and Purification Technology*, 2018. 200: p. 130–140.

[29] Gimbert, F. et al., Adsorption isotherm models for dye removal by cationized starch-based material in a single component system: Error analysis. *Journal of Hazardous Materials*, 2008. 157(1): p. 34–46.

[30] Yu, S. et al., Application of thin-film composite hollow fiber membrane to submerged nanofiltration of anionic dye aqueous solutions. *Separation and Purification Technology*, 2012. 88: p. 121–129.

[31] Zhang, Q. et al., Positively charged nanofiltration membrane based on cardo poly (arylene ether sulfone) with pendant tertiary amine groups. *Journal of Membrane Science*, 2011. 375(1–2): p. 191–197.

[32] Zhang, G., J. Li, and S. Ji, Self-assembly of novel architectural nanohybrid multilayers and their selective separation of solvent-water mixtures. *AIChE Journal*, 2012. 58(5): p. 1456–1464.

[33] Athanasekou, C., M. Pedrosa, T. Tsoufis, L.M. Pastrana-Martínez, G. Romanos, E. Favvas, F. Katsaros, A. Mitropoulos, V. Psycharis, and A.M.T. Silva, Comparison of self-standing and supported graphene oxide membranes prepared by simple filtration: Gas and vapor separation, pore structure and stability. *Journal. Membrane Science*, 2017. 522: p. 303–315.

[34] Jiang, Z., X. Zhao, Y. Fua, and A. Manthiram, Composite membranes based on sulfonated poly(ether ether ketone) and SDBS-adsorbed graphene oxide for direct methanol fuel cells. *Journal of Materials Chemistry*, 2012. 22: p. 24862–24869.

[35] Van der Bruggen, B. et al., Mechanisms of retention and flux decline for the nanofiltration of dye baths from the textile industry. *Separation and Purification Technology*, 2001. 22–23: p. 519–528.

Nanohybrid Graphene-Based Materials

[36] Huang, Z. et al., Multilayer poly(vinyl alcohol)-zeolite 4A composite membranes for ethanol dehydration by means of pervaporation. *Separation and Purification Technology*, 2006. 51(2): p. 126–136.

[37] Srivastava, V.C. et al., Adsorptive removal of phenol by bagasse fly ash and activated carbon: Equilibrium, kinetics and thermodynamics. *Colloids and Surfaces A: Physicochemical and Engineering Aspects*, 2006. 272(1–2): p. 89–104.

[38] Ho, Y.S., J.F. Porter, and G. McKay, Equilibrium isotherm studies for the sorption of divalent metal ions onto peat: Copper, nickel and lead single component systems. *Water, Air, and Soil Pollution*, 2002. 141(1–4): p. 1–33.

[39] Allen, S.J., G. McKay, and J.F. Porter, Adsorption isotherm models for basic dye adsorption by peat in single and binary component systems. *Journal of Colloid and Interface Science*, 2004. 280(2): p. 322–333.

[40] Limousin, G. et al., Sorption isotherms: A review on physical bases, modeling and measurement. *Applied Geochemistry*, 2007. 22(2): p. 249–275.

[41] Ncibi, M.C., Applicability of some statistical tools to predict optimum adsorption isotherm after linear and nonlinear regression analysis. *Journal of Hazardous Materials*, 2008. 153(1–2): p. 207–212.

[42] Bulut, E., M. Özacar, and I.A. Şengil, Adsorption of malachite green onto bentonite: Equilibrium and kinetic studies and process design. *Microporous and Mesoporous Materials*, 2008. 115(3): p. 234–246.

[43] Malek, A. and S. Farooq, Comparison of isotherm models for hydrocarbon adsorption on activated carbon. *AIChE Journal*, 1996. 42(11): p. 3191–3201.

[44] De Boer, J.H., *The Dynamical Character of Adsorption*, Oxford University Press, 1953, p. 239.

[45] Myers, A.L. and J.M. Prausnitz, Thermodynamics of mixed-gas adsorption. *AIChE Journal*, 1965. 11: p. 121–127.

[46] Ruthven, D.M., *Principles of Adsorption and Adsorption Processes*, Wiley & Sons, New York, 1984.

[47] Langmuir, I., The constitution and fundamental properties of solids and liquids. *Part I. Solids. The Journal of the American Chemical Society*, 1916. 38 (2): p. 2221–2295.

[48] Freundlich, H., Over the adsorption in solution. *Zeitschrift fur physikalische chemie*, 1906. 57: p. 385–470.

[49] Tien, C., *Adsorption Calculations and Modeling*, Boston, MA: Butterworth-Heinemann, 1994.

[50] Wu, H., B. Tang, and P. Wu, Development of novel SiO2–GO nanohybrid/polysulfone membrane with enhanced performance. *Journal of Membrane Science*, 2014. 451: p. 94–102.

[51] Wang, N. et al., Self-assembly of graphene oxide and polyelectrolyte complex nanohybrid membranes for nanofiltration and pervaporation. *Chemical Engineering Journal*, 2012. 213: p. 318–329.

[52] Akbari, A. et al., New UV-photografted nanofiltration membranes for the treatment of colored textile dye effluents. *Journal of Membrane Science*, 2006. 286(1–2): p. 342–350.

[53] Sun, H. et al., Surface-modified zeolite-filled chitosan membranes for pervaporation dehydration of ethanol. *Applied Surface Science*, 2008. 254(17): p. 5367–5374.

[54] Mosleh, S. et al., Zeolite filled polyimide membranes for dehydration of isopropanol through pervaporation process. *Chemical Engineering Research and Design*, 2012. 90(3): p. 433–441.

5 State of the Art and Perspectives in Membranes for Membrane Distillation/ Membrane Crystallization

Carmen Meringolo and Gianluca Di Profio
Institute on Membrane Technology of the
    National Research Council (ITM-CNR)
c/o University of Calabria (UNICAL)
Rende (CS), Italy

Efrem Curcio
Department of Environmental and Chemical Engineering
    (DIATIC)
University of Calabria (UNICAL)
Rende (CS), Italy

Elena Tocci, Enrico Drioli, and Enrica Fontananova
Institute on Membrane Technology of the
    National Research Council (ITM-CNR)
c/o University of Calabria (UNICAL)
Rende (CS), Italy

5.1 INTRODUCTION TO MEMBRANE DISTILLATION (MD) AND MEMBRANE CRYSTALLIZATION (MCr) TECHNOLOGIES

Water scarcity and depletion of raw materials are challenging problems that are estimated to increase in the future owing to population growth and climate changes [1].

Membrane processes, particularly Membrane Contactors (MCs), have been shown to have great potential as an environmentally friendly, low-cost, and sustainable water and wastewater treatment technology. Membrane Distillation (MD) and Membrane Crystallization (MCr) technologies belong to this class of processes, in which macroporous hydrophobic membranes are used to promote mass transfer between phases, based on the principles of phase equilibrium. MCs technology is still under development, and is of current interest, focusing on the selection and fabrication of suitable membranes and modules adequate for the process.

To address water scarcity problems, MD can be a promising alternative against traditional method such as flash evaporation and reverse osmosis (RO). Being energy intensive and having low recovery factor are main drawback of flash evaporation and RO, respectively [2]. Advantages of MD are high recovery factor, high purity water production and the possibility to utilize alternative energy sources, such as solar and geothermal energy [3].

MCr is an efficient process for production, purification, or recovery of solid materials [4]. Traditional crystallization techniques such as vapor diffusion, batch reactor, and antisolvent crystallization, while presenting positive characteristics in particular conditions, presents a series of disadvantages that make them less very effective [2]. The main challenge is the poor reproducibility of products' characteristics, which is associated with the limited control of supersaturation owing to imperfect mixing and/or to the heterogeneous distribution of solvent removal or antisolvent addition points over the plant. In addition, these processes require high energy for heating in conventional evaporators or in vacuum systems, which represents another drawback [5]. The main advantage of MCr lies in the acceleration of the crystallization process in terms of conventional crystallization techniques. Moreover, the proven ability to select a specific polymorphic form through the fine control of supersaturation rate and the promotion of specific molecule/membrane interactions is of huge interest in the pharmaceutical industry [6]. For example, membrane technology could beneficially be introduced as a valid and innovative option to improve industrial crystallization.

Although the applications of MD technology is considered in different fields, providing good and valuable results, e.g., desalination, its commercial and industrial implementation is limited by some factors [1]. The most significant limitation involves availability of commercial membranes and modules specifically designed for the MD to guarantee the long-term application of the process with stable permeability, high performance, low conductive heat transfer, and high resistance to scaling, fouling and wetting [1, 5, 7]. An area of emerging scientific research in membrane science and technology is the design and development of advanced membranes for MD. The key membrane characteristics of an adequate MD membrane are: high porosity, hydrophobicity, liquid entry pressure (LEP) in the pores, narrow pore size distribution, optimum thickness, small tortuosity of pores, low thermal conductivity, good and long-term thermal stability, and excellent chemical resistance. The performance of the processes not only depends on the membrane itself; it is also affected by the module design, the MD configuration, and the applied operating conditions. Various types of membrane modules have been proposed, namely plate and frame modules, shell-and-tube or tubular modules, and spiral-wound membrane modules. An adequate MD module should exhibit high membrane packing density, high permeate product rate, low temperature and concentration polarization effects, low fluid pressure drop, and low external heat loss [1].

This chapter addresses the fundamental concepts and the relevant mathematics related to the transport phenomena through porous hydrophobic membranes of interest in MD and MCr. Membrane material properties, configuration, structures, and fabrication techniques are discussed. Some of the most promising applications and new perspectives of these membrane operations are illustrated. Commercially available membranes and environmental issues are also presented. Finally, future trends and useful sources for more details are included.

5.1.1 PRINCIPLES OF MD TECHNOLOGY

MD is a promising thermally driven separation technology for desalting water [8, 9]. This membrane-based separation process belongs to the class of MCs, systems in which porous membranes are used to carry out mass transfer between phases (liquid/liquid or gas/liquid) contacted through the membrane without dispersion of one of the two phases in the other [10].

In MD, a microporous hydrophobic membrane is used as a non-selective interface placed between an aqueous heated solution (feed or retentate) and a condensing phase (permeate or distillate) [11]. The hydrophobic nature of the membrane, generally polymeric, prevents penetration of the pores by aqueous solutions and allows the establishment of a vapor – liquid interface at the entrance of each pore. The temperature gradient between the two streams leads to a vapor pressure difference causing volatile compounds (most commonly water) to evaporate on the hot feed solution – membrane interface, to transfer the vapor phase through the membrane pores, and to condense on the cold side membrane – permeate solution interface. Some of the key advantages of MD processes over conventional separation technologies are: relatively lower energy costs as compared to distillation, RO, and pervaporation; a considerable rejection of dissolved, nonvolatile species; much lower membrane fouling as compared with microfiltration, UF, and RO; reduced vapor space as compared to conventional distillation; lower operating pressure than pressure-driven membrane processes and lower operating temperature as compared with conventional evaporation [12–14].

Four main MD configurations exist: Direct Contact Membrane Distillation (DCMD); Air Gap Membrane Distillation (AGMD); Sweeping Gas Membrane Distillation (SGMD) and Vacuum Membrane Distillation (VMD). These configurations schematized in Figure 5.1 differ by the method used to activate the vapor pressure gradient across the microporous hydrophobic membrane.

DCMD is the most common MD configuration and is widely employed in desalination processes, concentration of aqueous solutions in food industries [15–17], and acids manufacturing [18]. In this case, the feed solution is directly in contact with the hot membrane side surface. Consequently, evaporation takes place at the feed membrane surface. The vapor is moved by the partial pressure difference across the membrane to the permeate side where it

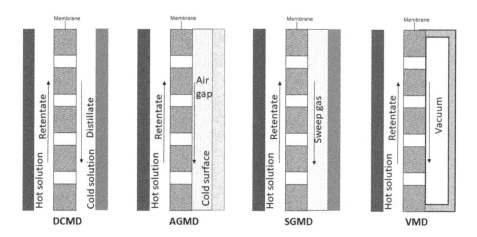

FIGURE 5.1 Scheme of the main membrane distillation configurations.

condenses. The feed cannot penetrate the membrane owing to the hydrophobic characteristic. The main drawback of this configuration is the heat lost by conduction [19].

The AGMD configuration can be employed with a particular interest for water desalination [7], as well as for separate volatile compounds such as alcohols [20, 21] from aqueous solutions, because there is no problem of membrane wetting at the distillate side [22].

In addition, AGMD was used for separation of oxygen isotopic water [23] to removing non-condensable gases [24], for enrichment of heavy-oxygen water ($H_2^{18}O$) in natural waters [25], and for removal of ethanol, butanol, and acetone–butanol–ethanol mixtures from water [26]. The AGMD design is similar than that of DCMD, but in this design, the feed solution is in direct contact with the hot side of the membrane surface only. An air gap is introduced between the membrane and the distillate side. The water vapor passes through the membrane and the air gap and then condenses over the cold surface inside the membrane module. The advantage of this configuration is the reduced heat lost by conduction. Additional resistance to mass transfer is generated, which is considered a disadvantage.

SGMD configuration is frequently useful for removal of volatile organic components from aqueous solutions, for direct separation of ethanol water [27], for water removal from dilute glycerol–water solutions [28], for ammonia removal from dilute and concentrated aqueous solution [29], for desalination [30], and for separating formic acid–water azeotropic mixtures [31]. In SGMD configuration, a cold inert gas is used to sweep the water vapor at the distillate membrane side to condense outside the membrane module. Compared to other MD configurations, the advantage of SGMD is that it has lower thermal polarization and no wetting from the distillate side. One of the disadvantages when operating MD in sweep gas configuration is the increasing sweep gas temperature along the module because of the transferred heat from the feed side through the membrane. Therefore, to avoid this, a cold wall in the permeate side can be used to remove the excess sweep gas heat. Another disadvantage of this configuration is that a small volume of permeate diffuses in a large sweep gas volume, requiring a large condenser.

VMD configuration is usually used to separate volatile organic compounds from aqueous solutions. Some specific applications include treatment of water containing different types of dyes [33], removal of ammonia from dilute aqueous solutions [32, 34], and seawater desalination [35]. In VMD, a vacuum pump is utilized to create a vacuum in the distillate side. To transport the water vapor, the vacuum pressure applied must be lower than the saturation pressure of the evaporating species. Condensation occurs outside of the membrane module. A great advantage of this configuration is that the heat lost by conduction is negligible [8].

Osmotic membrane distillation (OMD), which is also called isothermal MD, osmotic evaporation, gas membrane extraction, or osmotic concentration by a membrane, is an isothermal MD process; it uses similar modules (Figure 5.2) to those used in DCMD. However, the driving force generated in OMD results from the difference in concentration gradient across the membrane caused by a stripping solution at the permeate side [36]. In OMD, the temperatures of the feed and the stripping or absorbing liquids are kept almost at the same temperature, and transmembrane mass transfer takes place at much lower temperature than in MD, or even ambient temperature. Water vapor and volatile components are absorbed by a salt solution flowing continuously on the permeate side of the porous hydrophobic membrane. Owing to its low process temperature, OMD is preferred for the concentration of thermo-labile volatile components of fruits and aromatic extracts. Applications of OMD that have been investigated include: fruit juice concentration [37, 38], reconcentration of sugar solutions [39], and brine purification in the chloralkali industry [40].

In addition, extensive works have been carried out to develop new configurations with better permeation flux and improved energy efficiency; these include multieffect membrane distillation (MEMD), vacuum multieffect membrane distillation (V-MEMD), hollow-fiber multieffect membrane distillation MEMD, permeate gap membrane distillation (PGMD) and material gap membrane distillation (MGMD) [41–43].

MEMD shares similar concepts of multistage and multieffect distillation for seawater desalination. The cold feed solution is placed beneath the condensation surface as a coolant to condense the permeated vapors as well as to gain heat at the same time. The preheated feed solution is additionally heated before it enters the feed channel [44].

VMEMD is based on the principles of MEMD, except for the vacuum enhancement. The VMEMD module consists of a steam raiser, evaporation – condensation stages, and an external condenser [42]. A vacuum condition is employed at the air gap region to remove the excess air/vapor. Each stage recovers the heat of condensation, providing a multiple-effect design. Distillate

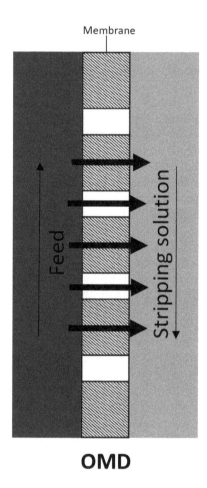

FIGURE 5.2 Osmotic membrane distillation OMD module configuration.

is produced in each evaporation – condensation stage, and in the condenser. The feed in each stage recovers the condensation heat and a multiple-effect characteristic.

Hollow-fiber MEMD involves a multieffect AGMD hollow-fiber module with internal heat recovery [45]. The feed solution is preheated to 90°C before entering the MD module. At the exit of the MD module, the concentrated feed solution is further cooled down by an external cooler at a reduced temperature. This cooled feed solution is feedback into the module and serves as the coolant to condensate vapor in the permeate side. As a multistage design, the effluent stream can serve as the feed solution for the next membrane module, to enhance heat recovery and efficiency.

PGMD is an enhancement of DCMD in which a third channel is introduced by an additional nonpermeable foil. Among the advantages of using PGMD with respect to DCMD are the direct use of feed water as a cooling liquid inside the module, and the consequent need for a single heat exchanger to heat the supply before entering the evaporator. A further significant advantage is the separation of the distillate from the coolant. In this way, the mass flow of the refrigerant in the condenser channel remains constant. However, a significant disadvantage of this configuration is the low flow rate of the distillate in the permeate gap because it leads to a poor heat conduction from the membrane surface to the condenser side. This problem has been partly solved with an alternative of PGMD called conductive gap membrane distillation CGMD, which adds thermally conductive spacers to the gaps [46].

The MGMD configuration was developed to address the problem of the lowest permeation flux of the AGMD configuration with respect to other configurations, owing to the presence of a layer of stagnant air between the membrane and the condensation surface [32]. A new MD membrane module was designed that fills the air gap with different materials like sponge (polyurethane) and polypropylene mesh.

5.1.1.1 TRANSPORT MECHANISMS IN MD

The mechanisms of transport in MD includes both heat and mass transfer across the membrane. Determination of the governing heat and mass transfer mechanisms is needed to be able to generate accurate and reliable predictions of interfacial membrane temperature and permeate flux from MD models [47, 48]. The average temperature inside the membrane is calculated from the predicted interfacial membrane temperature. This average membrane temperature is used to determine vapor pressure at the membrane interface, which is needed as input for estimating mass transfer in the MD process [49]. Thus, heat and mass transfer are interrelated in the MD process. Generally, for all MD configurations, estimation of heat transfer is a prerequisite for prediction of mass transfer. In all MD configurations except VMD, heat is transferred through membranes via conduction in the form of sensible heat and latent heat when vapor molecules diffuse through the pores of hydrophobic membranes. In VMD heat transfer by conduction is negligible [49–51]. It is noteworthy that heat transfer does not only depend on the membrane geometry, but also on the type of MD configuration and type of materials used in membrane preparation [52]. Various MD models [49] have described mass transfer of water vapor molecules through pores of hydrophobic membranes in the MD process. The existing mass transfer models differ, based on the configuration of pores of membrane [36, 49, 53]. Heat and mass transfer in the MD module create thermal and concentration boundary layers at the feed side membrane surface, and have different effects in the distillate side, depending on the configuration.

5.1.1.1.1 Heat Transfer in MD

Two main heat transfer mechanisms occur in the MD system: latent heat transfer and conduction heat transfer. There are three regions: the feed side, the membrane, and the permeate side through which heat is generally transferred in the MD process [47, 51, 54].

In the first region, heat is transferred from the feed solution by convection through the thermal boundary layer to the membrane surface on the feed side of the membrane module, imposing a resistance to mass transfer since a large quantity of heat must be supplied to the surface of the membrane to vaporize the liquid. The temperature at the membrane surface is lower than the corresponding value at the bulk phase. This negatively affects the driving force for mass transfer. This phenomenon is called temperature polarization [54]. The temperature polarization coefficient (TPC) is determined as the ratio of the transmembrane temperature to the bulk temperature difference:

$$TPC = \frac{T_{fm} - T_{pm}}{T_{fb} - T_{pb}} \tag{5.1}$$

where T_{fm}, T_{pm}, T_{fb} and T_{pb} are membrane surface temperatures and fluid bulk temperatures at the feed and permeate sides, respectively.

Heat transfer through the feed side boundary layer is given as:

$$Q_f = h_f \cdot (T_f - T_{fm}) \tag{5.2}$$

where h_f is the heat transfer coefficient of the feed side boundary layer.

The second region consists of a combination of heat transfer through the membrane by conduction (sensible heat Q_c) and heat transfer owing to water vapor transport through gas/air filled pores of the membrane (latent heat of vaporization Q_v) [36, 49]. The heat flux can be estimated by the following expressions:

$$Q_{mem} = Q_v + Q_c \tag{5.3}$$

$$Q_v = h_v \cdot (T_{fm} - T_{pm}) = N \cdot \lambda_v \tag{5.4}$$

$$Q_c = \frac{K_g \cdot \varepsilon + K_m (1 - \varepsilon)}{\delta} \cdot (T_{fm} - T_{pm}) = N \cdot \lambda_v \tag{5.5}$$

where N is the rate of mass transfer; λ_v is the heat of vaporization; ε is the membrane porosity; Kg is the thermal conductivity of the vapor within the membrane; Km is the thermal conductivity of the solid membrane material; δ is the membrane thickness; and T_{fm} and T_{pm} are the temperatures at membrane surface at feed side and permeate side, respectively.

In the third region, there is convective heat transfer from the membrane surface through the thermal boundary layer to the bulk permeate side.

$$Q_p = h_p \cdot (T_{pm} - T_p) \tag{5.6}$$

where h_p is the heat transfer coefficient of the permeate side boundary layer.

Membranes for MD/MCr 133

Both feed and permeate side boundary layers are functions of fluid properties and operating conditions, as well as hydrodynamic conditions. There are some convenient approaches in the literature to reduce the temperature polarization effects such as thorough mixing, working at high flow rates, or using turbulence promoters [55–59]. The boundary layer heat transfer coefficients are usually estimated from empirical correlations expressed in the form:

$$Nu = a \cdot Re^{\beta} \cdot Pr^{\gamma} \qquad (5.7)$$

where Nu is the Nusselt number; Re the Reynolds number; and Pr the Prandtl number [more details are available in References 55, 56].

With respect to the heat transferred by convection within the membrane pores, the aforementioned can also be considered, but are negligible because convection accounts for, at most, 6% of the total heat lost through the membrane and only 0.6% of the total heat transferred across the membrane [8].

5.1.1.1.2 Mass Transfer in MD

The mass transfer in MD is driven by the vapor pressure gradient imposed between two sides of the membrane. Mass transfer in MD consists of three consecutive steps: evaporation of water at the liquid – gas interface on the membrane surface of the feed side; water vapor transfer through the membrane pores; and condensation of water vapor at the gas/liquid interface on the membrane surface of the permeate side. The usual conceptual approach to the study of mass transfer in MD considers it in terms of serial resistances upon the transfer between the bulks of two phases contacting the membrane.

The permeability K is the overall mass transfer coefficient that is the reciprocal of the total resistance to the mass transport. This overall resistance is a combination of the mass transfer coefficients in the feed side (K_f), in the membrane (K_m), and in the distillate side (K_d):

$$K = \left[\frac{1}{K_f} + \frac{1}{K_m} + \frac{1}{K_d} \right]^{-1} \qquad (5.8)$$

In MD, only water vapor transport is allowed, owing to the hydrophobic character of the membrane. Therefore, the concentration of solute in feed solution becomes higher at the liquid/gas interface than that at the bulk feed as mass transfer proceeds. This phenomenon is called concentration polarization and results in reduction of the transmembrane flux by depressing the driving force for water transport. Concentration polarization coefficient (CPC) is defined to quantify the mass transport resistance within the concentration boundary layer at the feed side as the ratio of the solute concentration at the membrane surface (C_m) to that at the bulk feed solution (C_b):

$$CPC = \frac{C_m}{C_b} \qquad (5.9)$$

The concentration gradient between the liquid/gas interface and the bulk feed results a diffusive transfer of solutes from the surface of the membrane to the bulk solution. At steady state, the rate of convective solute transfer to the membrane surface is balanced by diffusion of solute to the bulk feed.

The mass balance across the feed side boundary layer yields the relationship between molar flux J, the diffusive mass transfer coefficient K_x through the boundary layer, the solution density ρ, and the solute concentrations C_{fm} and C_b at the interface and in the bulk, respectively:

$$\frac{J}{\rho} = K_x \ln\left(\frac{C_m}{C_b}\right) \quad (5.10)$$

In the literature, several empirical correlations of dimensionless numbers (Sherwood (Sh), Reynolds (Re) and Schmidt (Sc) numbers) can be used to estimate the value of the mass transfer coefficient K_x [60, 61].

These empirical relationships, often derived by analogy with those evaluated for the heat transport, are expressed in the form:

$$Sh = \alpha \cdot Re^\beta \cdot Sc^\gamma \quad (5.11)$$

Viscous resistance (resulting from the momentum transferred to the membrane), Knudsen diffusion resistance (owing to collisions between molecules and membrane walls), and/or ordinary diffusion (owing to collisions between diffusing molecules) affect mass transfer through the porous membrane, assuming negligible diffusion (Figure 5.3). The Knudsen number (K_n) is defined as the ratio between the mean free path of diffusing molecules and the average pore size of the membrane. It is responsible for mass transfer through the

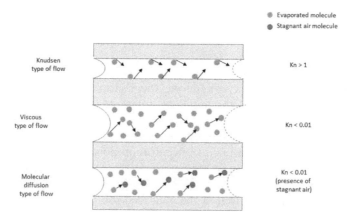

FIGURE 5.3 Scheme of the mass transport mechanisms trough a pore of an MD membrane.

Membranes for MD/MCr

membrane pore if the mean free path of the water molecules is much greater than the pore size of the membrane, and hence, the molecules tend to collide more frequently with the pore wall [19, 62, 63].

Still, when the pore size is relatively large, the molecule – molecule collisions are more frequent and molecular diffusion is responsible for mass transfer through the membrane pores [36].

Dusty gas model (DGM) is probably the best mathematical tool for describing gaseous molar fluxes through porous media; in the most general form (again neglecting surface diffusion), DGM [64] is expressed as:

$$\frac{J_i^D}{D_{ie}^k} + \sum_{j=1 \neq i}^{n} \frac{p_j J_i^D - p_i J_j^D}{D_{ije}^0} = -\frac{1}{RT} \nabla p_i \qquad (5.12)$$

$$J_i^v = -\frac{\varepsilon r^2 p_i}{8RT\tau\mu} \nabla P \qquad (5.13)$$

$$D_{ie}^k = \frac{2\varepsilon r}{3\tau} \sqrt{\frac{8RT}{\pi M_i}} \qquad (5.14)$$

$$D_{ije}^0 = \frac{\varepsilon}{\tau} D_{ij}^0 \qquad (5.15)$$

where J^D is the diffusive flux; J^V is the viscous flux; D^k is the Knudsen diffusion coefficient; D^0 is the ordinary diffusion coefficient; p_i is the partial pressure of the component i: P is the total pressure; M_i is the molecular weight of component i; r is the membrane pore radius; ε is the membrane porosity (assuming the membrane consists of uniform cylindrical pores); μ is the fluid viscosity; and τ is the membrane tortuosity. The subscript "e" is indicative of the effective diffusion coefficient function of the membrane structure. Although the DGM was derived for isothermal systems, it is successfully applied in MD, working under relatively small thermal gradients by assuming an average value of temperature across the membrane.

In most cases, simpler empirical correlation is preferred. The transmembrane flux (J) is expressed as a linear function of the vapor pressure difference across the membrane:

$$J = C\Delta P$$

where ΔP is the vapor pressure difference across the membrane (function of temperatures and compositions at the membrane surface); and C is the MD coefficient that can be obtained experimentally. This coefficient is a function of operative conditions temperature and pressure, membrane properties (pore size, thickness, porosity, and tortuosity), and properties of the vapor transported across the membrane (molecular weight and diffusivity) [65].

5.1.2 PRINCIPLES OF MCr TECHNOLOGY

Crystallization is a well-established unit operation for separation, purification, and for giving solid crystalline products from a liquid phase; it is employed in the chemical and pharmaceutical industries. MCr is an innovative crystallization process, put into operation by using membrane technology, by which crystals nucleation and growth is carried out in a well-controlled pathway, starting from an undersaturated solution. The large range of applications offered by the membranes and the future perspectives discussed in the literature present this technology as a promising and competitive alternative to conventional crystallizers for chemical production.

The working principle of a membrane crystallizer can be considered as an extension of the MD concept. The MCr system is based on the use of microporous and hydrophobic membranes, for the production of crystals starting from an unsaturated solution [4]. The membrane might be made by polymeric or inorganic materials or by a combination of both in a mixed matrix membrane. Hollow-fibers, as well as flat-sheet membranes, can be employed. In the typical configuration, the MCr is a device in which a solution containing a nonvolatile solute that is likely to be crystallized (defined as the crystallizing solution or feed), is contacted by means of a membrane, with a stripping solution, pure water, or the vacuum on the distillate side, based on the principle of the MCs. The physical-chemical properties of the membrane do not allow the aqueous solutions in contact to penetrate the pores in the liquid state and the establishment of a dual liquid – vapor interface on both sides of the membrane occurs (Figure 5.4).

The presence of these interfaces gives rise to a mechanism of evaporation – convection – condensation. The evaporation is established through the removal of solvent molecules in the vapor phase from the interface in contact with the solution containing the species to be crystallized. Following evaporation, the

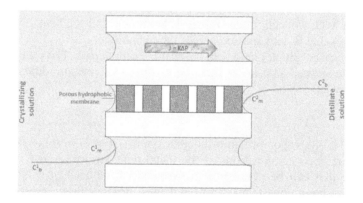

FIGURE 5.4 General principle of a membrane crystallizer: C_b, bulk concentration; C_m, concentration close to the membrane surface; J, transmembrane flux; K, phenomenological constant; and ΔP, gradient of partial pressure between the two sides of the membrane.

vapor is transported through the porous structure and finally, the vapor molecules recondensate in the liquid phase on the second interface. The driving force is a gradient of chemical potential that is generated by means of a concentration and/or temperature between the two sides of the membrane [66]. In the first case, the system is defined as a thermal membrane crystallizer, while in the second case, it is defined as an isothermal or osmotic membrane crystallizer. Under the action of the driving force, solvent or antisolvent molecules migrate, in the vapor phase through the porous membrane structure, from the site where their chemical potential is higher toward the lower chemical potential subsystem; thus, the crystallizing solution progressively concentrates, reaching thermodynamic conditions generating supersaturating, nucleation, and crystal growth.

The mechanism for mass transport depends on the operational configuration of the membrane crystallizer.

Solvent evaporation membrane crystallizer, thermal or osmotic, and antisolvent membrane crystallizer are two typical configurations. Both configurations are based on mass interexchange in the vapor phase, by which, in the antisolvent configuration, the mass transfer process also occurs by evaporation. For both solvent evaporation and antisolvent membrane crystallizers, static and dynamic process can be performed. In the static condition, the feed and the distillate solutions are quiescent, while in the dynamic process those two solutions are circulated in a forced solution flow environment, generally in a condition of laminar flow.

In the first configuration, the substances to be crystallized are dissolved in an undersaturated solution located at the feed side of the membrane. The distillate side of the membrane consists of a condensing fluid, often the pure solvent, at a temperature lower than the feed side in the case of thermal activation of the driving force, or in a stripping solution consisting of a hypertonic solution of inert inorganic salts such as $NaCl$, $MgCl_2$, and $CaCl_2$, in the case of the isothermal configuration. The gradient of chemical potential generates, at the first interface, a mechanism of evaporation of the solvent on the feed side, the transport of the vapor through the porous membrane under partial pressure gradient and its recondensation at the second interface on the permeate side. Therefore, solvent molecules are removed from the feed side, and thus, solute concentration increases up to supersaturation. This phenomenon results in concentration polarization layers adjacent to the two faces of the membrane. Consequently, next to the membrane surface, the solute concentration in the polarization layer is higher than the bulk solute concentration in the feed side. With the help of absorbed heat, solvent molecules transform to vapor molecules in the feed side, while this absorbed heat is released on the distillate side and condensation of vapor molecules occurs; thus, temperature polarization occurs.

Near the membrane surface, the degree of supersaturation could affect concentration and temperature polarization and the mechanism of crystallization can grow in a different approach with respect to the bulk of the solution. Therefore, by modifying the heat and mass profiles next to the membrane, it is possible to control the characteristics of the nucleated crystal and grown on/near the membrane surface.

The thermal configuration is appropriate for crystallizing inorganic substances or low molecular weight organic compounds, but in the case of heat sensitive molecules, such as proteins, the osmotic system is preferred, owing to its milder operating conditions [67–69].

Regarding antisolvent membrane crystallizers, two systems occur: solvent/antisolvent demixing configuration, and antisolvent addition configuration. In the case of solvent/antisolvent demixing configuration, a solute is dissolved in a mix that contains a solvent and an antisolvent. The solvent, with a higher vapor pressure than the antisolvent at the same temperature, evaporates at a higher flow rate owing to a gradient of chemical potential (e.g., temperature difference) generated between the two sides of the membrane, thus producing solvent/antisolvent mixing. The evaporation of solvent decreases the solubility of the solute in the mixture. Therefore, the antisolvent exceeds a certain volume fraction supersaturation and a phase separation occurs. For this configuration to occur, it is important that the antisolvent and the solvent are miscible; the initial solvent/antisolvent equilibrium in the mixture guarantees that the solute is under its solubility limit and the solvent evaporates at higher velocity with respect to the antisolvent.

While in the antisolvent addition configuration, a solute is dissolved in a solvent and to achieve a gradient of chemical potential, an antisolvent is gradually evaporated from the other side of the membrane. The composition of the mixture changes; the solute dilutes owing to the antisolvent and solvent mixing. At a specific moment, the excess of antisolvent creates supersaturation and solute crystallization. In this system, it is important that the solvent and antisolvent are miscible; the species to be crystallized can be soluble in aqueous solutions and poorly soluble inorganic low boiling liquids. Initially, on the permeate side, a certain quantity of solvent might be present to modulate the rate of antisolvent dosage in the crystallizing solution, thus controlling the process kinetics, or to avoid wetting of the membrane when using the pure antisolvent, for example, ethanol.

In a MCr, although the membrane put in contact the crystallizing solution with the distillate side, the operative parameters (e.g. flow rates) of the two solutions can be independently adjusted. The effect would be control of the rate and extent of nucleation over the crystal growth, thus investigating a broad set of kinetic trajectories for crystal nucleation and growth that are not readily achievable in conventional crystallization formats, and would lead to the production of specific crystalline morphologies and structures [70, 71]. The crystallizing solution is in direct contact with the membrane, and a heterogeneous contribution to the crystallization mechanism arising by solute – membrane interaction might occur. Consequently, the lowering of the activation energy for nucleation allows crystallization in conditions that would not be adequate for spontaneous nucleation and/or to enhance the crystallization kinetics with respect to conventional evaporative crystallization methodologies. Furthermore, the solute – membrane interaction can provide specific solute – solute interaction pathways that would lead to the formation of specific crystal forms. This effect can result from both the structural and chemical properties of the

Membranes for MD/MCr

membrane surface. The porous nature of the surface might achieve cavities where solute molecules are physically entrapped, leading to high supersaturation values suitable for nucleation; the nonspecific chemical interaction between the membrane and the solute molecules can lead to molecular orientation, and hence, to the facilitated effective interaction appropriate for crystallization.

5.1.3 FROM LABORATORY STUDIES TO LARGE-SCALE APPLICATIONS

MD and MCr are relatively new technologies, which have experienced keen academic interest and an increase in start-up businesses in the last years.

Particularly for MD, the ideas and first patents originated in the late 1960s; however, suitable membranes were not available until the late 1980s, and were created for biological and other purposes [8]. Research took off in the late 1990s and into the new millenium, and at this time, MD has become fairly well understood. Currently, most MD work is done in laboratories, although a number of test beds across the world for small-scale solar thermal MD have already been deployed, and a few other projects exist [7, 19, 72]. While increased research interest in MD is relatively recent [73], scaling under high-temperature conditions has been a key challenge in systems with water heating since the advent of the steam engine. MD has expanded into a niche of small-scale thermal desalination using solar and waste heat resources, owing to its fouling resistance, scalability, and acceptable efficiency. Recent studies indicate that MD could attain the efficiencies of state-of-the-art, mature thermal desalination technologies.

Temporarily, companies have begun using MD [74]. The first applications were in creating ultrapure water for computer manufacturing, becuase of the superior distillate quality with respect to the market-dominating technology – RO. Applications for fruit juice concentration and other food uses soon followed. More recently, companies began using MD for seawater desalination. So far, they have found a niche in small- to medium-scale desalination applications, especially where waste heat is available [75]. Interesting commercial systems based on the MD technology are MEMSTILL [76] and MEMSYS [77]. MEMSTILL is an MD system developed by a consortium of nine partners including Keppel Seghers (Singapore) and TNO (The Netherlands Organisation for Applied Scientific Research), who have established Aquastill (Netherlands) for commercialization of the technology. The MEMSTILL MD system is designed for internal heat recovery to be applied in applications having access to low-grade heat or industries with waste heat [78]. Three pilot plants have been tested since 2008: one in Johor, Malaysia for seawater desalination, and the other two in Rotterdam for brackish water distillation by using heat from a coal-fired power plant [79]. The pilot plants had design capacities of 5 m^3/day based on 24 h operation [80]. MEMSTILL relies on a modified AGMD process, while MEMSYS uses a multieffect vacuum MD process. MEMSYS is larger and has had many more installations. A particularly sustainable and clever application of MEMSYS has been in energy positive desalination in the

Maldives and in Singapore, where the desalination systems recover the heat from air-cooled engines that produce electricity for the grid, for example, diesel generators [36].

MEMSYS systems in industry reach a thermal efficiency ratio expressed as Gain Output Ratio (GOR – defined as the ratio of the output work to the input heat) of about 4 [42, 81].

A Fraunhofer spin-off company, Solarspring GmbH (Germany) has developed a solar- powered MD pilot test facility based on spiral-wound AGMD for desalination applications. The solar-driven MD desalination plants have capacities from 0.1 to 5 m^3/day and are located in areas with high solar intensity, such as Canary Islands in Spain, Pantelleria in Italy, and Amarika, Namibia. The modules can be configured to operate as DCMD, LGMD, or AGMD, and feature polytetrafluorethylene (PTFE) membrane inserts with PP backings; total membrane area ranges from 5–14 m^2 per module [82].

Recently Aquaver, a Dutch cleantech company commissioned and installed the first full-scale commercial MD facility for desalination in the Maldivian Island of Gulhi [83]. The MD system is supplied by memsys GmbH (Germany) and is based on Vacuum MultiEffect Membrane Distillation (V-MEMD) configuration. This module, called memDist, consists of PTFE membranes and is a modified form of VMD, so that thermal efficiency and GOR are improved through four or six stages modules in series with a total membrane area of 6.4 or 52 m^2, respectively [77]. The MD unit installed at Gulhi utilizes waste heat of a local power generator and has a production capacity of 10 m^3 drinking water per day. The first study with the objective of using a membrane unit properly as a crystallizer dates back to 1986 (3–6). Now, development of MCr technology is at laboratory scale or at pilot plants scale. In particular, the main applications in lab-scale of MCr relate to recovery of crystalline materials from the brine of nanofiltration (NF) and/or RO in integrated membrane operations for seawater desalination [84–87]. Sodium chloride, $CaCO_3$, and Epsom salts ($MgSO_4$ _7H_2O) are obtained as solid products from the NF retentate stream; NaCl is the product from the RO brine, while other products such as LiCl can be produced by further increasing the concentration factor [88]. In addition, MCr was employed in wastewater treatment for the recovery of high-purity silver [89] and sodium sulphate [90], CO_2 capture [91, 92], nanotechnology, such as the synthesis of $BaSO_4$ and $CaCO_3$ particles [93], the recovery of antibiotics [94], and microparticles of polystyrene [95]. In the studies dedicated to development of MCr and desalination systems, research has been conducted in lab-scale to crystallize inorganic [4, 84], small organic [96], and macromolecular [70, 71] materials. Some applications demonstrated that MCr is a versatile technique to control supersaturation, thus inducing polymorph selection in the crystallization of organic compounds [97–99]. Enhanced crystallization kinetics were obtained in the crystallization of proteins, owing to the heterogeneous contributions to nucleation provided by the membrane [6, 100]. As for all membrane operations, MCr techniques modularity might allow to scale up processes, for example, revolutionizing currently-used downstream practices in biopharmaceutical productions that are mainly based on expensive and cumbersome multistep batch chromatography platforms by developing an innovative continuous template-assisted MCr process as key unit. The Amecrys project [101], funded by the European Commission under the Horizon 2020 programme, combines two innovative concepts: membrane-assisted crystallization, a radical, new separation technology based on the use of porous hydrophobic membranes, which allows the accurate control of supersaturation by solvent removal in vapor phase; and template-assisted crystallization by engineered mesoporous silica 3D-nanotemplates to facilitate the crystallization of structurally complex therapeutic proteins, such as monoclonal antibodies (mAbs), from multicomponent solutions.

The activity is focused on the downstream processing of mAbs, one of the most important class of therapeutic proteins in modern medicine, which are used in a wide range of diseases, including cancer, as well as inflammatory, cardiovascular, and autoimmune disorders. Another interesting project involving MD/MCr units at pilot sacle is the IDEA-ERANETMED project [102]. The objective of the IDEA project is the development of a solar powered, zero liquid discharge, integrated desalination membrane system combining solar assisted MD/MCr unit with NF. The aim was to reduce the energy consumption, increase the water recovery factor, realize an environmental friendly with nearly zero-liquid discharge, reduce fouling and scaling problems using novel nanostructured membranes combined with an innovative process design.

5.1.4 ADVANTAGES AND STILL EXISTING LIMITATIONS

MD possesses unique advantages over other desalination technologies, including pressure-driven methods such as RO, and thermally driven methods such as flash distillation. MD is free of the specialized requirements of high-pressure RO systems, which include heavy gauge piping, complex pumps, and maintenance demands. Another advantage of MD, compared to RO, is the absence of osmotic pressure limitation [75].

Since MD is not a pressure-driven process and only vapor is allowed to cross through the membrane, MD is more fouling-resistant than RO [102] and has a 100% potential rejection of ions and macromolecules. MD can be run at lower temperatures than other thermal systems making untapped sources of waste heat usable; it requires significantly fewer parts, and can have a much smaller footprint as result of reduced vapor space [8]. Additionally, recent theoretical and computational work claims potential multistage DCMD configurations with efficiencies greater than that of other thermal technologies [103, 104], assuming very large available heat exchanger areas. The comparative simplicity makes MD more competitive for small scale applications such as solar-driven systems for remote areas, especially in the developing world [105–107]. However, significant advancements are needed in membrane technology for MD to reach theoretical cost competitiveness and develop market share growth [108]. In addition, MD processes have some clear advantages over the traditional distillation processes and over other well established membrane processes (Figure 5.5). Phase mixing and associated problems such as emulsification, bobbling, flooding and unloading are not present in MD operations. The phases are well separated, independently of their density. The modularity of the technology made the scale-up process linear with potential. Compared to conventional distillation processes, such as multistage flash (MSF) and multiple-effect distillation (MED), lower operative temperatures are required. This permits the possible use of alternative renewable energy sources (solar, wind, or geothermal). With respect to the RO process, lower operating pressure is necessary. The MD process can be achieved at operating pressures that are usually near the atmospheric pressure. This permits using apparatus made of plastic material to reduce or avoid corrosion problems. In addition, the concentration polarization effect is not a limitation. The concentrate can reach higher concentrations than in RO. Even though this technology presents many advantages, there are still some relevant drawbacks which are a matter of study. In the case of MD, it is possible to have 100% (theoretical) rejection of nonvolatile solute.

FIGURE 5.5 Advantages of MD with respect to conventional distillation (MSF and MED) and reverse osmosis (RO).

Performance that is not limited by high osmotic pressure allows MD to be used whenever elevated permeate recovery factors or high retentate concentrations are demanded. Less demanding membrane mechanical properties are characteristic of the MD process in comparison to RO and other pressure driven membrane operations because of the lower operating pressure for the first one. The materials of membranes can be made from almost any chemically resistant polymers, with hydrophobic intrinsic properties thus increasing membrane life. Membrane fouling in MD is less problematic than in other membrane-based operation because the pores are comparatively large with respect to RO and UF, as well as the low operating pressure of the method; the accumulation of aggregates on the membrane surface is less compact and slightly influence the mass transport. Some disadvantages of MD include the limitations of membranes and modules designed specifically for MD, risk of wetting the pores of the membrane if the liquid penetrates into the pores of the membrane, and the effect of temperature polarization resulting from the transfer of heat through the membrane.

The advantages of MCr with respect to traditional crystallization techniques are many (Figure 5.6) and include

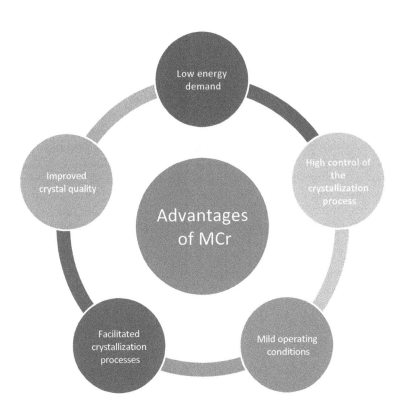

FIGURE 5.6 Advantages of MCr with respect to traditional crystallization techniques.

- **High control of the crystallization process.** Morphological (crystal size, size distribution, shape, and habit) and structural (polymorphism) crystal properties can behave by acting on supersaturation. In a membrane crystallizer, control of supersaturation is feasible and allows modulating the final crystals' properties with important effects in terms of production costs and product effectiveness.
- **Low energy demand.** MCr technique requires low energy input for its operation. A membrane crystallizer can be driven by using heat generated by other industrial processes and that would, in normal cases, be lost in the surrounding with high operating costs. On these bases, also nonconventional alternative energy sources, like wind, solar, geothermal, etc., might be potentially used to operate a MCr installation.
- **Facilitated crystallization processes.** In a membrane crystallizer, the membrane surface acts as a promoter of crystallization in conditions of supersaturation that would not be adequate for spontaneous nucleation. The specific effect of the heterogeneous contribution allows crystals to be grown faster and/or by using lower initial amount of substance with respect to usual crystallization techniques. Moreover, particular surface-assisted pathways for solute – solute interaction, can allow specific crystalline habits of the same structure or specific polymorphs to be obtained. This would be of extreme importance, for example, in pharmaceutics, microelectronics, and other nanotechnology areas in which crystals' structure and morphology have an important impact on the properties of the crystals produced in terms of their specific applications.
- **Mild operating conditions.** In a thermal MCr, low feed temperature is sufficient to operate at high transmembrane fluxes. In an osmotic system, the supersaturation inside the crystallizing solution can be generated by using a stripping agent, which normally consists in a solution of an inert inorganic salt. In both cases, the concentration of the feed is carried out in gentle conditions, without the solution experiencing thermal or mechanical stress. This means that with the membrane-based system, labile or thermal-sensitive molecules, such as proteins, viruses, or other macromolecules, can be crystallized in large amounts and in very mild conditions, thus avoiding degradation and/or denaturation.
- **Improved crystal quality.** MCr allows generating high ordered crystals by improving first stage aggregation and aggregate growth. This would be a fundamental improvement in the field of for example protein crystallization, in which high diffracting single crystal is required for X-ray diffraction analysis for medical advancement [109].

An advantage of both membrane processes, MD and MCr, is that they can be scaled-up and scaled-down. As for every membrane operations, modularity might allow one to scale-up or scale-down a process by simply adding or removing membrane modules.

One principal limitation for MD and MCr techniques is membrane fouling. Fouling is defined as the accumulation of unwanted material on membrane surfaces with an associated detriment of function. In particular, fouling increases costs of energy consumption, downtime, cleaning, required membrane area, and required membrane replacement, and creates problems with product water contamination from pore-wetting [110]. For many years, it was believed that the relatively high mean pore size of the membranes and the low feed pressure in the process are sufficient to prevent the significant fouling and scaling. In contrast, MD/MCr membranes are relatively vulnerable to these unwanted phenomena and often require well-engineered designs and operating methods to avoid and mitigate the membranes by fouling. These design choices, especially in the case of inorganic scaling, are related to maintaining the concentration of ions and the temperature at the membrane interface within limits in which crystallization is not favored. Understanding temperature and concentration polarization effects (relative reduction in temperature and increase in solute concentration at the membrane interface, compared to the feed bulk, owing to the removal of energy and water mass through the membrane) therefore becomes very important. The types of fouling that occur in MD and MCr processes can be divided into four categories: inorganic salt scaling, particulate fouling and biological fouling. Inorganic scaling generally falls into one of three categories: alkaline, non-alkaline, and uncharged molecule scale [110, 111]. Inorganic scaling risk varies greatly based on the salts present. Biofouling of hydrophobic membranes has been a key research interest in the food, beverage, and wastewater industries. Many of the membranes used in these processes originated in these industries. Biofouling relevant to desalination includes bacteria, fungi, and biofilm studies. The operating conditions of the processes, especially the high temperatures and salinity, can restrict the microbial growth to a great extent [111]. As a consequence, the problems caused by biofouling in membrane processes, including NF, UF, or RO, should not occur to such a high degree in MD and MCr systems [112]. However, organic fouling can play a more important role in membrane wetting. Particulate and colloidal fouling risk is common in many feedwater solutions. Larger particles can often be addressed with modern filtration technology but smaller particles can be an issue in fouling. Particulate fouling in MD has proven difficult to remove, but it can largely be prevented by ultra- or microfiltration pre-treatment. Chemical degradation and damage to the membrane has proven to be a concern as well, but can be mitigated by selecting operating conditions that avoid fouling, extreme pH, and certain salts. The choice of membrane material and properties can also help to avoid chemical degradation; for example PTFE membranes may be less susceptible to damage than Polyvinylidene fluoride (PVDF) [113]. Studies with extreme susceptibility to fouling have almost exclusively been performed with hollow-fiber capillary membranes, with the feed internal to the capillary tubes [114]. Membrane materials that are somewhat less hydrophobic show relative resistance to fouling. Modifying pH in the feed or with cleaning may prevent or remove certain types of fouling very effectively as well. In MD, keeping feed temperature above 60°C has proven very effective in mitigating biofouling. Rinsing with a basic solution such as NaOH may resolve fouling for some substances, including humic acid. Mildly effective fouling prevention methods include boiling for removal of carbonate, ultrasonic cleaning, magnetic water treatment, flocculation, covering the membrane surface with a less porous smaller pore-size layer, and for humic acid, oscillating the feed temperature. System design characteristics also influence fouling. Concentration polarization, closely related to feed Reynolds number and rate of permeate production, is critical in causing fouling, and can be mitigated by increasing the feed flow rate [115]. Finally, stagnation zones or high residence times in the module may contribute to fouling as well [116].

5.2 MEMBRANES FOR MD AND MCR

The core of the MCs processes is the membrane itself, since it defines the range of the operative conditions, and in many cases, it is the most important element to consider for the mass transport operation. In these membrane processes, it is essential to have suitable membranes with well-controlled properties, because the final performance of the full process depends on the morphological and physical-chemical characteristics of the employed membranes. In particular, the structure of the film in terms of thickness, porosity, mean pore size, pore distribution, and geometry influences the successful outcome of the membrane process. The membrane must have the capability to interface two media without dispersing one phase into another, and to combine high volumetric mass transfer with high resistance to liquid intrusion in the pores. Thus, it is fundamental to know the characteristics of the membrane, to have a safe operation with the maximum potentialities of the membranes and the process. The membranes for MCs applications must be hydrophobic and porous, and must have good thermal stability and excellent chemical resistance to a wide variety of feed solutions, as well as adequate mechanical stability. In MCs processes, mass and heat transport depend directly on the membrane properties. Therefore, suitable membranes must meet specific characteristics that maximize mass transport and minimize heat transfer [36, 52, 75, 107, 117, 118]. Regarding the morphological characteristics, the membranes must have specific characteristics such as sharp pore size distribution, high porosity, low tortuosity and thickness. The pore size may be between several nanometers and a few micrometers. This parameter influences membrane performance by increasing the flux with increasing pore size. Depending on the application, if high fluxes are desired then high pore sizes are required. The pore size distribution should be as narrow as possible to minimize the wetting risk of the membrane. The liquid entry pressure (LEP) should be higher than the transmembrane pressure applied to prevent liquid feed from penetrating into the pores. This is realized by employing hydrophobic membrane materials and keeping the membrane pores relatively small [119]. Membrane porosity refers to the void volume fraction of the membrane (defined as the volume of the pores divided by the total volume of the membrane). Membranes with high porosity have a larger evaporation surface area. Therefore, high porosity membranes usually have higher permeate fluxes and lower conductive heat loss.

The tortuosity factor refers to the deviation of the pore structure from the cylindrical shape. A higher tortuosity value results in a lower permeate flux, therefore the tortuosity factor should be small.

The membrane thickness has great influence on the membrane performance. An inversely proportional relationship exists between the permeate flux and the membrane thickness. As the membrane becomes thicker, the permeate flux reduces, owing to an increased mass transfer resistance, while heat loss is reduced because of a smaller amount of transferred heat. A compromise should be made between the heat and mass transfer by properly choosing the membrane thickness [119]. The membrane material and its interaction with solutes and solvents primarily determine the physical-chemical characteristics.

The membrane should be highly wetting-resistant, and have LEP, the minimum hydrostatic pressure that must be applied onto the feed solution before it overcomes the hydrophobic forces of the membrane and penetrates into the membrane pores. LEP is characteristic of each membrane, and to achieve high LEP, the membrane should have high hydrophobicity and a small maximum pore size; its material should have low surface energy or, still better, its surface should be omniphobic (i.e., large contact angle with both, high and low surface tension liquids).

$$LEP_w = \frac{B\gamma_L \cos\theta}{d_{max}} \tag{5.16}$$

Where B is a geometric factor related to the pore structure, γ_L is the interfacial tension, and θ is the liquid – solid contact angle. However, as the maximum pore size decreases, the mean pore size of the membrane decreases and the permeability of the membrane becomes low.

In addition, the membranes used for MCs application should have high permeability. The flux increases with an increase in the membrane pore size and porosity, and with a decrease of the membrane thickness and pore tortuosity. Molar flux through a pore is related to the membrane's average pore size ($\langle r^a \rangle$), the membrane porosity (ε), the membrane tortuosity (τ) and the membrane thickness (δ) parameters by:

$$J \propto \frac{\langle r^a \rangle \cdot \varepsilon}{\tau \cdot \delta}$$

In fact, to obtain a high permeability, the surface layer that governs the membrane transport must be as thin as possible, and its surface porosity and pore size must be as large as possible.

The thermal conductivity of the membrane should be as low as possible. To meet this requirement, the membrane material must have low thermal conductivity and the membrane have high porosity. The thermal conductivity coefficients of the gases entrapped in the pores are an order of magnitude smaller than that of the majority of the polymers used in membrane preparation. Most of the hydrophobic polymers have similar thermal conductivities [120]. Therefore, high porosities are required to decrease the heat transfer by conduction through the membrane. It is advisable that the feed surface of the membrane be highly fouling-resistant. Fouling is one of the major problems in the application of porous membranes. MCs' fouling effect is not as strong as in pressure-driven membrane processes. However, membrane surface modification could be necessary if certain types of feed solutions must be treated. Another important aspect of these processes is that the membrane should have a long operating life with stable performance, for example, permeate flux and selectivity. The membranes for MCs applications must have high chemical stability and high thermal stability. The chemical stability of the membrane material has a substantial consequence on the long-term stability of the membrane. The membrane material may not be able to resist to degradation when high temperatures are employed. In fact, the proprieties of the polymer depend on the transition temperature that influenced the nature of the membranes in terms of long-term stability, for example, glass transition temperature (T_g) in the case of amorphous polymers, and the melting point (T_m) in the case of crystalline polymers,

The suitable membranes for MD and MCr applications should fulfill all the aforementioned features to achieve optimal morphological structures that lead to high permeate fluxes and rejection factors, together with high thermal efficiencies.

5.2.1 Membranes Materials: Traditional and Emerging

Selection of membrane materials and the geometry are crucial factors in membrane process design. As mentioned previously, a membrane employed for MCs applications must be macroporous and hydrophobic. These characteristics are key parameters in the design of a successful membrane for a MCs applications. In addition, the materials used for membranes fabrication should meet the mechanical, thermal, and chemical stability requirements, combined with low surface tension, low thermal conductivity, low absorptivity with respect to water, appropriate process ability, and eventually, low cost.

Owing to the broad spectrum of membrane features required for an appropriate membrane, several materials have been tried for synthesizing membranes suitable for MCs applications.

A wide variety of traditional membranes, including polymeric and inorganic membranes of hydrophobic nature, can be utilized in MD and MCr processes. Because of the possibility to modulate their proprieties such as mechanical, thermal, and chemical stability, selectivity, and permeability, polymeric membranes have attracted much more attention and many researchers have concentrated on their use in MD and MCr applications. Polymers such as polypropylene (PP), PVDF, and PTFE (Figure 5.7) are generally employed in polymeric membrane preparation for MCs applications, owing to their low surface tension values.

These nonpolar polymers naturally feature excellent hydrophobic characteristics with respect to the polar water properties. The hydrophobic nature of the membrane prevents solution from penetrating into the pores, thus creating a vapor – liquid interface at each pore entrance. This value may be quantified by a contact angle measurement. For a hydrophobic surface, a water droplet has a contact angle θ>90° with respect to the solid surface, and larger contact angles indicate greater hydrophobicity.

PP is a thermoplastic linear polymer with structure consisting of – CH_2 $CH(CH_3)$ as the repeating unit. PP can exist in semicrystalline as well (isotactic PP and syndiotactic polypropylene) as amorphous forms (atactic polypropylene) depending on the tacticity, arrangement of pendant groups along the backbone chain. Glass transition temperature of this form is −10°C and it melts round about 155°C–166°C. Generally, in membrane preparation at commercial scale the most used form is Isotactic PP owing to very good solvent resistance still can be dissolved in strong solvents such as 1,2,4-trichlorobenzene, halogenated hydrocarbons, decalin, aliphatic ketones, and xylene at relatively high temperatures.

Another polymer largely used for membrane preparation suitable for MD and MCs processes is PTFE. Its backbone structure comprises of carbon chains with two fluorine atoms for each carbon atom. Complete fluorination of carbon atom, combined with the bonding strength of carbon and fluorine, makes PTFE a highly stable, biologically inert, and chemically nonreactive material. With respect to PVDF and PP, PTFE polymer displays the highest hydrophobicity because the surface energy of this polymer is the lowest. PTFE has a very high melting point (342°C) and it does not flow, even above its melting point. In phase of preparation present issues because is insoluble in any solvent at room temperature. Commonly used phase inversion or melt spinning techniques for membrane synthesis do not work for PTFE.

Therefore, the most research on membrane preparation has been carried out by using the PVDF polymer. PVDF is a semicrystalline polymer with typical

FIGURE 5.7 Chemical formula of some of the most used polymers in membrane preparation for MCs.

crystallinity between 35% and 70%. This polymer has glass transition temperature (Tg) and melting point in the range of −40°C to −30°C and 155°C to 192°C, respectively. PVDF show excellent chemical and thermal resistances, good mechanical strength, and in addition, exhibits an excellent process ability in different common solvents such as N-methyl-2-pyrrolidone (NMP), N,N-dimethyl acetamide (DMAc), dimethyl formamide (DMF) and dimethyl sulfoxide (DMSO). Both the easy processing and the excellent properties of this polymer makes the PVDF the most applied material for membrane fabrication. In addition, the strong hydrophobic character of PVDF provides the motivation for its use for MD and MCr membranes.

Polymeric membranes are selected for MCs applications based on their morphological characteristics, higher hydrophobicity and LEP of water inside the membrane pores, although PP and PTFE, can only be processed at temperatures higher than their melting temperature. In addition, they are often selected because they can be manufactured in various configurations: flat-sheet nanofibers, tubular, and in the form of hollow-fibers, which is the configuration allowing the highest compaction because it presents the highest surface/volume ratio permitting a very large evaporation surface in a small volume [121].

Regarding membrane fabrication, PP membranes are prepared by stretching and track-etching techniques. Research on PP membrane fabrication is investigating the ways improved hydrophobicity, high porosity, and pore size distribution by using methods such as thermally induced phase separation (TIPS) and electrospinning. In addition, incorporation of several materials such as carbon nanotubes (CNTs) and other fillers was also explored to improve membrane performance. While PTFE membranes are generally synthetized by complicated extrusion, stretching, and sintering procedures, PVDF membranes are usually prepared by the phase separation, particularly through nonsolvent-induced phase separation (NIPS). Future research directions for PVDF membranes for MD applications are expected to focus on better control of membrane morphology, pore size distribution, and enhancement and stabilization of hydrophobic character over a longer duration of time by such using techniques as TIPS and electrospinning [72, 122]. The use of other materials such as polymers and nanoparticles with PVDF to achieve desired properties is also expected to gain more attention. These manufacturing techniques are discussed in detail in the next section. The polymer type influences the membrane hydrophobicity. In addition, another important factor in membrane preparation affecting the membrane proprieties is the polymer concentration in the casting solution [73, 123, 124]. Furthermore, it was observed that the reduction of the polymer concentration enhanced the porosity, surface pore size and permeate flux. On the other hand, low polymer concentration reduced the membrane mechanical properties and the LEP in the membrane pores. The increase of the polymer concentration in the solution owing to the reduction of the coagulation rate led to a change of the resulting membrane structure altered from a fingerlike structure to a spongelike structure [123].

Generally, mixed matrix membranes (MMMs) were prepared by introducing organic or inorganic additives in the polymer solutions. This allows changing the membrane morphology, inducing pores in the membrane matrix, improving pore interconnections and, as a result, increasing the performance of the processes. Additionally, increases the viscosity of the polymer solutions, and changes both the kinetic and the thermodynamic properties of the coagulation process affecting the final membrane structure [125, 126]. Different types of additives have been utilized for preparation of membranes for MCs. For example, inorganic salts such as lithium chloride and lithium perchlorate, polymeric additives such as polyvinylpyrrolidone, poly(ethylene glycol), or any other type of nonsolvent additives capable of increasing both the pore size and the porosity of the prepared membranes, have been employed [127, 128].

Recently, several fluoro-copolymers, such as polyvinylidenefluoride-co-hexafluoropropylene (PVDF-HFP), polyvinylidenefluoride-tetrafluoroethylene (PVDF-TFE), poly(vinylidene fluoride-co-chlorotrifluoroethylene) (P(VDF-co-CTFE)), poly(tetrafluroethylene-co-hexafluoropropylene) (FEP), poly(ethylene chlorotrifluoroethylene) (ECTFE), Hyflon AD, and Teflon AF have received consideration as novel materials to prepare suitable membranes for MCs applications to improve wetting resistance, thermal stability, chemical resistance and mechanical strength of MD membranes [129].

PVDF-HFP shows higher hydrophobicity than PVDF owing to the hexafluoropropylene (HFP) group, which increases the fluorine content of the copolymer [130]. Different PTFE, P(VDF-co-TFE) can be dissolved in common solvents and can thus be utilized for the preparation of phase inversion membranes [131]. ECTFE and FEP are easier to process than PTFE. In addition, ECTFE is more hydrophobic than PTFE, and exhibits excellent chemical and thermal stabilities, compared to PVDF. These copolymers have been employed to prepare membranes by the phase inversion technique. High hydrophobicity, narrow pore size distribution, and good mechanical strength [132] also characterize PVDF membranes coated with Hyflon.

Although the polymeric membranes are those most used in membrane contactor applications, great attention has also been given to use of inorganic membranes in these processes. Among the polymeric membranes, inorganic membranes can tolerate aggressive environments because of their excellent chemical, structural, and thermal stabilities. Ceramic oxides, including alumina (γ-Al$_2$O$_3$ and α-Al$_2$O$_3$), zirconia (ZrO$_2$), titania (TiO$_2$), and silica (SiO$_2$), rather than such oxides as silicon carbide (SiC), are the most commonly used materials for the fabrication of membranes. Ceramic materials show a natural hydrophilic character, owing to the presence of hydroxyl groups on the surface. This has limited the popularity of ceramic membranes for MD and MCr applications.. The use of ceramic membranes for these membrane processes requires the modification of the membrane surface, using hydrophobic materials [109, 110]. Reactive groups such as methoxy, ethoxy, or active chlorine are used for the surface modification process [111]. The reaction of the reactive groups with the hydroxyl creates a stable covalent bond that forms a hydrophobic molecular layer. An interesting surface modification on ceramic membranes was performed by modifying the silica surface with tetraethoxysilane (TEOS) [133]. To obtain a greater hydrophobicity on ceramic membranes, zirconia surfaces were modified with hexadecyltrimethoxysilane (HDTMS) [134]. The grafting of the silane on the membrane surface has, in general. little effect on the membrane morphology and pore structure. Perfluoroalkylsilanes (PFAS), possessing reactive grouping, are the most popular molecules to add a hydrophobic character to ceramic membranes [135–137]. In addition, tubular zirconia and titania ceramic membranes were imparted hydrophobic property by grafting them with ethanolic solution of triethoxy-1H, 1H, 2H, 2H-perfluorodecylsilane [138]. The modified membranes was investigated in three configurations of MD in particular DCMD, VMD and AGMD.

Another interesting choice over ceramic materials, generally brittle, are metals because possess higher mechanical strength. Additionally, metals have distinct properties such as thermal and electrical conductivities. Several methods have been adapted to confer hydrophobicity to metallic substrates. Some of the recently used methods are immersion in stearic acid solution $(CH_3(CH_2)_{16}COOH)$ [139], perfluorinated octyltrichlorosilane treatment to a stainless-steel surface [140], fluoropolymer deposition by inductively coupled radio frequency plasma technique [141], treatment with fluoroalkylsilane (FAS) dissolved in silver nitrate solution for electroless deposition of Ag (silver) structures [142], and immersion of metallic substrates in a solution containing low-surface-energy molecules such as 1H, 1H, 2H, 2H – perfluorooctyltrichlorosilane [143]. As explained previously, the hollow-fiber configuration provides a very large area/volume ratio enabling the reduction of membrane system volumes, increasing the surface of exchange, and hence the manufacturing and process costs. Thus, the combination of metallic material and hollow-fiber geometry offers an interesting prospect for future MD and MCr development [144, 145].

Hydrophobicity is a significant parameter in the design of a suitable membrane for membrane contactor application. In fact, maintaining dry pores throughout the process is necessary for a successful MD and/or MCr process. This is the reason that hydrophobic polymers are most often the choice for design of membrane materials. However, pore wetting by vapor condensation and liquid penetration may happen during continuous MD operation even with membranes made of very well-known hydrophobic polymers such as PTFE [146]. To enhance the hydrophobic character, many approaches have been adapted. Generally, there are two approaches to enhance the hydrophobicity of membranes; one is the construction of rough structures on the membrane surface and the other is surface-modification to lower its surface energy.

Membrane surface modification using various technologies such as grafting, differential chemical etching, coating of different low energy fluoropolymers and copolymers [147, 148], plasma treatments, formation of various hierarchical structures, treatment with different types of silane compounds, incorporation of nanoparticles into the dope solution during membrane synthesis, and making the surface rough have been applied [149–151]. Surface roughness is an interesting technique to render the super hydrophobicity to the membrane surface; however, its further effects on surface scaling/fouling and thermal polarization still need to be addressed [135]. While doing so, the most important factor to be considered is not to alter the porous characteristics of the native membrane.

More recently, other strategies have been investigated to improve hydrophobic properties of the membranes, such as blending between two or more polymeric materials [8], incorporation of inorganic and/or additives in casting solutions, composite hydrophobic/hydrophilic layered membrane synthesis [94, 152–155].

Various types of fluorinated surface modifying macromolecules (SMMs) were blended into hydrophilic polymers such as polyetherimide, PEI, polysulfone, polyethersulfone, and PES to prepare porous composite hydrophobic/hydrophilic membranes suitable for MCs applications by phase inversion. During membrane formation, SMMs migrate to the polymer – air interface, rendering it more hydrophobic. This type of membrane showed good performances in DCMD because they combine the low resistance with mass transport, achieved by the reduction of the water vapor transport path and the low conductive heat loss through the whole membrane owing to the thicker hydrophilic bottom layer [156].

Additionally, thermally rearranged polymers have appeared as a class of materials for membrane preparation with tunable cavity size, narrow cavity size distribution, and good mechanical and chemical resistance [157]. These polymers can be very suitable contenders for water purification applications including candidates for water [158].

Advanced functional membranes with tailored properties became a widely researched area over the last few years to improve the performances in membrane contactor applications of polymeric membranes [159, 160]. The realization of systems composed of polymer and nanomaterials (organic and an inorganic phase coexist) has a considerable role in membrane-based processes, including MD and MCr. Recently, electrospun nanofiber membranes have been investigated in many studies and have shown very high porosity, excellent hydrophobicity, very good interconnectivity, and very high surface to volume ratio, making them interesting candidates for MCs applications. Several functional materials can be incorporated into the nanofibers during or after their spinning, thus incorporating multifunctionality into the fibers. Lately, some experiments were performed with membrane prepared by the electrospinning process. More recently, graphene oxide (GO) is becoming an interesting material used to fabricate high-performance membranes for membrane contactor applications, owing to its very high strength-to-weight ratio that exhibits amazing selective permeability toward various components [161]. GO can be used in membrane preparation, either as the main constituent material, or as an additive to other matrix materials. In addition, owing to its abundant functional groups, GO can be easily functionalized with other molecules that can improve its properties [162]. Functionalization with silanes, for example, have been shown to improve the performance and mechanical strength of membranes [18]. The facile fabrication of PVDF mixed-matrix membranes incorporating GO and APTS-functionalized graphene oxide as nanofillers to investigate their effects on the properties of the PVDF polymer for MD applications. The obtained results show the great potential of using small additions of GO and related materials to increase the productivity of the highly promising, low-cost MD technology [163–165]. Furthermore, in preparation of membranes for water desalination, carbon nanotubes are becoming more popular in small-scale applications. Nanoscale dimensions of CNTs appear to be like cylindrical rolled up graphene. They have outstanding mechanical strength, chemical resistance, and thermal properties. The high transport rate of water molecules inside the CNTs, together with their potential to change the water – membrane interaction (able to stop the permeation of water molecules while favoring the preferential transport of vapors through the pores) has encouraged their incorporation into the membrane matrix [160, 166]. In addition, presence of CNTs in the membrane matrix optimized porosity and hydrophobicity [167–169].

The next-generation of materials to prepare membrane with extraordinarily high permeability are two-dimensional (2D) materials of atomic thickness, which include graphene, and molybdenum disulfide. 2D membranes show well-defined transport channels and ultralow thicknesses; they were employed with exceptional performance for liquid and gas separation applications. The unique atomic thickness of the membrane confers ultralow resistance to mass transport. The current main challenges for the commercial scale implementation of these membranes include limited available techniques for exfoliating the high aspect ratio and intact nanoporous monolayers from bulk crystals; drilling of the pores with required characteristics such as being uniform, high-density, large-area, sub-nanosized in membrane matrices; and scaling up these atomic scale membranes into real-scale separation devices.

Another important aspect to consider in MD and MCr when dealing with complex feed solution is antiwetting proprieties of the membranes. For this reason, the concept of omniphobicity has attracted increasing attention from membrane researchers, and several attempts have been done to fabricate omniphobic membranes for membrane contactor applications [170, 171].

By forming a stable solid – liquid – vapor interface, omniphobic membranes exhibit strong wetting resistances toward liquids with both high and low surface tensions. The most widely adopted strategy to design omniphobic surfaces is through the combination of a re-entrant surface texture and a low-surface-energy coating. All the omniphobic MD membranes show superior wetting resistance in treating feed solutions containing salts and surfactants.

5.2.2 Membrane Configurations and Membrane Modules

The performance of a MCs process not only depends on the membrane itself, but is also affected by the membrane module design, the membrane configuration, and the applied operating conditions. Membrane geometry and position determine the relation of the membrane, the feed flow and the permeate. Depending on the modular design, it is possible to arrange compactness, fluid dynamics, easiness of cleaning and maintenance. An adequate membrane contactor module should exhibit high membrane packing density, high permeate product rate, low temperature and concentration polarization effects, low fluid pressure drop, and low external heat loss [172]. Once the particular membrane has been selected for a particular application, the configuration on which these membranes are going

Membranes for MD/MCr

to be arranged should be defined. A good module configuration can provide compactness and robustness, and through the appropriate fluid dynamics, it might reduce pressure drops, enhance the mass transfer, and impart a positive impact on the process [118]. As a result, thermal/concentration polarization, fouling, and energy consumption would be reduced. The high-density packing, interface area and high mass transfer coefficient make the height of the transfer unit lower than conventional unit design for the same operation. A benefit of the membrane process is a straightforward scale-up because the available surface area between gas and liquid phase is known [173]. Assembling modules on series and parallels is a practical way to increase capacity, reach desired performance, and control pressure drops, with the aim of avoiding liquids' entry into the membrane pores. Various types of membrane modules were proposed and tested in membrane contactor applications. A membrane module is a complete unit composed of the membranes, a housing, feed inlet, concentrate outlet, and permeate outlet. Membrane modules may contain different membrane configurations, including plate-and-frame, spiral-wound, tubular, capillar and hollow–fiber. The membrane geometry is planar in the first two configurations and cylindrical in the two others. In the plate-and-frame configuration (Figure 5.8), flat-sheet membranes are disposed together with spacers between two plates. The membranes may be square or circular, and arranged in vertical or horizontal stacks. The membranes in this configuration are easy to clean and replace. This type of module cannot withstand very high pressure. The surface area to volume ratio of plate-and-frame modules is not high. This module configuration is used mainly on laboratory scale applications.

In the spiral-wound configuration (Figure 5.9), flat-sheet membranes and spacers are rolled around a perforated central collector tube. One of the phases flows in the axial direction of the modules, while the second flows radially to the

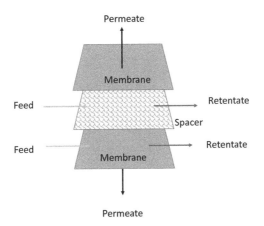

FIGURE 5.8 Plate-and-frame configuration.

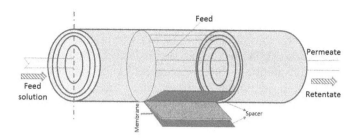

FIGURE 5.9 Spiral-wound configuration.

center. The packing density on this configuration is high, but the flow pattern reduces the fouling effects and the pressure drops.

Tubular membranes can be distinguished in: hollow fibers (fiber diameter below 0.5 mm), capillary (fiber diameter comprised between 0.5 and 10 mm) and tubular (fiber diameter > 10mm). A very large number of them are connected to perforated end plates and the entire bundle is inserted in a vessel or jacket. Flow direction may be inside-out or outside-in (Figure 5.10).

The bundle is set inside a shell ensuring phase separation through proper potting sealing. Accordingly with process requirements one of the phases flows through the tubes and the second flows on the shell side, parallel or perpendicular to the tubes if baffles are used.

Capillary and hollow fiber modules have the highest membrane surface area per element; however, due to the high density packaging, these modules appear to be very sensitive to the feed stream quality in terms of fouling potential. The main advantage of hollow–fiber modules is their compactness; the high packing density gives this module configuration a great membrane area. Their disadvantage is their high susceptibility to fouling and clogging, which limits their use to clear fluids of relatively low viscosity. The compactness of the bundle made the substitution of single membranes impossible and cleaning process is more difficult than in other configurations.

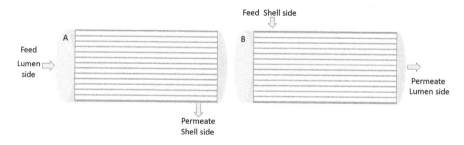

FIGURE 5.10 Tubular, hollow-, or capillary configurations.

Membranes for MD/MCr 157

The tubular configurations are similar to the hollow–fiber set-up, but the tubes diameter is larger. The membrane is cast on the inner wall of rigid porous tubes, made of polymer or ceramic. The tubes are connected to end-plates and installed as parallel bundles inside a shell. The tubes may have diameters in the range of 10 to 25 mm. Flow direction is usually inside-out, the retentate flows inside the tubes and the permeate is collected at the shell-side. It is often possible to reverse the flow for cleaning and unclogging of the membrane. Tubular configurations provide the possibility of maintaining high tangential velocity in the feed stream and are therefore particularly suitable for applications where the feed contains a high proportion of suspended solids or must be strongly concentrated. Because of their relatively large diameter, tubular membranes are easy to clean. The surface area to volume ratio of tubular modules is not high. For many applications, it is preferred as the packing density is higher than plates and frame but the drawbacks phenomena in hollow fibers are reduced.

5.2.3 Membrane Fabrication Techniques

Different membrane fabrication techniques were applied to prepare synthetic membranes suitable for MCs application. Membranes can be prepared by sintering, stretching, track-etching, electrospinning and phase inversion [1, 4, 85, 86]. Additionally, some types of membranes are prepared combining different techniques. For example, PTFE hydrophobic hollow-fiber membranes were prepared by a cold pressing method including extrusion, stretching, and sintering [87].

Sintering is a simple technique that allows obtaining porous membranes from both organic and inorganic materials. The method involves compressing a powder composed of particles of known size and sintering at high temperatures. The required temperature depends on the material used. During sintering, the interface between the particles in contact disappears. Different materials can be used such as polymer (polyethylene (PE), PTFE, PP), metals (stainless steel, tungsten), ceramics (aluminum oxide, zirconium oxides), graphite (carbon) and glass (silicates). The pore size of the resulting membrane is determined by the particle size and its distribution in the powder. This technique allows obtaining pores between 0.1 and 10 μm. Sintering is a technique suitable for preparing membranes from polytetrafluoroethylene, because this polymer, which is chemically and thermally very resistant, is insoluble. In fact, all the materials mentioned above as basic materials for the sintering process have the common characteristic of possessing considerable chemical, thermal and mechanical stability, particularly inorganic materials. However, only microfiltration membranes can be prepared by sintering.

Stretching is a solvent-free technique, a film extruded from a partially crystalline polymeric material (polytetrafluoroethylene, polypropylene, polyethylene) is stretched perpendicularly to the extrusion direction, so that the crystalline regions are located parallel to the extrusion direction. Membranes are made at a temperature close to its melting point coupled with a rapid stretching to form micropores [88]. When a mechanical stress is applied, small breaks occur and a porous structure is obtained with pores of minimum size of about 0.1 μm and maximum of 3 μm. This technique is suitable when using highly crystalline polymers.

The final porous membrane structure and its properties depend both on the physical properties of the materials used such as crystallinity, melting point, and tensile strength and the applied processing parameters. With this technique, membranes with a relatively uniform porous structure and porosity of about 90% are prepared.

In the track-etching technique, a polymeric film (often polycarbonate) is irradiated with energetic heavy ions applied perpendicularly to the film. The particles damage the polymer matrix and create trajectories. The film is then immersed in an acid or alkaline bath and the polymeric material is etched along these trajectories to form cylindrical pores with a narrow pore distribution [92, 93]. Pore sizes may vary from 0.02 to 10 μm but surface porosity is low (about 10% maximum). The choice of material depends above all on the thickness of the film used and on the energy of the particles applied (usually around 1 MeV). The maximum penetrating thickness of the particles with this energy is about 20 μm. With the increase in energy of the particles, the film thickness also increases, making it possible to use inorganic materials. The membrane porosity is determined by the duration of radiation, while the pore size is related to the etching time and temperature [94]. It was observed that the membranes prepared with this technique have low porosities.

Electrospinning is a simple and versatile technology to produce nanofibers and has high potential for commercialization [174]. A high electric field applied on a polymeric solution generate nanofibers, which can be attached on a collector [159]. The electrospinning device consists of three parts the high voltage electric source (0–30 kV), the syringe containing polymeric solution on a pressure syringe pump with a metallic tip and rotating drum collector collecting fibers [175]. A syringe containing polymeric solution is pressurized by a syringe pressure pump and the high voltage supply is applied on the metallic tip. When the applied electric field is high enough to overcome the surface tension of the solution in the metallic tip, Taylor cone shape droplet of the solution form on the metallic tip and fine nanofibers will be produced [176]. The produced nanofibers are ejected from the tip and deposited on the rotating drum collector. The produced fibers can be greatly elongated by electrostatic repulsion of Coulombic forces and the solvent evaporate during the process. The morphology of the produced nano/microfibers changes by varying the solution viscosity, electrical conductivity, surface tension, environmental conditions, applied electric potential, flow rate of the polymer solution, gap between the needle and the collector, etc [174].

Most commercially available membranes for MCs application are polymeric prepared by phase separation (often called phase inversion) technique [177]. This is a very versatile technique that allows to obtain several types of morphology.

Membranes for MD/MCr

Phase inversion is a demixing process in which the initially homogeneous polymer solution is transformed, in a controlled way, from a liquid to a solid phase. The solidification process begins with the transition of a single liquid to two liquids (liquid-liquid demixing). At a certain stage of demixing one of the liquid phases (the phase with the highest concentration of polymer) will solidify to form a solid matrix. By controlling the initial stage of the phase transition, it is possible to check the morphology of the membrane. This technique can be used to prepare both asymmetric and symmetric porous membranes via different methods, immersion precipitation (IP) or nonsolvent induced phase separation (NIPS), thermally induced phase separation (TIPS), vapor induced phase separation (VIPS), and evaporation induced phase separation (EIPS). Largely, NIPS, TIPS and VIPS are the most commonly employed methods for MD and MCr membrane preparation [178].

The NIPS technique has three major components: polymer, nonsolvent and solvent. In NIPS technique, the polymer is dissolved in an appropriate solvent and the formed solution is then cast on a flat support. After a partial evaporation of the solvent (dry-wet phase inversion) or without allowing any solvent evaporation to occur, the cast film is immersed into a bath of coagulant (non-solvent) media. The polymer solution is separated spontaneously into two phases, a polymer-rich phase and a liquid-rich phase, via solvent nonsolvent exchange phenomenon. Demixing takes place and the liquid phase becomes a solid polymeric film. During the evaporation step, a thin skin layer of solid polymer is formed at the top of the cast film, owing to the loss of solvent and the increase in the polymer concentration. The fabricated membranes by dry-wet NIPS typically shows a dense surface active layer with an asymmetric structure [179].

The TIPS technique has only two components, which are polymer and solvent. In the TIPS method, a homogenous polymer solution is prepared at high temperature and then this is cooled in a single solvent or in a mixture of solvents to make phase separation possible. The precipitation step takes place via cooling the solution to a given demixing point, which depends on the type of the polymer solution. The evaporation of the solvent results in the formation of pores in the solid matrix. This technique is applicable to a wide range of polymers, including those that cannot be used for formation of membranes via NIPS and semicrystalline polymers [45, 91]. The fabricated membranes via the TIPS technique indicates a highly porous and symmetric morphology that leads to increased water flux performance [180]. In addition this procedure are easy to be reproduced, owing to the simplicity of its characteristics [181]. Even via the technique needs high energy consumption, it allows a different selection of polymers and solvents due to the high operating temperature. Despite high energy consumption, TIPS technique is capable of fabricating PVDF membranes with high porosity and narrower pore size distribution, which are desired properties for MD application compared with NIPS method.

In the VIPS method, a film consisting of a polymer and a solvent is placed in a vapor phase composed of a nonsolvent saturated with a solvent. The high concentration of solvent in the vapor phase prevents the solvent from evaporating from the film. The formation of the membrane is due to the diffusion of the non-solvent in the film. A porous membrane is obtained without a dense skyn layer. The beginning of the membrane formation process is crucial and largely determines the final properties of the separation [182].

The simplest technique for preparing membranes by phase inversion is precipitation by evaporation of the solvent. In this method a polymer is dissolved in a solvent and the polymeric solution is poured onto an appropriate support, for example a glass plate or another type of support, which can be porous (e.g. polyester) or nonporous (metal, glass or polymers such as polymethylmethacrylate or Teflon). The solvent is allowed to evaporate in an inert gas for example nitrogen to exclude water vapor, obtaining a homogeneous dense membrane. Instead of pouring it, it is possible to deposit the polymer solution on a substrate by immersion coating or vaporization, followed by evaporation.

In the precipitation by controlled evaporation, the polymer is dissolved in a mixture of solvent and nonsolvent (the mixture acts as a solvent for the polymer). Since the solvent is more volatile than the non-solvent, the composition changes during evaporation to a higher content of non-solvent and polymer. This leads to the precipitation of the polymer with the formation of a membrane having a denser skin layer.

5.2.4 Advances in Membrane Fabrication and Modification

The membrane structure plays a major role in the selective and the permeability of the membrane-based processes [169]. However, appropriate membranes specifically designed for membrane contactor application is still lacking. Currently, most of the membranes used for MD and MCr studies are membranes made of either PVDF, PFTE or PP developped for MF or UF [183]. However, some properties must be enhanced. Higher efficiency in MD and MCr, require better hydrophobicity, higher porosity, adequate pore size, and narrower pore size distribution. In addition, these membranes have low permeability and some wetting issues with more complex feed solutions during the MD and MCr processes. Thus, a need to develop new membranes specifically designed for MCs application is imperative. Although phase inversion is the major technique used to fabricate membranes suitable for MCs application, in more recent years different methods have been employed for the synthesis and modification of polymeric membranes to improve their performances.

Electrospinning and 3D printing technique have been introduced to fabricate membranes for MCs application [156, 184, 185]. However, the 3D printing technology is still at its infancy and present several limitations, especially on control of membrane pore size [186]. Electrospun nanofiber membranes have garnered wide attention in recent years as potential membrane for MD owing to their interesting characteristics [187]. These membranes fabricated through an electrospinning process show many advantages such as very high porosity, excellent hydrophobicity, very good interconnectivity and very high surface to volume ratio making them interesting candidates for desalination applications. Eletrospinning can be performed with polymer solution or melt and the properties can be modulate by changing the process parameters, material used and the post treatment step applied [34–36]. This new technique give the opportunities to make the membranes with vast variety of polymers due to the possibility to use polymer melt instead solution. Several functional materials can be incorporated into the nanofibers during or after their spinning thus incorporating multi functionality into the fibers. Some lab scale applications of electrospun nanofiber membranes have also been reported in recent literature [35–37].

Composite membranes fabricated with different materials allowed to obtained membranes with characteristics suitable for MD. Bilayer hydrophobic/hydrophilic porous composite membranes have been used in direct contact MD. The hydrophilic layer is brought into contact with the permeate liquid while the hydrophobic layer is maintained in contact the feed aqueous solution.

Composite membranes were prepared using different strategies either by modifying the already existing membranes such as coating, grafting, modification of the plasma surface, etc. or in situ modification during membrane formation [156, 188, 189]. Good performance in MD, were obtained employed a triple layer membrane composed of a thin hydrophobic electrospun nanofibrous layer coated on PVDF microporous layer, a middle layer was made by immersion precipitation method, in addition to the bottom hydrophilic polyethylene terephthalate (PET) support layer [190]. Triple-layered membrane showed hydrophobic character of the nanofibrous layer, higher water contact angle and higher LEP values as compared to the bilayer membrane.

Recently, interest has also shifted to the preparation of composite hollow-fiber membranes suitable for MD application. In particular, dual-layer hollow-fiber membranes were prepared using the dry/wet spinning process by a triple-orifice spinneret. The bore fluid was extruded through the central channel of the spinneret and both spinning solutions for example hydrophobic and hydrophilic solutions, were circulated through its annular channels, middle and the outer. The inner solution was composed of PVDF/polyacrylonitrile (PAN)/hydrophilic Cloisite NАþ/EG/NMP and the effect of adding like methanol (MeOH) and fluorinated silica particles into the outer dope solutions composed of PVDF/NMP was studied [191]. In another work, PTFE particles were added to the PVDF/NMP/EG solution to prepare the outer layer of the hollow-fiber membranes, while the inner layer was composed of the same polymer solution without particles [192]. In this case, a macrovoid-free morphology and a thin outer layer were obtained when 30% by weight of PFTE particles was added in the outer layer. A number of research groups started to look into nanotechnology specifically the electrospinning technique to prepare novel membranes [193]. Nanofibers composed of two layers were prepared by electrospinning and tested for desalination by DCMD. The use of a dual nanofibers membrane using PVDF-HFP as the active membrane surface layer and polyacrylonitrile (PAN), or Nylon-6 (N6) or polyvinyl alcohol (PVA) as a support layer showed interesting results in MD. The performance in desalination by AGMD of dual-layer nonwoven nanofiber membranes have been evaluated and compared with a single layer nanofibers membrane and a commercial PVDF flat-sheet membrane [194]. Although there has been good progress in the preparation of superhydrophobic electrospun membranes for MD application, there are still concerns on the membrane durability and robustness, mechanical stability and simplicity of preparation.

Membrane surface coating is applied to improve the surface hydrophobicity of the membranes. The main drawback of this technique is the instability of the coated layer, which could be removed during the operation process owing to the weak physical interaction between the membrane and the coated layer, and the high risk to close the pores and/or to reduce their size. In some cases, chemical treatment or solid-vapor interaction crosslinking was applied. Surface grafting is advisable because the membrane surface is modified through the covalent attachment between the grafted chains and the membrane. In contrast to the physically surface coating method, covalent bonding interaction of graft chains on the membrane surface avoids their delamination, offering a long-term chemical stability of grafted chains.

Recently, several surface-modified membranes have been designed and applied for MD technology. Self-supporting bucky-paper membranes were prepared using CNTs and coated on both sides with a thin layer of PTFE by sputtering to increase hydrophobicity and mechanical stability of the membrane, without altering pore size and porosity [166]. This composite membrane composed of four layers: PTFE layer, CNT BP, PTFE layer and PE support showed good performances in MD tests.

A great interest has been addressed to prepare superhydrophobic membranes suitable for MCs application [195–197]. Superhydrophobic membranes for DCMD were fabricated by developing a hierarchical structure with multilevel roughness. TiO$_2$ nanoparticles were deposited on PVDF membranes using low temperature hydrothermal (LTH) technique [195]. TiO$_2$ precursor was prepared by mixing anhydrous ethanol with 2,4-pentanedione, perchloric acid, and titanium (IV) iso-propoxide and ultrapure water. Then a solution of Pluronic F127 as a templating agent in anhydrous ethanol was added. Then, the resulting coated membranes were fluorosilanized using 1H,1H,2H,2H-perfluorododecyltrichlorosilane (FTCS). The functionalized membrane showed good thermal and mechanical resistances. Other superhydrophobic PVDF membranes were prepared by coating using different templating agents such as polyethylene glycololyethylene glycol (PEG), Wacker IM-22 and cetyltrimethylammonium bromide [197]. The structure, wettability and MD performance of the obtained coatings membranes depended on the physical and chemical properties of the templating agent. In particular, the PEG showed the best properties, with optimal LEP and DCMD performance evaluated in terms of flux and rejection. Other techniques for preparing superhydrophobic membranes have been employed. PVDF nanofibrous membranes integrally modified (I-PVDF) and surface-modified (S-PVDF) were prepared by electrospinning. Surface modification with polydopamine (PDA) improved the adhesive force, silver nanoparticle deposition during chemical reduction optimized the morphology and roughness of the membrane, and hydrophobic treatment was conducted with 1-dodecanethiol. After modification, the obtained I-PVDF nanofibrous membrane was not wetted in MD.

Another method to fabricate superhydrophobic PVDF membranes for MD application is via spraying a mixture of polydimethylsiloxane and a hydrophobic silica (SiO_2) nanoparticles on membranes using an airbrush [198]. The resulting membrane was successfully tested in DCMD and excellent antifouling property was observed compared to the polymeric membrane.

The carbon tetrafluoride (CF_4) plasma surface modification technique allowed to change the membrane surface wettability without changing the bulk properties. This method was employed to prepare PVDF hollow-fiber membranes for MD using plasma and a chemical modification procedure by hydroxylation. Greater hydrophobicity, better mechanical strengths, higher LEP values and smaller maximum pore sizes was obtained respect to unmodified PVDF hollow-fiber membrane [199].

This technique was also used to improve the membrane hydrophobicity owing to the strong fluorination of the membrane surface by fluorine functional groups. For example, an asymmetric hydrophilic polyethersulfone (PES) flat-sheet membrane has been made hydrophobic and suitable for DCMD application [200].

To improve hydrophobicity self-synthesized fluorinated silica particles [55], calcium carbonate particles [84] and clay particles [49, 52, 83] were used. Clay particles blending improved the tensile modulus and the long-term stability of PVDF hollow-fiber membranes.

Thermally conductive nanoparticles were dispersed in the hydrophilic layer of dual-layer membranes to evaluate the effects on the thermal efficiency, heat transport by conduction through the hydrophilic layer, and permeate flux. A significant increase of the permeate flux in MD was observed when graphite particles and multiwall carbon nanotubes (MWCNTs) were dispersed into PVDF/PAN/Cloisite NA^+ [201].

To improve the membrane properties and performance CNT was used to prepare membranes for MCs application owing to its exceptional mechanical, electrical, and thermal properties, high aspect ratio and lightweight [202]. Initially CNTs were extensively studied as stand-alone or filler materials of polymer composites for different application [203].

Self-supporting CNTs membranes prepared by vacuum filtration [188] were suitable for MD application owing to the high hydrophobicity, high void volume fraction, high specific surface area and relatively low heat transfer by conduction.

However, stand-alone CNTs membranes during MD testing are prone to damage.

The CNTs coated on polymer membrane should be robust enough to remain in the membrane matrix and at the same time confer new characteristics to the membrane. Other research has been carried out to try to immobilize the CNTs on support without being carried away by fluid flow.

Composites membranes were prepared by immobilizing CNTs in the pores of hydrophobic PVDF membranes [160]. CNTs were dispersed in a polymer solution and under a hydrostatic pressure the solution was injected into the lumen of a conventional hollow fiber [204]. The resulting membranes were thermally stable, robust, and possessed high selectivity and prevented liquid penetration into the membrane pores. Regarding performance in MD, an increment in permeate flux and salt reduction was observed [160].

Membrane technology has dramatically advanced over the past 50 years by combining advances in materials with innovative module design and an increasingly deeper understanding of the associated transport phenomena. More recently, 2D materials, typically represented by graphene and its derivatives, have attracted vast interest and are generally believed to have great potential.

Graphene particles and CNTs were introduced into PVDF-HFP membrane, altering the membrane's thermal properties [179]. Changes in membrane morphology were small with the introduction of these particles, although CNTs and graphene acted as nodules within the composite structure. The addition of these particles may increase the surface roughness, the contact angle, and effective heat transfer surface, and may stiffen the material and thereby reduce membrane compressibility [205].

Hydrophobic macroparticles have been used in polymer blends to change the contact angle and surface energy of polymer membranes [206, 207]. The resultant membranes had improved MD performance compared to polymer only membranes. Such changes allow increased surface hydrophobicity and contact angles, and reduced thermal conductivity in that the area of contact between the liquid and the membrane is reduced at the microscopic level.

Recently, other techniques have been used to modify the surface of the hydrophilic membranes and make them superhydrophobic. The surfaces of commercial membranes such as polyamide (PA) and polyethersulfone PES with different pore size were modified using a fluorolinkperfluoropolyether (PFPE) by dip-coating and in situ polymerization methods [208]. The resulting hydrophobic/hydrophilic membranes were tested in DCMD, confirming the stability of the coating and the good performance of the membranes.

5.2.5 Environmental Issues in Membrane Fabrication

In recent years, the hazard linked to chemical compounds and processes has become a more and more important social and environmental issue, requiring a comprehensive regulation. The initiatives and regulations in force have driven the interest to protect the environment and to reduce pollution and the consequent risks for human health. The regulation REACH (Registration, Evaluation, Authorisation and Restriction of Chemicals) was approved by the European Parliament on June 1 2007 [209]. REACH focuses greater responsibility on industries to assess the risks that chemicals may pose to the health and the environment. In this way, REACH regulation requires that substances of very high concern (new and existing) are ruled out or adequately controlled, and progressively substituted by safer substances or technologies, or only used where there is an overall benefit for the society.

Although membrane-based operations are commonly considered by themselves green and sustainable, usually the membrane fabrication method is quite far to be green. Nowadays, designing, manufacturing membranes to be applied in membrane processes without using or generating hazardous substances has become very interesting. Green Chemistry principles should be applied to all the aspects of the membrane life cycle, from its invention to disposal, including the environmental fate after the use. The risk related to membrane fabrication is a crucial factor being an indicator of the potential harm, which can be done by the process itself. Although several works have been aimed at developing high performance membranes suitable for MCs application, to date only a few studies have focused on a more sustainable membrane preparation. The most widespread industrial processes to manufacture polymeric membranes by phase inversion (NIPS, VIPS or TIPS) use large solvent quantity hazardous to human health and to the environment [210]. The most employed solvents used of the major companies are NMP, DMF, DMA [211], or mixed solvents such as NMP/tetrahydrofuran, NMP/acetone and trimethyl phosphate/DMA, and TEP/DMA [212]. Moreover, these traditional organics solvents have been identified as substances of very high concern (SVHC) [213]. It has been estimated that every year more than 50 billion liters of wastewater contaminated with SVCH solvents are produced in membranes manufacture at industrial scale [210].

Many organic solvents are volatile organic compounds (VOCs), which means that their high volatility, very useful for industrial applications, contributes both to increase the risks of fire and explosion, and to facilitate the release in the atmosphere in which these solvents can act as air pollutants causing ozone depletion, photochemical smog and global warming. Moreover, the conventional solvents are highly toxic for human beings, animals and plants, and often their chronic toxicological properties are not completely known [214]. Limiting the use and exposure to toxic hazardous solvents, have become very important in membrane manufacture. Precautions to minimise the effects of these solvents by recycling have limited success and cannot avoid some losses into the environment. Moreover, the risk connected to potential accidents is still present. For these reasons, the replacement of these hazardous compounds with innocuous substitutes, as the fifth principle of Green Chemistry requires, seems to be the only valid alternative for a sustainable use of solvents. In membrane production, two main routes towards green solvents have been developed. The substitution of petrochemically fabricated solvents with biosolvents from renewable resources or the substitution of hazardous solvents with those that show better EHS (Environmental, Health and Safety) properties [209].

The choice of the proper alternative should be focused on workers safety, including carcinogenicity, mutagenicity, reproduction hazards, skin and respiration absorption/sensitisation, and toxicity. Process safety, including flammability, explosiveness, potential for emissions through high vapor pressure, static charge, potential for peroxide formation and odour issues. Environmental safety and safety for population, including ecotoxicity, persistence, ground water contamination, ozone depletion potential, photo-reactive potential, global warming potential [215]. However, in the development of innovative production protocols, in which less toxic substances replace harmful solvents, to make the production of industrial membranes more environmentally friendly [216, 217], care must be taken to maintain the unaltered performance of the membranes or even to improve them, preserving the objective chemical-physical-structural parameters.

Many research activities for discovering new solvents to replace conventional toxic ones have been conducted. The solvent-polymer compatibility, and phase inversion kinetics were systematically investigated to optimize the membrane performance, with particular focus also on environmental sustainability of the membrane fabrication process for MD and MCr applications. Dihydrolevoglucosenone (Cyrene), a dipolar aprotic biosolvent [218] was employed in PVDF membrane fabrication via VIPS coupled with NIPS [219]. This bioderivated-green solvent was selected, owing to its similarity with traditionally used organic aprotic solvents in terms of polarity, density and miscibility with water [220]. Acetyl tributyl citrate (ATBC) was used for membrane preparation by TIPS to take full advantage of ATBC's low toxicity [221]. The effect of different citrated-based solvents for TIPS membrane preparation was thoroughly investigated [222]. In the field of the N-TIPS process, the effects of γ-butyrolactone (GBL) [223, 224] and glycerol triacetate (triacetin) as less toxic solvents in membrane preparation was investigated [225]. More recently, an environmentally friendly solvent, methyl-5-(dimethylamino)-2-methyl-5-oxo pentanoate (PolarClean), with a good biodegradability and no reported potential health issues, was employed to fabricate highly permeable membranes with strong mechanical strength [226, 227]. However, in all cases, environmentally-friendly alternatives were not entirely free from the limitations that conventional TIPS and N-TIPS solvents face. Another solvent selected as nonhazardous solvents for PVDF membrane preparation was DMSO. In particular, DMSO was employed to produce PVDF and PVDF-HFP membranes suitable for contactors applications in a combination of VIPS and LIPS stages and without using any chemical additive as pore forming [130, 228]. Triethyl phosphate (TEP) has a been used for membrane fabrication by liquid–induced phase separation (LIPS) [229]. TEP was also investigated to prepare (polyvinylidenefluoride-*co*-hexafluoropropylene) (PVDF-HFP) hydrophobic porous membranes for MD operations, also providing evidence in this case that less-toxic solvents can replace hazard chemicals in membrane preparation for water treatment [230]. Table 5.1 shows some of the most recent nontoxic or less toxic solvents employed as novel solvents in membrane preparation through phase inversion.

5.2.6 An Overview of Commercially Available Membranes Used in MD

The principal requirement for MD process is that the membrane must not be wetted by the feed solutions in contact with and only vapor and non-condensable gases should be present within the membrane pores. In Tables 5.2 and 5.3 are reported some flat-sheet, capillary, and hollow-fiber commercial

Membranes for MD/MCr

TABLE 5.1

Nontoxic or less toxic solvents used in membrane preparation. [Part of the table was reprinted from reference *Journal of Membrane Science* 580 (2019) 224–234 Copyright (2019), with permission from Elsevier.]

Polymer	Name	Solvent			Membrane Preparation technique	Membrane Configuration	Reference
		Molecular weight (g/mol)	Boiling point (°C)	Solubility in water (20°C)			
PVDF	Dihydrolevoglucosenone (Cyrene)	128.13	227	Totally miscible	VIPS-NIPS	Flat sheet	[219]
PVDF	Dimethyl isosorbide	174.2	93–95	Totally miscible	NIPS VIPS-NIPS	Flat sheet	[231]
PVDF	Acetyl tributyl citrate (ATBC)	402.48	331	Slightly miscible	TIPS	Hollow-fiber	[221]
PVDF	Ethyl-lactate (ethyl 2-hydroxy-propanoate)	118.132	155	Miscible	NIPS	Flat-sheet	[232]
PVDF	1,2-butylene carbonate (BC)	116.116	238.7	Miscible	NIPS	Flat-sheet	[233]
PVDF-HFP	Tamisolve NXG	141.21	241	Totally miscible	NIPS	Flat-sheet	[234]
PVDF	Dimethyl carbonate (DMC)	90.08	90	Totally miscible	NIPS	Flat sheet	[233]
PVDF	Diethyl carbonate (DMC)	126–128	118.132	Totally miscible	NIPS	Flat-sheet	[233]
PVDF	methyl-5-(dimethylamino)-2-methyl-5-oxo-pentanoate (Polar Clean)	187.8	278–282	Totally miscible	N-TIPS TIPS	Hollow-fiber	[226, 227]
PVDF	Dimethyl sulfoxide (DMSO)	78.13	189	Totally miscible	LIPS-NIPS NIPS	Flat-sheet	[228 235,229]
PVDF-HFP					LIPS-NIPS	Flat- sheet	[130]
PVDF	Diethylene glycol monoethyl ether acetate (DCAC)	176.21	218–219	Totally miscible	TIPS	Hollow-fiber	[236]
PVDF	Glycerol triacetate (TRIACETIN)	218.21	258	Slightly miscible	TIPS	Hollow-fiber	[225]
PVDF	Triethylene glycol diacetate (TEGDA)	234.35	286	Totally miscible	TIPS	Flat-sheet	[237]
PVDF		182.15	215		NIPS	Flat-sheet	

(Continued)

TABLE 5.1 (Cont.)

| Polymer | Name | Solvent | | | Membrane Preparation technique | Membrane Configuration | Reference |
		Molecular weight (g/mol)	Boiling point (°C)	Solubility in water (20°C)			
PVDF-HFP	Triethyl phosphate (TEP)			Totally miscible	VIPS-NIPS	Hollow-fiber	[235, 238–241]
PVDF	Triethyl citrate (TEC)	278.28	294	Totally miscible	TIPS	Flat-sheet	[222]
PVDF	Acetyl triethyl citrate (ATEC)	318.32	228–229	Partially miscible	TIPS	Flat-sheet	[222]
PVDF	Maleic acid dibutyl ester (DBM)	228.28	281	Not miscible	TIPS	Flat-sheet	[242]
PVDF	Dibutyl sebacate	314.46	178–179	Not miscible	TIPS	Flat-sheet	[223]
PVDF	Propylene carbonate	102.09	240	Totally miscible	TIPS	Flat-sheet	[223]
PVDF	γ-Butyrolactone (γ-BL)	86.09	204–205	Totally miscible	TIPS	Flat-sheet	[223, 224]

macroporous hydrophobic membranes made of different polymers such as PP, PVDF, PTFE and PE used in MD studies together with their principal characteristics and performances in MD experiments. The choice of an appropriate membrane for a given MD application is a crucial factor. A low-thermal conductivity achieved by thicker membranes, a high permeate flux achieved by thin membrane, a large pore size, as well as a high porosity and a high separation factor are important characteristics for suitable membrane selection.

Table 5.4 reports some commercial and patented MD membrane modules. The major application of these MD modules is desalination.

In addition to the commercial membranes, numerous laboratory researches have been carried out on the preparation and/or modification of synthetic membranes designed specifically for MD process. Recently, a number of research studies have been geared on membrane surface modification to enhance hydrophobicity and antifouling proprieties of MD membranes. Membranes were modified by different techniques for MD application, which were reported in literature, are outlined in Table 5.5.

A general summary for polymeric flat-sheet- and hollow-fiber membranes fabricated using various phase separation techniques in laboratory for MD lab-scale applications is given in Table 5.6.

A list of electrospun nanofiber membranes prepred at laboratory scale for MD is reported in Table 5.7.

TABLE 5.2

Some flat-sheet commercial membranes used in MD. [Part of the table was reprinted from reference *Advances in Colloid and Interface Science* **164 (2011) 56–88 Copyright (2010), with permission from Elsevier]**

Membrane trade name	Manufacturer	Membrane materials	LEP_w (bar)	δ (μm)	d_p (μm)	ε (%)	Operating Conditions: MD configuration; Feed solution composition; Feed inlet temperature T_f; Distillate inlet temperature T_p	Flux (L/ m2·h)	Rejection (%)	Reference
GVHP	Millipore	PVDF	204	110	0.22	75	DCMD; Distilled water; $T_f = 90.7°C$ $T_p = 20°C$	48.7	—	[243]
HVHP (Durapore)	Millipore	PVDF	105	140	0.45	75	Deaeration DCMD; distilled water/ 14–25 wt% NaCl $T_f = 80°C$ $T_p = 20°C$	67 50-40	—	[244]
PV22	Millipore	PVDF	2.29	126	0.22	62	—	—	—	[245]
PV45	Millipore	PVDF	1.10	116	0.45	66	—	—	—	
GVSP	Millipore	PVDF	—	108	0.22	80	—	—	—	
FGLP	Millipore	PTFE/PE*	280	130	0.25	70	DCMD; Distilled water; NaCl (0.5 to 2.0 M) Tf = 57.2°C Tp = 20°C	8.56	—	[54]
FHLP	Millipore	PTFE/PE*	124	175	0.5	85	—	—	—	—
PTS20	Gore	PTFE/PP*	4.63	184	0.2	44	—	—	—	[245]

(*Continued*)

TABLE 5.2 (Cont.)

Membrane trade name	Manufacturer	Membrane materials	Properties				Operating Conditions: MD configuration; Feed solution composition; Feed inlet temperature T_f; Distillate inlet temperature T_p	Flux (L/m²·h)	Rejection (%)	Reference
			LEP_w (bar)	δ (μm)	d_p (μm)	ε (%)				
PT20	Gore	PTFE/PP*	3.68	64	0.2	90	—	—	—	[245]
PT45	Gore	PTFE/PP*	2.88	77	0.45	89	—	—	—	[245]
TF200	Gelman	PTFE/PP*	282	178	0.2	80	DCMD; Distilled water; $T_f = 80.1°C$	67.3	—	[243]
TF450	Gelman	PTFE/PP*	138	178	0.45	80	AGMD; 30 g/L NaCl; $T_f = 71°C$	51.1	99.64-99.97	[30]
TF1000	Gelman	PTFE/PP*	48	178	1	80	—	—	—	—
Taflen	Gelman	PTFE/PP*	—	60	0.8	50	—	—	—	—
Metricel	Gelman	PP	—	90	0.1	55	—	—	—	—
PFTE Sartorius	Sartorius	PTFE	—	70	0.2	70	DCMD; Distilled water; $T_f = 70°C$; $T_p = 20°C$	14.00	—	[53]
Enka	Sartorius	PP	—	100	0.1	75	—	—	—	—
Enka	Sartorius	PP	—	140	0.2	75	—	—	—	—
Celgard 2500	Hoechst Celanese Co	PP	—	28	0.05	45	—	—	—	—
Celgard 2400	Hoechst Celanese Co	PP	—	25	0.02	38	—	—	—	—
TS22	Osmonics	PTFE/PP*	—	175	0.22	70	DCMD; 0.6 g/L NaCl; $T_f = 40°C$; $T_p = 20°C$	80	99.8	[57]

TS45	Osmonics	PTFE/PP*	—	175	0.45	70	DCMD 0.6 g/L NaCl $T_f = 60°C$ $T_p = 20°C$	80	99.8	[57]
TS1.0	Osmonics	PTFE/PP*	—	175	1.0	70	DCMD 0.6 g/L NaCl $T_f = 40°C$ $T_p = 20°C$	40	99.9	[57]
PP22	Osmonics	PP	—	150	0.22	70	DCMD 0.6 g/L NaCl $T_f = 40°C$ $T_p = 20°C$	25	99.8	[57]
Membrana	Membrana, Germany	PP	—	91	0.2	—	Cross flow VMD; distilled water; $T_f =$ 59.2°C; $Pp=10^3$ Pa; 1.1 kW/kg h^{-1}. Longitudinal VMD; distilled water; $T_f =$ 59°C; $T_p = 14.3°C$; 3.55 kW/kg h^{-1}	15.61 7.06	—	[246]
Vladipore	Vladipore	—	—	120	0.25	70	—	—	—	[247]
3MA	3M Corporation	PP	—	91	0.29	66	VMD; Distilled water; $T_f = 74°C$ $P_p = 3·10^3$ Pa	102.6	—	[248]
3MB	3M Corporation	PP	—	81	0.40	76	VMD; distilled water; $T_f = 74°C$; $P_p = 3·10^3$ Pa.	117.0	—	[248]
3MC	3M Corporation	PP	—	76	0.51	79	VMD; distilled water; $T_f = 74$ °C; $P_p = 3·10^3$ Pa.	160.2	—	[248]

(Continued)

TABLE 5.2 (Cont.)

Membrane trade name	Manufacturer	Membrane materials	LEP$_w$ (bar)	δ (µm)	d$_p$ (µm)	ε (%)	Operating Conditions: MD configuration; Feed solution composition; Feed inlet temperature T$_f$; Distillate inlet temperature T$_p$	Flux (L/m2·h)	Rejection (%)	Reference
						Properties				
3MD	3M Corporation	PP	—	86	0.58	80	DCMD; Distilled water; T$_f$ = 70 °C; T$_p$ = 20 °C.	27	—	[249]
3ME	3M Corporation	PP	—	79	0.73	85	DCMD; distilled water T$_f$ = 80 C; T$_p$ = 20 °C;	40.5	—	[249]
Teknokrama	Teknokrama	PTFE	—	—	0.2	80	DCMD; distilled water ΔT = 10°C	—	—	[250]
Teknokrama	Teknokrama	PTFE	—	—	0.5	80	DCMD; distilled water ΔT= 10°C	—	—	[250]
Teknokrama	Teknokrama	PTFE	—	—	1.0	80	DCMD; distilled water ΔT= 10°C	—	—	[250]
M09G0020H500	GVS SPA	PVDF	—	—	0.2	54	DCMD 1 g/L NaCl 60°C 20°C	—	—	[228]

* Flat- sheet polytetrafluoroethylene (PTFE), membranes supported by polypropylene (PP), or polyethylene (PE). Active layer of PTFE/PP membranes purchased from Osmonics Corp. are between 5 and 10 µm.
LEP$_w$ liquid entry pressure of water; δ membrane thickness; d$_p$ mean pore size,; ε porosity.

TABLE 5.3

Capillary and hollow-fiber commercial membranes commonly used in MD [Reprinted from reference *Advances in Colloid and Interface Science 164* (2011) 56–88 Copyright (2010), with permission from Elsevier]

| Membrane trade name | Manufacturer | Membrane materials | Properties* | | | | | Operating Conditions: MD configuration; Feed solution composition; Feed inlet temperature (°C); Distillate inlet temperature (°C) | Flux (L/m2·h) | Rejection (%) | Reference |
			LEP_w (bar)	δ (μm)	d_i (μm)	d_p (μm)	ε (%)				
AccurelS6/2	AkzoNobel	PP	1.4	450	1.8	0.2	70	DCMD Tap water 90 20	33.3	99.8	[251]
MD080CO2N	Enka Microdyn	PP	—	650	—	—	70	DCMD; 35 g/L NaCl; $T_f = 55°C$; $T_p = 15°C$.	$0.83 \cdot 10^{-3}$	—	[252]
MD020TP2N	Enka Microdyn	PP	—	1550	—	—	70	DCMD; 35 g/L NaCl; $T_f = 70°C$; $T_p = 15°C$.	$0.97 \cdot 10^{-3}$	—	[252]
Accurel	Enka A.G.	PP	—	400	—	—	74	—	—	—	—
Accurel	Enka A.G.	PP	—	200	0.33	0.2	70	—	—	—	—
Accurel	Enka A.G.	PP	—	150	—	0.43	70	DCMD; 0.05 M NaCl; $T_f = 42.5°C$; $T_p = 22.5°C$	$4.63 \cdot 10^{-3}$	—	[253]

(Continued)

TABLE 5.3 (Cont.)

Membrane trade name	Manufacturer	Membrane materials	Properties* LEP$_w$ (bar)	δ (µm)	d$_i$ (µm)	d$_p$ (µm)	ε (%)	Operating Conditions: MD configuration; Feed solution composition; Feed inlet temperature (°C); Distillate inlet temperature (°C)	Flux (L/ m$_2$·h)	Rejection (%)	Reference
Accurel	Enka A.G.	PP	—	150	0.33	0.5	74	—	—	—	—
Celgard X-20	Hoechst Celanese Co	PP	—	25		0.03	35	—	—	—	—
Capillary membranes	Membrana GmbH	PP	—	510	1.79	0.2	75	VMD; dye solution (Remazol brillant blue R, 50 ppm in water); T$_f$ = 60°C; P$_p$ = 10^3 Pa.	15.83· 10^{-3}	—	[33]
EHF270FA-16	Mitsubishi	PE	—	55	—	—	70	—	—	—	[254]
UPE test fibered	Millipore	PE	—	250	0.2	0.2	—	—	—	—	—
PTFE	Sumitomo Electric	PORE-FLON	—	550	0.9	0.8	62	SGMD	—	100	[255]

* LEP$_w$ liquid entry pressure of water; δ membrane thickness; d$_p$ mean pore size; ε porosity; d$_i$ internal diameter.

TABLE 5.4

Commercial and patented MD modules using different configurations [Part of the table was reprinted from reference *Desalination* 356 (2015) 56–84 Copyright (2014), with permission from Elsevier.]

Manufacturer	Membrane material	Pore size (μm)	Effective membrane area (m²)	MD configuration	Permeate flux (L/ m²h)	MD conditions: Feed solution composition; ΔT: Feed inlet temperature (°C)- Distillate inlet temperature (°C); Flow rate	Other module specifications	Reference
PLATE AND FRAME MEMBRANE MODULES								
GE Osmonics	PTFE GE	0.45	0.014	DCMD	22.3	10,000 mg/L NaCl; 400 mL/min ΔT = 25°C	—	[256]
SEPA CF	PTFE Membrane solutions	0.45	0.014	DCMD	23.4	10,000 mg/L NaCl; ΔT = 25°C; 400 mL/min	—	[256]
Keppel Seghers	—	—	9	—	3.25	35 g/L NaCl; ΔT = 30°C; 26 L/min	Compact single module	[257]
Scarab development AB	PTFE	0.2	2.8	AGMD	6.5	1 g/L NaCl; ΔT = 65°C	Solar desalination pilot	[258]
Memsys	PTFE	0.2	1.88 (7 frames)	V-MEMD	8.7	Tap water; ΔT = 25°C; 1.5 L/min	2 stage system	[42]
Memsys/ memDist module	PP	—	—	V-MEMD	—	—	—	[77]
Aquastill	—	—	—	AGMD	—	—	membrane surface area 24 m²	[259]

(Continued)

TABLE 5.4 (Cont.)

Manufacturer	Membrane material	Pore size (μm)	Effective membrane area (m²)	MD configuration	Permeate flux (L/m²h)	MD conditions: Feed solution composition; ΔT: Feed inlet temperature (°C)- Distillate inlet temperature (°C); Flow rate	Other module specifications	Reference
SPIRAL-WOUND MEMBRANE MODULES								
Fraunhofer ISE	PTFE	0.2	10	LGMD	1–2.5	Tap water; ΔT = 25°C 200–500 kg/h	Solar desalination Spiral coil diameter = 0.28 m	[41]
Fraunhofer ISE	PTFE	0.2	10	LGMD (with Deaeratiom)	1.2–3.0	Tap water E T = 80°C; CD T = 25°C F FR = 200–500 kg/h VP = 40 mbar	Spiral coil diameter = 0.28 m	[260]
SEP GmbH	PTFE	0.2	4	DCMD	2.5–12.5	LLLW T_f = 30°–80°C; T_p = 5°–30°C 300–1500 kg/h	G-4.0-6-7 module Dimensions (φ x h) = 0.45 x 0.62 m	[261]
Patent (US4545862)	PTFE 5C.2 W. L. Gore & Associates, Inc.	0.45	5	DCMD	2.65	40 g/L salt water; 170 L/h	L = 19.5 m Diameter = 0.5 m	[262]
TUBULAR MEMBRANE MODULES								
Enka-Microdyn	PP	0.2	0.036	DCMD	4	35 g/L NaCl ΔT=55°C 0.055 kg/s	Countercurrent flow module: MD020TP2N (3 tubes) PF = 0.5; L = 0.75 m	[252]

CAPILLARY AND HOLLOW-FIBER MEMBRANE MODULES

Microdyn	PP	0.2	0.1	DCMD	13	35 g/L NaCl $\Delta T=55°C$ 0.055 kg/s	Countercurrent flow module: MD020CP2N (40 HF)	[252]
Microdyn	PP	0.2	0.1	VMD	3.6	Distilled water; $T_f= 40°C$ F V = 0.4 m/s; Vacuum pressure = 40 mbar	Capillary module: MD020CP2N 40 capillary fibers;	[168]
Microdyn	PP	0.2	0.1	VMD	6.1	Distilled water; $T_f = 50°C$ Vacuum pressure = 60 mbar		
Microdyn	PP	0.2	0.1	VMD	9.1	Distilled water; $T_f = 60°C$ Vacuum pressure = 90 mbar		
Memstill	—	—	—	LGMD	3.53	Seawater; $T_{top} = 74.3°C$ Channel V = 0.035 m/s	M28 module (BS)	[44]
	—	—	—		2.45	Seawater; $T_{top} = 51.2°C$ Channel V = 0.027 m/s	M31 module (BS)	
	—	—	—		3.78–1.94	Seawater; $T_{top} = 77-69°C$ Channel V = 0.034–0.023 m/s	M32 module (BS)	
Enka Microdyn	PP	0.45	0.1	DCMD	1.38	465 g/L dilute apple juice; $T_f = 28°C$; $\Delta T = 19°C$ 300 L/h.	Tube and shell module: MD-020-2N-CP (41 HF)	[263]
Enka Microdyn	PP	0.2	2	DCMD	5.5	35 g/L NaCl F T = 70°C; P T = 15°C 0.055 kg/s.	Countercurrent flow module: MD080CO2N (467 HF)	[252]

TABLE 5.5

Examples of Modified membranes using various modification techniques for MD applications

Modification technique	Membrane material and configuration	MD application; Feed inlet temperature T_f; Distillate inlet temperature T_p	Flux (L/ m²h)	Salt rejection (%)	Reference
Filtration coating	PVDF Hollow-fiber	DCMD; T_f = 60°C; T_p =20°C	28	99.99	[264]
Filtration coating	Hollow-fiber	DCMD; T_f = 70°C; T_p =20°C.	29.9	—	[265]
Grafting	Alumina ceramic Hollow-fiber	VMD; T_f = 60°C; T_p =20°C	18	—	[266]
Grafting	PVDF-PTFE Flat-sheet	DCMD; T_f = 60°C; T_p =20°C	25.2	99.99	[267]
Grafting	PVDF Flat-sheet	VMD; Tf = 60°C; Tp =20°C Vacuum pressure: 9kPa	31.5	99.99	[268]
Cross-linking	PVDF Hollow-fiber	DCMD; T_f = 60°C; T_p =20°C	10.5	99.99	[199]
Spraying using an airbrush	PVDF Flat-sheet	DCMD; T_f = 69.85°C; T_p =20°C	8	99.99	[198]
Initiated chemical vapor deposition (iCVD)	Flat-sheet	AGMD; T_f = 60°C; T_p =20°C	8	—	[269]
Dip-coating	PVDF Hollow-fiber	DCMD; T_f = 60°C; T_p =20°C	21	99.99	[270]
Dip-coating	PVDF-co-HFP Flat-sheet	DCMD; T_f = 50°C; T_p =20°C	12	99.99	[271]
Dip-coating	Hydrophobic Flurolink AD 1700 on commercial hydrophilic polyamide (PA) membranes	DCMD; T_f = 60°C; T_p =20°C	13	99.3	[208]

(Continued)

Membranes for MD/MCr

TABLE 5.5 (Cont.)

Modification technique	Membrane material and configuration	MD application; Feed inlet temperature T$_f$; Distillate inlet temperature T$_p$	Flux (L/ m^{2}h)	Salt rejection (%)	Reference
Dip-coating	Flat-sheet PVDF Flat-sheet	DCMD; T$_f$ = 60°C; T$_p$ =20°C	31.8	99.9	[196]
Heat-press	PVDF-co-HFP Flat-sheet	DCMD; T$_f$ = 60°C; T$_p$ =20°C	20–22	99.99	[272]
Heat-press	PVDF Flat-sheet	DCMD; T$_f$ = 60°C; T$_p$ =20°C	20.6	99.99	[159]
Heat-press	PVDF-co-HFP Flat-sheet	DCMD; T$_f$ = 60°C; T$_p$ =20°C	29	99.99	[185]
Annealing	PVDF-co-HFP Flat-sheet	DCMD; T$_f$ = 60°C; T$_p$ =20°C	35	99.99	[194]

5.3 Conclusions and Perspectives

The growth and progress in MD and MCr applications have greatly evolved. However, the availability of suitable membranes and modules for MCs technology is still an issue today for several applications. Frequently, membranes and modules developed for microfiltration and ultrafiltration are adapted for these processes. However, these systems are not fully appropriate for scale-up applications of MD and MCr, owing to limited performance, wetting (over long periods), fouling, and scaling problems.

The use of nanotechnology to prepare membranes can help address the above-mentioned problems. Moreover, the incorporation of functional additives in the membrane structure is a valuable method to enhance the membrane performance. Module design is also a key aspect for these technologies.

Future perspectives for the fabrication of suitable membrane should also focus on the use of sustainable materials and methods to make industrial membrane production more environmentally friendly.

ACKNOWLEDGEMENT

This work was partially supported by the Italian Ministry of Education University and Research (MIUR) within the project "Development of a solar powered, zero liquid discharge Integrated desalination membrane system to

TABLE 5.6

Examples of Flat-sheet and hollow-fiber laboratory-made membranes prepared by phase separation techniques for MD applications

Membrane fabri-cation technique	Membrane configuration	Membrane materials	Flux (L/m²·h)	Rejection (%)	Operating Conditions: MD configuration; Feed solution composition; Feed inlet temperature T_f; Distillate inlet temperature T_p	Reference
VIPS+LIPS	Flat-sheet	PVDF	12	99.7	DCMD; distilled water; NaCl 1g/L; NaCl 0.5 M $T_f = 60°C$ $T_p = 20°C$	[228]
VIPS+LIPS	Flat-sheet	PVDF/PVDF-HFP	11	99.7	DCMD; distilled water; NaCl 1g/L $T_f = 60°C$ $T_p = 20°C$	[130]
VIPS	Flat-sheet	Polysulfone with silica nanoparticle	12.5–14	99.9	DCMD; distilled water; $T_f = 73°C$ $T_p = 25°C$ D	[273]
VIPS	Flat-sheet	Polysulfone	10.9	99.99	DCMD; Feedwater: 3.5 wt% NaCl $T_f = 73°C$ $T_p = 25°C$	[274]
Solvent phase separation	Flat-sheet	PVDF	26	99.99	DCMD; Feedwater: 3.5wt% NaCl $T_f = 60°C$	[240]

Phase separation	Flat-sheet	PVDF with MWNTs	36.8	99.99	$T_p = 20°C$ DCMD; Feed water: 1 wt% NaCl $T_f = 70°C$	[169]
NIPS or solvent phase separation	Flat-sheet	PVDF	16–39	99.99	$T_p = 20°C$ DCMD; Feedwater: 3.5wt% NaCl $T_f = 60°C$	[275]
NIPS or solvent phase separation	Flat-sheet	PVDF with MWNTs	32.4	—	$T_p = 20°C$ DCMD; Feedwater: 3.5 wt% NaCl $T_f = 60°C$	[276]
NIPS	Flat-sheet	PVDF with copper oxide nanoparticle	4.5	99.99	$T_p = 20°C$ VMD; Distilled water; $T_f = 60°C$ Vacuum pressure = 1.2 KPa	[277]
NTIPS	Flat-sheet	PVDF	40.9	99.99	DCMD; Feedwater: 3.5 wt% NaCl $T_f = 60°C$ $T_p = 20°C$	[278]
NIPS	Flat-sheet	PVDF	13	99.27	DCMD; Feedwater: 3.5 wt% NaCl $T_f = 69.85°C$ $T_p = 20°C$	[198]
NIPS	Flat-sheet	Polyvinyl chloride (PVC)	37.5	—	VMD; Distilled water; $T_f = 60°C$	[279]

(*Continued*)

TABLE 5.6 (Cont.)

182

Membrane fabri-cation technique	Membrane configuration	Membrane materials	Flux (L/m²·h)	Rejection (%)	Operating Conditions: MD configuration; Feed solution composition; Feed inlet temperature T_f; Distillate inlet temperature T_p	Reference
NIPS	Flat-sheet	Polyethersulfone (PES) with titanium oxide nanotubes	5.5	96.7	Vacuum pressure = 2 bar VMD; distilled water; $T_f = 60°C$	[280]
NIPS	Flat-sheet	PVDF	3.8 4.8	—	Vacuum pressure = 300 mbar DCMD; VMD Feedwater: 3.5 wt% NaCl $T_f = 60°C$ $T_p = 20°C$	[281]
NIPS	Flat-sheet	PVDF	2–12	99.0	VMD; distilled water; $T_f = 70°C$	[282]
NIPS	Flat-sheet	PVDF	5.0	99.22	Vacuum pressure = 3 kPa DCMD; Feedwater: 3.5 wt% NaCl $T_f = 60°C$ $T_p = 20°C$	[205]
NIPS	Flat-sheet	PVDF with fluorin-ated silica nanoparticles	15.23	—	VMD; Distilled water; $T_f = 70°C$	[283]
NIPS	Flat-sheet	PVDF	12.45	—	Vacuum pressure = 3 kPa DCMD; Feedwater: 3.5 wt% NaCl $T_f = 60°C$	[284]

Method	Configuration	Material	Value	Rejection (%)	Conditions	Ref.
NIPS	Flat-sheet	PVDF on hydrophilic support layer	20	—	$T_p = 20°C$ DCMD; Feedwater: 3.5 wt% NaCl $T_f = 60°C$ $T_p = 20°C$	[285]
NIPS	Flat-sheet	PVDF with graphene nanoparticle	20.5	99.99	AGMD	[286]
NIPS	Hollow-fiber	PVDF	19.8	—	DCMD; Feedwater: 3.5 wt% NaCl $T_f = 60°C$ $T_p = 20°C$	[126]
NIPS	Hollow-fiber	PVDF/PVA (hydrophobic/hydrophilic dual layer)	8.2	99.99	DCMD; Feedwater: 3.5 wt% NaCl $T_f = 60°C$ $T_p = 20°C$	[287]
VIPS/NIPS	Flat-sheet	PVDF	10.8–18	99.99	DCMD; Feedwater: 3.5 wt% NaCl $T_f = 60°C$ $T_p = 20°C$	[288]
TIPS	Flat-sheet	PP with ethylene vinyl acetate co-blending polymer	27.6	—	VMD; Distilled water; $T_f = 70°C$ Vacuum pressure = 3 kPa	[180]
TIPS	Hollow-fiber	PVDF	27–29	99.9	DCMD; Feedwater: 3.5 wt% NaCl $T_f = 60°C$ $T_p = 25°C$	[289]
TIPS	Hollow-fiber	PVDF	8–10	99.99	DCMD; Feed water: 3.5 wt% NaCl	[181]

(Continued)

TABLE 5.6 (Cont.)

Membrane fabri-cation technique	Membrane configuration	Membrane materials	Flux (L/m²·h)	Rejection (%)	Operating Conditions: MD configuration; Feed solution composition; Feed inlet temperature T_f; Distillate inlet temperature T_p	Reference
NIPS (Dry-wet phase inversion and spinning technique)	Flat-sheet and hollow-fiber	PVDF-CTFE	Flat-sheet 26.5 Hollow-fiber 22.5	99.99	$T_f = 60°C$ $T_p = 20°C$ DCMD; Feedwater: 3.5 wt% NaCl	[290]
NIPS with solvent phase separation	Flat-sheet	PVDF	17	99.99	$T_f = 60 °C$ $T_p = 20°C$ DCMD; Feedwater: 5 wt% NaCl	[291]
NIPS (Dry-jet wet phase inversion)	Hollow-fiber	PVDF	16	99.99	$T_f = 70°C$ $T_p = 20°C$ DCMD; Feedwater: 3.5 wt% NaCl	[289]
NIPS (Dry-jet wet phase inversion)	Hollow-fiber	PVDF/PFTE composite	18–20	99.99	$T_f = 70°C$ $T_p = 20°C$ DCMD; Feedwater: 3.5 wt% NaCl	[192]
NIPS (Dry-jet wet phase inversion)	Hollow-fiber	PVDF/PAN dual layer (hydrophobic/ hydrophilic)	38	99.99	$T_f = 60°C$ $T_p = 20°C$ DCMD; Feedwater: 3.5 wt% NaCl	[179]
	Hollow-fiber	PVDF	14.5	99.99	$T_f = 60°C$ $T_p = 20°C$ DCMD; Feedwater: 3.5 wt% NaCl	[292]

NIPS (Dry-jet wet phase inversion)					$T_f = 60°C$ $T_p = 20°C$	
NIPS (Dry-jet wet phase inversion)	Hollow-fiber	PVDF	8–18	—	DCMD; VMD Feedwater: 3.5 wt% NaCl $T_f = 60°C$ $T_p = 20°C$	[293]
NIPS (Dry-jet wet phase inversion)	Hollow-fiber	PVDF	32	—	DCMD; Feedwater: 3.5 wt% NaCl $T_f = 60°C$ $T_p = 20°C$	[264]
NIPS (Dry-jet wet phase inversion)	Hollow-fiber	PVDF with modified graphene oxide	28.2	99.99	DCMD; Feedwater: 3.5 wt% NaCl $T_f = 60°C$ $T_p = 20°C$	[294]
NIPS (Co-extrusion spinning process)	Hollow-fiber	PVDF	36.0	—	VMD; Feedwater: 3.5 wt% NaCl $T_f = 60°C$	[295]
NIPS (Dry-jet wet phase inversion)	Hollow-fiber	PVDF	8.2 41.8	—	DCMD; Feedwater: 3.5 wt% NaCl $T_f = 50°C$ $T_p = 20°C$ VMD; Feedwater: 3.5 wt% NaCl $T_f = 50°C$	[296]
	Hollow-fiber		7.5	99.99	DCMD;	[297]

TABLE 5.6 (Cont.)

Membrane fabrication technique	Membrane configuration	Membrane materials	Flux (L/m²·h)	Rejection (%)	Operating Conditions: MD configuration; Feed solution composition; Feed inlet temperature T_f; Distillate inlet temperature T_p	Reference
NIPS (Dry-jet wet phase inversion)		PVDF/PVA (hydrophobic/ hydrophilic dual layer)			Feedwater: 3.5 wt% NaCl $T_f = 60°C$ $T_p = 20°C$	
NIPS (Dry-jet wet phase inversion)	Hollow-fiber	PVDF	25–30	—	DCMD; Feedwater: 3.5 wt% NaCl $T_f = 60°C$ $T_p = 20°C$	[189]
NIPS (Dry-jet wet phase inversion)	Hollow-fiber	PVDF with CaCO₃ particles	20	—	DCMD; Feedwater: 3.5 wt% NaCl $T_f = 60°C$ $T_p = 20°C$	[270]
Immersion-induced phase inversion with sintering	Hollow-fiber	PVDF with alumina nanoparticle	42.5	—	VMD; Feedwater: 3.5 wt% NaCl $T_f = 80°C$	[266]

TABLE 5.7
Electrospun nanofiber membrane fabricated at laboratory scale for MD application

Layer format	Active layer	Properties*			Operating Conditions: MD configuration; Feed solution; Feed inlet temperature T_f; Distillate inlet temperature T_p	Flux (L/m²·h)	Salt rejection (%)	Reference
		CA (°)	d_p (μm)	ε (%)				
Single-layer	PVDF	130	—	—	AGMD; Feed solution: 3.5wt % NaCl; T_f = 80°C; T_p = 20°C.	11–12	98.7–99.9	[298]
Single-layer	PVDF-co-HFP with benzyltriethylammonium chloride (BTEAC)	130	0.42	—	DCMD; Feed solution: 3.5 wt % NaCl; T_f = 60°C; T_p = 20°C.	36	99.99	[271]
Single-layer	PVDF	138	0.31	39	DCMD; Feed solution: 3.5 wt % NaCl; T_f = 60°C; T_p = 20°C.	31.6	99.9	[196]
Single-layer	PVDF-co-HFP	134	0.73	91	DCMD; Feed solution: 1%, 3% and 6 % NaCl; T_f = 80°C; T_p = 20°C.	11–12	98.7–99.9	[272]
Single-layer	Matrimid 5218	130	2.15	—	DCMD; Feed solution: 3.5 wt % NaCl; T_f = 60°C; T_p = 20°C.	24	99.99	[299]
Single-layer	PVDF	138	0.18	71.4	DCMD; Feed solution: 3.5 wt % NaCl; T_f = 60°C; T_p = 20°C.	10	—	[159]

(Continued)

TABLE 5.7 (Cont.)

Layer format	Active layer	Properties*			Operating Conditions: MD configuration; Feed solution; Feed inlet temperature T$_f$; Distillate inlet temperature T$_P$	Flux (L/ m2·h)	Salt rejection (%)	Reference
		CA (°)	d$_p$ (μm)	ε (%)				
Single-layer	PTFE/PVA composite membrane	156.7	—	81.5	VMD; Feed solution: 3.5 wt % NaCl; T$_f$ = 60°C; Vacuum pressure: 30 kPa.	10	99.11	[300]
Single-layer	PVDF-co-HFP with nanocrystalline cellulose (NCC)	127	0.2	70	DCMD; Feed solution: 3.5 wt % NaCl; T$_f$ = 60°C; T$_p$ = 20°C.	10.2 11.5	99	[301]
Single-layer	PVDF/SiO$_2$	152	0.23	79	DCMD; Feed solution: 3.5 wt % NaCl; T$_f$ = 60°C; T$_p$ = 20°C.	25	99.99	[201]
Single-layer	PVDF with clay particles	154.2	0.64	81	DCMD; Feed solution: 3.5 wt % NaCl; T$_f$ = 60°C; T$_p$ = 20°C.	2	99.97	[302]
Single-layer	PTFE/PVA with PAN	150	1.0	90	DCMD; Feed solution: 3.5 wt % NaCl; T$_f$ = 60°C; T$_p$ = 20°C.	14.59	99.8	[303]
Single-layer	PVDF-co-HFP	153	1.4	91	DCMD; Feed solution: 3.5 wt % NaCl; T$_f$ = 60°C; T$_p$ = 20°C.	18	—	[194]

Single-layer	PVDF/SiO$_2$	121.5	0.22	55	VMD; Feed solution: 3.5 wt % NaCl; T$_f$ = 60°C; Vacuum pressure: 9 kPa.	15	—	[268]
Single-layer by dual-nozzle	PVDF-co-HFP with functionalized TiO$_2$ nanoparticles	153.4	0.76	89.8	DCMD; Feed solution: 7 wt % NaCl; T$_f$ = 60°C; T$_p$ = 20°C.	40	99.99	[304]
Single-layer	PVDF-co-HFP with fluorosilane coated TiO$_2$	149	0.75	91.6	DCMD; Feed solution: 7 wt % NaCl; T$_f$ = 60°C; T$_p$ = 20°C.	37.8	99.99	[186]
Single-layer	PVDF	142.8	0.68	85	DCMD; Feed solution: 3.5 wt % NaCl; T$_f$ = 60°C; T$_p$ = 20°C.	12	99.9	[305]
Dual-layer	PVDF/SiO$_2$	156.3	0.69	82		19	99.99	
Dual-layer	PVDF/silica	150	0.36	64	DCMD; Feed solution: 3.5 wt % NaCl; T$_f$ = 60°C; T$_p$ = 20°C.	16.27	99.99	[187]
Dual-layer	Polystyrene (PS) on PET support layer	150.2	0.76	69	DCMD; Feed solution: 3.5 wt % NaCl; T$_f$ = 75°C; T$_p$ = 20°C.	58	—	[306]
Dual-layer	PVDF with SiO$_2$ on nonwoven support (hydrophobic/hydrophilic dual-layer)	150	0.36	62	DCMD; Feed solution: 3.5 wt % NaCl; T$_f$ = 60°C; T$_p$ = 20°C.	21	99.99	[305]

(Continued)

TABLE 5.7 (Cont.)

Layer format	Active layer	Properties* CA (°)	dₚ (µm)	ε (%)	Operating Conditions: MD configuration; Feed solution; Feed inlet temperature Tᶠ; Distillate inlet temperature Tₚ	Flux (L/ m₂·h)	Salt rejection (%)	Reference
Dual-layer	PVDF with surface modifying macro-molecules (SMMs)/ PVDF phase-inversion support layer	148.4	0.06	-	DCMD; Feed solution: 3.5 wt % NaCl $T_f = 60°C$; $T_p = 20°C$.	10	99.98	[307]
Dual-layer	PVDF-co-HFP/PAN (hydrophobic/ hydrophilic)	150	1.0	90	DCMD; Feed solution: 3.5 wt % NaCl $T_f = 60°C$; $T_p = 20°C$.	30	98.5	[308]
Triple-layer (casting with electrospinning)	PVDF	145	0.10	-	AGMD; Feed solution: 3.5 wt % NaCl; $T_f = 80°C$; $T_p = 20°C$.	15.2	99.9	[190]
Triple-layer	PVDF nanofiber (top)/PVDF phase separation (middle)/ PET support (bottom)	145.02	0.1	-	AGMD; Feed solution: 3.5 wt % NaCl; $T_f = 80°C$; $T_p = 20°C$.	10	99.99	[190]

* Contact angle CA; δ membrane thickness; d_p mean pore size; ε porosity.

address the needs for water of the Mediterranean region," (IDEA-ERANETMED2-72-357; prot. MIUR N. 10912 del 06/06/2016; concession grant decree n. 3366 of 12/18/2018).

REFERENCES

[1] M. Kummu, P.J. Ward, H. De Moel, O. Varis, Is physical water scarcity a new phenomenon? Global assessment of water shortage over the last two millennia, *Environ. Res. Lett.* 5 (2010). https://doi.org/10.1088/1748-9326/5/3/034006.

[2] V.G. Gude, Energy consumption and recovery in reverse osmosis, Desalin, *Water Treat.* 36 (2011) 239–260. https://doi.org/10.5004/dwt.2011.2534.

[3] X. Yang, A.G. Fane, R. Wang, Membrane distillation: Now and future, Desalin, *Water from Water* 9781118208, 373–424, 2014. https://doi.org/10.1002/9781118904855.ch8.

[4] E. Curcio, A. Criscuoli, E. Drioli, Membrane crystallizers, *Ind. Eng. Chem. Res.* 40 (2001) 2679–2684. https://doi.org/10.1021/ie000906d.

[5] E. Drioli, G. Di Profio, E. Curcio, Progress in membrane crystallization, *Curr. Opin. Chem. Eng.* 1 (2012) 178–182. https://doi.org/10.1016/j.coche.2012.03.005.

[6] G. Di Profio, E. Curcio, A. Cassetta, D. Lamba, E. Drioli, Membrane crystallization of lysozyme: Kinetic aspects, *J. Cryst. Growth.* 257 (2003) 359–369. https://doi.org/10.1016/S0022-0248(03)01462-3.

[7] F.A. Banat, J. Simandl, Desalination by membrane distillation: A parametric study, *Sep. Sci. Technol.* 33 (1998) 201–226. https://doi.org/10.1080/01496399808544764.

[8] K.W. Lawson, D.R. Lloyd, Membrane distillation, *J. Memb. Sci.* 124 (1997) 1–25. https://doi.org/10.1007/s00216-011-4733-9.

[9] Z. Lei, B. Chen, Z. Ding, *Membrane distillation*, 1st ed., Elsevier, Amsterdam, 2005.

[10] E. Drioli, E. Curcio, G. Di Profio, State of the art and recent progresses in membrane contactors, *Chem. Eng. Res. Des.* 83 (2005) 223–233. https://doi.org/10.1205/cherd.04203.

[11] E. Curcio, E. Drioli, Membrane distillation and related operations – A review, *Sep. Purif. Rev.* 34 (2005) 35–86. https://doi.org/10.1081/SPM-200054951.

[12] A.L. Quoc, M. Mondor, F. Lamarche, J. Makhlouf, Optimization of electrodialysis with bipolar membranes applied to cloudy apple juice: Minimization of malic acid and sugar losses, *Innov. Food Sci. Emerg. Technol.* 12 (2011) 45–49. https://doi.org/10.1016/j.ifset.2010.12.007.

[13] L. Bazinet, C. Cossec, H. Gaudreau, Y. Desjardins, Production of a phenolic antioxidant enriched cranberry juice by electrodialysis with filtration membrane, *J. Agric. Food Chem.* 57 (2009) 10245–10251. https://doi.org/10.1021/jf9021114.

[14] D.S. Couto, M. Dornier, D. Pallet, M. Reynes, D. Dijoux, S.P. Freitas, L.M. C. Cabral, Evaluation of nanofiltration membranes for the retention of anthocyanins of açai (euterpe oleracea mart.) juice, Desalin. *Water Treat.* 27 (2011) 108–113. https://doi.org/10.5004/dwt.2011.2067.

[15] S.T. Hsu, K.T. Cheng, J.S. Chiou, Seawater desalination by direct contact membrane distillation, *Desalination* 143 (2002) 279–287. https://doi.org/10.1016/S0011-9164(02)00266-7.

[16] M.P. Godino, L. Peña, C. Rincón, J.I. Mengual, Water production from brines by membrane distillation, *Desalination* 108 (1997) 91–97. https://doi.org/10.1016/S0011-9164(97)00013-1.

[17] V.D. Alves, I.M. Coelhoso, Orange juice concentration by osmotic evaporation and membrane distillation: A comparative study, *J. Food Eng.* 74 (2006) 125–133. https://doi.org/10.1016/j.jfoodeng.2005.02.019.

[18] M. Tomaszewska, M. Gryta, A.W. Morawski, Study on the concentration of acids by membrane distillation, *J. Memb. Sci.* 102 (1995) 113–122. https://doi.org/10.1016/0376-7388(94)00281-3.

[19] S. Srisurichan, R. Jiraratananon, A.G. Fane, Mass transfer mechanisms and transport resistances in direct contact membrane distillation process, *J. Memb. Sci.* 277 (2006) 186–194. https://doi.org/10.1016/j.memsci.2005.10.028.

[20] F.A. Banat, J. Simandl, Membrane distillation for dilute ethanol: Separation from aqueous streams, *J. Memb. Sci.* 163 (1999) 333–348. https://doi.org/10.1016/S0376-7388(99)00178-7.

[21] M.C. García-Payo, M.A. Izquierdo-Gil, C. Fernández-Pineda, Air gap membrane distillation of aqueous alcohol solutions, *J. Memb. Sci.* 169 (2000) 61–80. https://doi.org/10.1016/S0376-7388(99)00326-9.

[22] G.W. Meindersma, C.M. Guijt, A.B. de Haan, Desalination and water recycling by air gap membrane distillation, *Desalination* 187 (2006) 291–301. https://doi.org/10.1016/j.desal.2005.04.088.

[23] J. Kim, D.S. Chang, Y.Y. Choi, Separation of oxygen isotopic water by using a pressure-driven air gap membrane distillation, *Ind. Eng. Chem. Res.* 48 (2009) 5431–5438. https://doi.org/10.1021/ie900277r.

[24] A.S. Alsaadi, L. Francis, H. Maab, G.L. Amy, N. Ghaffour, Evaluation of air gap membrane distillation process running under sub-atmospheric conditions: Experimental and simulation studies, *J. Memb. Sci.* 489 (2015) 73–80. https://doi.org/10.1016/j.memsci.2015.04.008.

[25] R. Moradi, J. Karimi-Sabet, M. Shariaty-Niassar, Y. Amini, Air gap membrane distillation for enrichment of H218O isotopomers in natural water using poly (vinylidene fluoride) nanofibrous membrane, *Chem. Eng. Process. Process Intensif.* 100 (2016) 26–36. https://doi.org/10.1016/j.cep.2015.11.015.

[26] A. Kujawska, J.K. Kujawski, M. Bryjak, M. Cichosz, W. Kujawski, Removal of volatile organic compounds from aqueous solutions applying thermally driven membrane processes. 2. Air gap membrane distillation, *J. Memb. Sci.* 499 (2016) 245–256. https://doi.org/10.1016/j.memsci.2015.10.047.

[27] M.M.A. Shirazi, A. Kargari, M. Tabatabaei, Sweeping gas membrane distillation (SGMD) as an alternative for integration of bioethanol processing: Study on a commercial membrane and operating parameters, *Chem. Eng. Commun.* 202 (2015) 457–466. https://doi.org/10.1080/00986445.2013.848805.

[28] M.M.A. Shirazi, A. Kargari, M. Tabatabaei, A.F. Ismail, T. Matsuura, Concentration of glycerol from dilute glycerol wastewater using sweeping gas membrane distillation, *Chem. Eng. Process. Process Intensif.* 78 (2014) 58–66. https://doi.org/10.1016/j.cep.2014.02.002.

[29] Z. Xie, T. Duong, M. Hoang, C. Nguyen, B. Bolto, Ammonia removal by sweep gas membrane distillation, *Water Res.* 43 (2009) 1693–1699. https://doi.org/10.1016/j.watres.2008.12.052.

[30] M. Khayet, C. Cojocaru, Artificial neural network model for desalination by sweeping gas membrane distillation, *Desalination* 308 (2013) 102–110. https://doi.org/10.1016/j.desal.2012.06.023.

[31] M.C. García-Payo, C.A. Rivier, I.W. Marison, U. Von Stockar, Separation of binary mixtures by thermostatic sweeping gas membrane distillation II. Experimental results with aqueous formic acid solutions, *J. Memb. Sci.* 198 (2002) 197–210. https://doi.org/10.1016/S0376-7388(01)00649-4.

[32] F. Banat, S. Al-Asheh, M. Qtaishat, Treatment of waters colored with methylene blue dye by vacuum membrane distillation, *Desalination* 174 (2005) 87–96. https://doi.org/10.1016/j.desal.2004.09.004.

[33] A. Criscuoli, J. Zhong, A. Figoli, M.C. Carnevale, R. Huang, E. Drioli, Treatment of dye solutions by vacuum membrane distillation, *Water Res.* 42 (2008) 5031–5037. https://doi.org/10.1016/j.watres.2008.09.014.

[34] T. Duong, Z. Xie, D. Ng, M. Hoang, Ammonia removal from aqueous solution by membrane distillation, *Water Environ. J.* 27 (2013) 425–434. https://doi.org/10.1111/j.1747-6593.2012.00364.x.

[35] W. Cao, Q. Liu, Y. Wang, I.M. Mujtaba, Modeling and simulation of VMD desalination process by ANN, *Comput. Chem. Eng.* 84 (2016) 96–103. https://doi.org/10.1016/j.compchemeng.2015.08.019.

[36] M. Khayet, T.T. Matsuura, Introduction to membrane distillation, in *Membrane distillation*, Elsevier, Oxford, UK, 2011, pp. 1–16.

[37] J. Kujawa, E. Guillen-Burrieza, H.A. Arafat, M. Kurzawa, A. Wolan, W. Kujawski, Raw juice concentration by osmotic membrane distillation process with hydrophobic polymeric membranes, *Food Bioprocess Technol.* 8 (2015) 2146–2158. https://doi.org/10.1007/s11947-015-1570-4.

[38] W. Kujawski, A. Sobolewska, K. Jarzynka, C. Güell, M. Ferrando, J. Warczok, Application of osmotic membrane distillation process in red grape juice concentration, *J. Food Eng.* 116 (2013) 801–808. https://doi.org/10.1016/j.jfoodeng.2013.01.033.

[39] J. Warczok, M. Gierszewska, W. Kujawski, C. Güell, Application of osmotic membrane distillation for reconcentration of sugar solutions from osmotic dehydration, *Sep. Purif. Technol.* 57 (2007) 425–429. https://doi.org/10.1016/j.seppur.2006.04.012.

[40] D. Bessarabov, Z. Twardowski, New opportunities for osmotic membrane distillation, *Membr. Technol.* 2006 (2006) 7–11. https://doi.org/10.1016/S0958-2118(06)70744-3.

[41] D. Winter, J. Koschikowski, M. Wieghaus, Desalination using membrane distillation: Experimental studies on full scale spiral wound modules, *J. Memb. Sci.* 375 (2011) 104–112. https://doi.org/10.1016/j.memsci.2011.03.030.

[42] K. Zhao, W. Heinzl, M. Wenzel, S. Büttner, F. Bollen, G. Lange, S. Heinzl, N. Sarda, Experimental study of the memsys vacuum-multi-effect-membrane-distillation (V-MEMD) module, *Desalination* 323 (2013) 150–160. https://doi.org/10.1016/j.desal.2012.12.003.

[43] L. Francis, N. Ghaffour, A.A. Alsaadi, G.L. Amy, Material gap membrane distillation: A new design for water vapor flux enhancement, *J. Memb. Sci.* 448 (2013) 240–247. https://doi.org/10.1016/j.memsci.2013.08.013.

[44] A.E. Jansen, J.W. Assink, J.H. Hanemaaijer, J. van Medevoort, E. van Sonsbeek, Development and pilot testing of full-scale membrane distillation modules for deployment of waste heat, *Desalination* 323 (2013) 55–65. https://doi.org/10.1016/j.desal.2012.11.030.

[45] Y. Qin, L. Liu, F. He, D. Liu, Y. Wu, Multi-effect membrane distillation device with efficient internal heat reclamation function and method, China Patent 201010570625, 2013.

[46] J. Swaminathan, H.W. Chung, D.M. Warsinger, F.A. AlMarzooqi, H.A. Arafat, J. H. Lienhard, Energy efficiency of permeate gap and novel conductive gap membrane distillation, *J. Memb. Sci.* 502 (2016) 171–178. https://doi.org/10.1016/j.memsci.2015.12.017.

[47] Ó. Andrjesdóttir, C.L. Ong, M. Nabavi, S. Paredes, A.S.G. Khalil, B. Michel, D. Poulikakos, An experimentally optimized model for heat and mass transfer in

direct contact membrane distillation, *Int. J. Heat Mass Transf.* 66 (2013) 855–867. https://doi.org/10.1016/j.ijheatmasstransfer.2013.07.051.

[48] H.J. Hwang, K. He, S. Gray, J. Zhang, I.S. Moon, Direct contact membrane distillation (DCMD): Experimental study on the commercial PTFE membrane and modeling, *J. Memb. Sci.* 371 (2011) 90–98. https://doi.org/10.1016/j.memsci.2011.01.020.

[49] I. Hitsov, T. Maere, K. De Sitter, C. Dotremont, I. Nopens, Modelling approaches in membrane distillation: A critical review, *Sep. Purif. Technol.*, 142 (2015) 48–64. https://doi.org/10.1016/j.seppur.2014.12.026.

[50] B.B. Ashoor, S. Mansour, A. Giwa, V. Dufour, S.W. Hasan, Principles and applications of direct contact membrane distillation (DCMD): A comprehensive review, *Desalination* 398 (2016) 222–246. https://doi.org/10.1016/j.desal.2016.07.043.

[51] L.M. Camacho, L. Dumée, J. Zhang, J. de Li, M. Duke, J. Gomez, S. Gray, Advances in membrane distillation for water desalination and purification applications, *Water (Switzerland)* 5 (2013) 94–196. https://doi.org/10.3390/w5010094.

[52] M. Khayet, Membranes and theoretical modeling of membrane distillation: A review, *Adv. Colloid Interface Sci.* 164 (2011) 56–88. https://doi.org/10.1016/j.cis.2010.09.005.

[53] J. Phattaranawik, R. Jiraratananon, A.G. Fane, Effect of pore size distribution and air flux on mass transport in direct contact membrane distillation, *J. Memb. Sci.* 215 (2003) 75–85. https://doi.org/10.1016/S0376-7388(02)00603-8.

[54] M. Qtaishat, T. Matsuura, B. Kruczek, M. Khayet, Heat and mass transfer analysis in direct contact membrane distillation, *Desalination* 219 (2008) 272–292. https://doi.org/10.1016/j.desal.2007.05.019.

[55] M.S. El-Bourawi, Z. Ding, R. Ma, M. Khayet, A framework for better understanding membrane distillation separation process, *J. Memb. Sci.* 285 (2006) 4–29. https://doi.org/10.1016/j.memsci.2006.08.002.

[56] M.N. Chernyshov, G.W. Meindersma, A.B. de Haan, Comparison of spacers for temperature polarization reduction in air gap membrane distillation, *Desalination* 183 (2005) 363–374. https://doi.org/10.1016/j.desal.2005.04.029.

[57] T.Y. Cath, V.D. Adams, A.E. Childress, Experimental study of desalination using direct contact membrane distillation: A new approach to flux enhancement, *J. Memb. Sci.* 228 (2004) 5–16. https://doi.org/10.1016/j.memsci.2003.09.006.

[58] F. Macedonio, Encyclopedia of membranes, *Encycl. Membr.* (2015). https://doi.org/10.1007/978-3-642-40872-4.

[59] A. Ali, F. Macedonio, E. Drioli, S. Aljlil, O.A. Alharbi, Experimental and theoretical evaluation of temperature polarization phenomenon in direct contact membrane distillation, *Chem. Eng. Res. Des.* 91 (2013) 1966–1977. https://doi.org/10.1016/j.cherd.2013.06.030.

[60] V. Gekas, B. Hallström, Mass transfer in the membrane concentration polarization layer under turbulent cross flow. I. Critical literature review and adaptation of existing sherwood correlations to membrane operations, *J. Memb. Sci.* 30 (1987) 153–170. https://doi.org/10.1016/S0376-7388(00)81349-6.

[61] R.B. Bird, W.E. Stewart, E.N. Lightfoot, *Transport phenomena*, 2nd ed., Wiley, New York, 2007. https://doi.org/10.1017/CBO9781107415324.004.

[62] N. Nagaraj, G. Patil, B.R. Babu, U.H. Hebbar, K.S.M.S. Raghavarao, S. Nene, Mass transfer in osmotic membrane distillation, *J. Memb. Sci.* 268 (2006) 48–56. https://doi.org/10.1016/j.memsci.2005.06.007.

[63] N.N. Li, A.G. Fane, W.S.W. Ho, T. Matsuura, *Advanced membrane technology and applications*, Wiley, New York, 2008.

[64] E.A. Mason, A.P. Malinauskas, *Gas transport in porous media: The dusty-gas model*, Elsevier, New York, 1983.

Membranes for MD/MCr

[65] L. Martínez-Díez, M.I. Vázquez-González, A method to evaluate coefficients affecting flux in membrane distillation, *J. Memb. Sci.* 173 (2000) 225–234. https://doi.org/10.1016/S0376-7388(00)00362-8.

[66] E. Drioli, E. Curcio, G. Di Profio, Experimental and theoretical analysis of a membrane crystallizer, *Chem. Eng. Trans.* 1 (2002) 927–932. https://doi.org/10.1016/j.desal.2010.11.024.

[67] E. Curcio, G. Di Profio, E. Drioli, A new membrane-based crystallization technique: Tests on lysozyme, *J. Cryst. Growth* 247 (2003) 166–176. https://doi.org/10.1016/S0022-0248(02)01794-3.

[68] E. Curcio, G. Di Profio, E. Drioli, Membrane crystallization of macromolecular solutions, *Chem. Eng. Trans.* 145 (2002) 173–177. https://doi.org/10.1016/S0011-9164(02)00404-6.

[69] M. Gryta, Concentration of NaCl solution by membrane distillation integrated with crystallization, *Sep. Sci. Technol.* 37 (2002) 3535–3558. https://doi.org/10.1081/SS-120014442.

[70] G. Di Profio, E. Curcio, E. Drioli, Trypsin crystallization by membrane-based techniques, *J. Struct. Biol.* 150 (2005) 41–49. https://doi.org/10.1016/j.jsb.2004.12.006.

[71] G. Di Profio, G. Perrone, E. Curcio, A. Cassetta, D. Lamba, E. Drioli, Preparation of enzyme crystals with tunable morphology in membrane crystallizers, *Ind. Eng. Chem. Res.* 44 (2005) 10005–10012. https://doi.org/10.1021/ie0508233.

[72] F. Liu, N.A. Hashim, Y. Liu, M.R.M. Abed, K. Li, Progress in the production and modification of PVDF membranes, *J. Memb. Sci.* 375 (2011) 1–27. https://doi.org/10.1016/j.memsci.2011.03.014.

[73] M. Essalhi, M. Khayet, Self-sustained webs of polyvinylidene fluoride electrospun nano-fibers: Effects of polymer concentration and desalination by direct contact membrane distillation, *J. Memb. Sci.* 454 (2014) 133–143. https://doi.org/10.1016/j.memsci.2013.11.056.

[74] A.M. Alklaibi, N. Lior, Membrane-distillation desalination: Status and potential, *Desalination* 171 (2005) 111–131. https://doi.org/10.1016/j.desal.2004.03.024.

[75] A. Alkhudhiri, N. Darwish, N. Hilal, Membrane distillation: A comprehensive review, *Desalination* 287 (2012) 2–18. https://doi.org/10.1016/j.desal.2011.08.027.

[76] www.keppelseghers.com/en/content.aspx?sid=3023.

[77] www.memsys.eu/technology/membrane-distillation-technology.html

[78] J.H. Hanemaaijer, J. van Medevoort, A.E. Jansen, C. Dotremont, E. van Sonsbeek, T. Yuan, L. De Ryck, Memstill membrane distillation – A future desalination technology, *Desalination* 199 (2006) 175–176. https://doi.org/10.1016/j.desal.2006.03.163.

[79] R. Schwantes, A. Cipollina, F. Gross, J. Koschikowski, D. Pfeifle, M. Rolletschek, V. Subiela, Membrane distillation: Solar and waste heat driven demonstration plants for desalination, *Desalination* 323 (2013) 93–106. https://doi.org/10.1016/j.desal.2013.04.011.

[80] A. Jansen, J. Hanemaaijer, J. Assink, E. van Sonsbeek, C. Dotremont, M.J. Van, Pilot plants prove feasibility of a new desalination technique, *Asian Water* 26 (2010) 22–26.

[81] Q. He, P. Li, H. Geng, C. Zhang, J. Wang, H. Chang, Modeling and optimization of air gap membrane distillation system for desalination, *Desalination* 354 (2014) 68–75. https://doi.org/10.1016/j.desal.2014.09.022.

[82] F. Macedonio, A. Ali, T. Poerio, E. El-Sayed, E. Drioli, M. Abdel-Jawad, Direct contact membrane distillation for treatment of oilfield produced water, *Sep. Purif. Technol.* 126 (2014) 69–81. https://doi.org/10.1016/j.seppur.2014.02.004.

[83] www.aquaver.com/knowledge-center-articles/membrane-distillation-a-lowcost-breakthrough-technology-for-water-treatment/

[84] E. Drioli, E. Curcio, A. Criscuoli, G. Di Profio, Integrated system for recovery of CaCO3, NaCl and MgSO4·7H2O from nanofiltration retentate.pdf, *J. Memb. Sci.* 239 (2004) 27–38.

[85] C.M. Tuna, A.G. Fane, J.T. Matheickal, R. Sheikholeslami, Membrane distillation crystallization of concentrated salts – Flux and crystal formation.pdf, *J. Memb. Sci.* 257 (2005) 144–155.

[86] E. Drioli, E. Curcio, G. Di Profio, F. Macedonio, A. Criscuoli, Integrating membrane contactors technology and pressure-driven membrane operations for seawater desalination: Energy, Exergy and costs analysis.pdf, *Chem. Eng. Res. Des.* 84 (2006) 209–220.

[87] L. Mariah, C.A. Buckley, C.J. Brouckaert, E. Curcio, E. Drioli, D. Jaganyi, D. Ramjugernath, Membrane distillation of concentrated brines – Role of water activities in the evaluation of driving force, *J. Memb. Sci.* 280 (2006) 937–947. https://doi.org/10.1016/j.matlet.2016.12.100.

[88] Z.X.P. Zhang, K. Wei, Y. Wang, R. Ma, file.pdf, *Desalination* 219 (2008) 101–117.

[89] B. Tang, G. Yu, J. Fang, T. Shi, Recovery of high-purity silver directly from dilute effluents by an emulsion liquid membrane-crystallization process, *J. Hazard Mater.* 177 (2010) 377–383.

[90] W. Li, B. Van der Bruggen, P. Luis, Integration of reverse osmosis and membrane crystallization for sodium sulphate recovery, *Chem. Eng. Process. Process Intensif.* 85 (2014) 57–68.

[91] R. Salmon, L. Alconero, R. Salmon, L. Alconero, P. Membrane, Membrane crystallization of sodium carbonate in CO2capture scenario: Mass and heat transfer study, Référence bibliographique, (n.d.) 1–2.

[92] P. Luis, D. Van Aubel, B. Van der Bruggen, Technical viability and exergy analysis of membrane crystallization: Closing the loop of CO2 sequestration, *Int. J. Greenh. Gas Control* 12 (2013) 450–459. https://doi.org/10.1016/j.ijggc.2012.11.027.

[93] Z. Jia, Z. Liu, F. He, Synthesis of nanosized BaSO4 and CaCO3 particles with a membrane reactor: Effects of additives on particles, *J. Colloid Interface Sci.* 266 (2003) 322–327.

[94] S.Z. Li, X.Y. Li, D.Z. Wang, Membrane (RO-UF) filtration for antibiotic wastewater treatment and recovery of antibiotics, *Sep. Purif. Technol.* 34 (2004) 109–114.

[95] E. Drioli, M.C. Carnevale, A. Figoli, A. Criscuoli, Vacuum membrane dryer (VMDr) for the recovery of solid microparticles from aqueous solutions, *J. Memb. Sci.* 472 (2014) 67–76.

[96] E. Curcio, G. Di Profio, E. Drioli, Recovery of fumaric acid by membrane crystallization in the production of l-malic acid, *Sep. Purif. Rev.* 33 (2003) 63–73.

[97] S. Simone, E. Curcio, G. Di Profio, M. Ferraroni, E. Drioli, Polymeric hydrophobic membranes as a tool to control polymorphism and protein-ligand interactions, *J. Memb. Sci.* 283 (2006) 123–132. https://doi.org/10.1016/j.memsci.2006.06.028.

[98] G. Di Profio, S. Tucci, E. Curcio, E. Drioli, Selective glycine polymorph crystallization by using microporous membranes, *Cryst. Growth Design* 7 (2007) 526–530. https://doi.org/10.1021/cg800838b.

[99] G. Di Profio, S. Tucci, E. Curcio, E. Drioli, Controlling polymorphism with membrane-based crystallizers: Application to form I and II of paracetamol, *Chem. Mater.* 19 (2007) 2386–2388. https://doi.org/10.1021/cm0701005.

[100] E. Curcio, E. Fontananova, G. Di Profio, E. Drioli, Influence of the structural properties of poly(vinylidene fluoride) membranes on the heterogeneous nucleation rate of protein crystals, *J. Phys. Chem. B.* 110 (2006) 12438–12445. https://doi.org/10.1021/jp061531y.

[101] www.amecrys-project.eu/

[102] www.idea-eranetmed.eu/

[103] H. Lee, F. He, L. Song, J. Gilron, K.K. Sirkar, Desalination with a cascade of cross-flow hollow fiberhollow-fiber membrane distillation devices integrated with a heat exchanger, *AIChE J.* 57 (2011) 1780–1795. https://doi.org/10.1002/aic.

[104] J. Gilron, L. Song, K.K. Sirkar, Design for cascade of crossflow direct contact membrane distillation, *Ind. Eng. Chem. Res.* 46 (2007) 2324–2334. https://doi.org/10.1021/ie060999k.

[105] R.B. Saffarini, E.K. Summers, H.A. Arafat, J.H. Lienhard V, Economic evaluation of stand-alone solar powered membrane distillation systems, *Desalination* 299 (2012) 55–62. https://doi.org/10.1016/j.desal.2012.05.017.

[106] M. Khayet, Solar desalination by membrane distillation: Dispersion in energy consumption analysis and water production costs (a review), *Desalination* 308 (2013) 89–101. https://doi.org/10.1016/j.desal.2012.07.010.

[107] E.K. Summers, J.H. Lienhard V, A novel solar-driven air gap membrane distillation system, Desalin. *Water Treat.* 51 (2013) 1344–1351. https://doi.org/10.1080/19443994.2012.705096.

[108] A.S. Hassan, H.E.S. Fath, Review and assessment of the newly developed MD for desalination processes, Desalin. *Water Treat.* 51 (2013) 574–585. https://doi.org/10.1080/19443994.2012.697273.

[109] G. Di Profio, S.M. Salehi, E. Curcio, E. Drioli, 3.11 membrane crystallization technology, *Compr. Membr. Sci. Eng.* 3 (2017) 297–317. https://doi.org/10.1016/b978-0-12-409547-2.12247-4.

[110] S. Shirazi, C.J. Lin, D. Chen, Inorganic fouling of pressure-driven membrane processes – A critical review, *Desalination* 250 (2010) 236–248. https://doi.org/10.1016/j.desal.2009.02.056.

[111] A.S. Al-Amoudi, Factors affecting natural organic matter (NOM) and scaling fouling in NF membranes: A review, *Desalination* 259 (2010) 1–10. https://doi.org/10.1016/j.desal.2010.04.003.

[112] M. Gryta, The assessment of microorganism growth in the membrane distillation system, *Desalination* 142 (2002) 79–88. https://doi.org/10.1016/S0011-9164(01)00427-1.

[113] E. Guillen-Burrieza, R. Thomas, B. Mansoor, D. Johnson, N. Hilal, H. Arafat, Effect of dry-out on the fouling of PVDF and PTFE membranes under conditions simulating intermittent seawater membrane distillation (SWMD), *J. Memb. Sci.* 438 (2013) 126–139. https://doi.org/10.1016/j.memsci.2013.03.014.

[114] F. He, J. Gilron, H. Lee, L. Song, K.K. Sirkar, Potential for scaling by sparingly soluble salts in crossflow DCMD, *J. Memb. Sci.* 311 (2008) 68–80. https://doi.org/10.1016/j.memsci.2007.11.056.

[115] G. Chen, X. Yang, R. Wang, A.G. Fane, Performance enhancement and scaling control with gas bubbling in direct contact membrane distillation, *Desalination* 308 (2013) 47–55. https://doi.org/10.1016/j.desal.2012.07.018.

[116] D.M. Warsinger, K.H. Mistry, K.G. Nayar, H.W. Chung, J.H.V. Lienhard, Entropy generation of desalination powered by variable temperature waste heat, *Entropy.* 17 (2015) 7530–7566. https://doi.org/10.3390/e17117530.

[117] L. Eykens, K. De Sitter, C. Dotremont, L. Pinoy, B. Van Der Bruggen, How to optimize the membrane properties for membrane distillation: A review, *Ind. Eng. Chem. Res.* 55 (2016) 9333–9343. https://doi.org/10.1021/acs.iecr.6b02226.

[118] E. Drioli, A. Ali, F. Macedonio, Membrane distillation: Recent developments and perspectives, *Desalination* 356 (2015) 56–84. https://doi.org/10.1016/j.desal.2014.10.028.

[119] M. Khayet, T. Matsuura, *Membrane distillation principles and applications*, 1st ed., Elsevier, Amsterdam, 2011.

[120] C. Huang, X. Qian, R. Yang, Thermal conductivity of polymers and polymer nanocomposites, *Mater. Sci. Eng. R Reports* 132 (2018) 1–22. https://doi.org/10.1016/j.mser.2018.06.002.

[121] M. Mulder, Basic principles of membrane technology (1991). https://doi.org/10.1524/zpch.1998.203.part_1_2.263.

[122] L.D. Tijing, J.S. Choi, S. Lee, S.H. Kim, H.K. Shon, Recent progress of membrane distillation using electrospun nanofibrous membrane, *J. Memb. Sci.* 453 (2014) 435–462. https://doi.org/10.1016/j.memsci.2013.11.022.

[123] M.C. García-Payo, M. Essalhi, M. Khayet, Effects of PVDF-HFP concentration on membrane distillation performance and structural morphology of hollow fiber membranes, *J. Memb. Sci.* 347 (2010) 209–219. https://doi.org/10.1016/j.memsci.2009.10.026.

[124] M. Khayet, C. Cojocaru, M. Essalhi, M.C. García-Payo, P. Arribas, L. García-Fernández, Hollow fiber spinning experimental design and analysis of defects for fabrication of optimized membranes for membrane distillation, *Desalination* 287 (2012) 146–158. https://doi.org/10.1016/j.desal.2011.06.025.

[125] A.C. Sun, W. Kosar, Y. Zhang, X. Feng, A study of thermodynamics and kinetics pertinent to formation of PVDF membranes by phase inversion, *Desalination* 309 (2013) 156–164. https://doi.org/10.1016/j.desal.2012.10.005.

[126] D. Hou, J. Wang, D. Qu, Z. Luan, X. Ren, Fabrication and characterization of hydrophobic PVDF hollow fiber membranes for desalination through direct contact membrane distillation, *Sep. Purif. Technol.* 69 (2009) 78–86. https://doi.org/10.1016/j.seppur.2009.06.026.

[127] E. Fontananova, J.C. Jansen, A. Cristiano, E. Curcio, E. Drioli, Effect of additives in the casting solution on the formation of PVDF membranes, *Desalination* 192 (2006) 190–197. https://doi.org/10.1016/j.desal.2005.09.021.

[128] S. Wongchitphimon, R. Wang, R. Jiraratananon, L. Shi, C.H. Loh, Effect of polyethylene glycol (PEG) as an additive on the fabrication of polyvinylidene fluoride-co-hexafluropropylene (PVDF-HFP) asymmetric microporous hollow fiber membranes, *J. Memb. Sci.* 369 (2011) 329–338. https://doi.org/10.1016/j.memsci.2010.12.008.

[129] Z. Cui, E. Drioli, Y.M. Lee, Recent progress in fluoropolymers for membranes, *Prog. Polym. Sci.* 39 (2014) 164–198. https://doi.org/10.1016/j.progpolymsci.2013.07.008.

[130] C. Meringolo, T. Poerio, E. Fontananova, T.F. Mastropietro, F.P. Nicoletta, G. De Filpo, E. Curcio, G. Di Profio, Exploiting fluoropolymers immiscibility to tune surface properties and mass transfer in blend membranes for membrane contactor applications, *ACS Appl. Polym. Mater.* 1 (2019) 326–334. https://doi.org/10.1021/acsapm.8b00105.

[131] C. Feng, B. Shi, G. Li, Y. Wu, Preparation and properties of microporous membrane from poly(vinylidene fluoride-co-tetrafluoroethylene) (F2.4) for membrane distillation, *J. Memb. Sci.* 237 (2004) 15–24. https://doi.org/10.1016/j.memsci.2004.02.007.

[132] A. Gugliuzza, E. Drioli, PVDF and HYFLON AD membranes: Ideal interfaces for contactor applications, *J. Memb. Sci.* 300 (2007) 51–62. https://doi.org/10.1016/j.memsci.2007.05.004.

[133] T. Okubo, H. Inoue, Single gas permeation through porous glass modified with tetraethoxysilane, *AIChE J.* 35 (1989) 845–848. https://doi.org/10.1002/aic.690350515.

[134] Y. Gao, J. Qin, Z. Wang, S.W. Østerhus, Backpulsing technology applied in MF and UF processes for membrane fouling mitigation: A review, *J. Memb. Sci.* 587 (2019) 117136. https://doi.org/10.1016/j.memsci.2019.05.060.

[135] J. Kujawa, W. Kujawski, S. Koter, A. Rozicka, S. Cerneaux, M. Persin, A. Larbot, Efficiency of grafting of Al2O3, TiO2 and ZrO2 powders by perfluoroalkylsilanes, *Colloids. Surfaces A. Physicochem. Eng. Asp.* 420 (2013) 64–73. https://doi.org/10.1016/j.colsurfa.2012.12.021.

[136] S.R. Krajewski, W. Kujawski, M. Bukowska, C. Picard, A. Larbot, Application of fluoroalkylsilanes (FAS) grafted ceramic membranes in membrane distillation process of NaCl solutions, *J. Memb. Sci.* 281 (2006) 253–259. https://doi.org/10.1016/j.memsci.2006.03.039.

[137] S.R. Krajewski, W. Kujawski, F. Dijoux, C. Picard, A. Larbot, Grafting of ZrO2 powder and ZrO2 membrane by fluoroalkylsilanes, *Colloids. Surfaces A. Physicochem. Eng. Asp.* 243 (2004) 43–47. https://doi.org/10.1016/j.colsurfa.2004.05.001.

[138] J. Kujawa, S. Cerneaux, W. Kujawski, K. Knozowska, Hydrophobic ceramic membranes for water desalination, *Appl. Sci.* 7 (2017) 1–11. https://doi.org/10.3390/app7040402.

[139] Y. Wang, W. Wang, L. Zhong, J. Wang, Q. Jiang, X. Guo, Super-hydrophobic surface on pure magnesium substrate by wet chemical method, *Appl. Surf. Sci.* 256 (2010) 3837–3840. https://doi.org/10.1016/j.apsusc.2010.01.037.

[140] R. Jagdheesh, B. Pathiraj, E. Karatay, G.R.B.E. Römer, A.J.H. In'T Veld, Laser-induced nanoscale superhydrophobic structures on metal surfaces, *Langmuir.* 27 (2011) 8464–8469. https://doi.org/10.1021/la2011088.

[141] A. Delimi, E. Galopin, Y. Coffinier, M. Pisarek, R. Boukherroub, B. Talhi, S. Szunerits, Investigation of the corrosion behavior of carbon steel coated with fluoropolymer thin films, *Surf. Coatings Technol.* 205 (2011) 4011–4017. https://doi.org/10.1016/j.surfcoat.2011.02.030.

[142] J. Song, X. Liu, Y. Lu, L. Wu, W. Xu, A rapid two-step electroless deposition process to fabricate superhydrophobic coatings on steel substrates, *J. Coatings Technol. Res.* 9 (2012) 643–650. https://doi.org/10.1007/s11998-012-9431-9.

[143] J. Ou, W. Hu, M. Xue, F. Wang, W. Li, One-step solution immersion process to fabricate superhydrophobic surfaces on light alloys, *ACS Appl. Mater. Interfaces* 5 (2013) 9867–9871. https://doi.org/10.1021/am402303j.

[144] S.M. Lee, I.H. Choi, S.W. Myung, J. young Park, I.C. Kim, W.N. Kim, K.H. Lee, Preparation and characterization of nickel hollow fiber membrane, *Desalination* 233 (2008) 32–39. https://doi.org/10.1016/j.desal.2007.09.024.

[145] M.W.J. Luiten-Olieman, L. Winnubst, A. Nijmeijer, M. Wessling, N.E. Benes, Porous stainless steel hollow fiber membranes via dry-wet spinning, *J. Memb. Sci.* 370 (2011) 124–130. https://doi.org/10.1016/j.memsci.2011.01.004.

[146] R.B. Saffarini, B. Mansoor, R. Thomas, H.A. Arafat, Effect of temperature-dependent microstructure evolution on pore wetting in PTFE membranes under membrane distillation conditions, *J. Memb. Sci.* 429 (2013) 282–294. https://doi.org/10.1016/j.memsci.2012.11.049.

[147] B.S. Lalia, I. Janajreh, R. Hashaikeh, A facile approach to fabricate superhydrophobic membranes with low contact angle hysteresis, *J. Memb. Sci.* 539 (2017) 144–151. https://doi.org/10.1016/j.memsci.2017.05.071.

[148] A.K. An, J. Guo, E.J. Lee, S. Jeong, Y. Zhao, Z. Wang, T.O. Leiknes, PDMS/PVDF hybrid electrospun membrane with superhydrophobic property and drop impact dynamics for dyeing wastewater treatment using membrane distillation, *J. Memb. Sci.* 525 (2017) 57–67. https://doi.org/10.1016/j.memsci.2016.10.028.

[149] Z. Jin, D.L. Yang, S.H. Zhang, X.G. Jian, Hydrophobic modification of poly (phthalazinone ether sulfone ketone) hollow fiber membrane for vacuum

membrane distillation, *J. Memb. Sci.* 310 (2008) 20–27. https://doi.org/10.1016/j.memsci.2007.10.021.

[150] S. Cerneaux, I. Struzyńska, W.M. Kujawski, M. Persin, A. Larbot, Comparison of various membrane distillation methods for desalination using hydrophobic ceramic membranes, *J. Memb. Sci.* 337 (2009) 55–60. https://doi.org/10.1016/j.memsci.2009.03.025.

[151] M. Khayet, T. Matsuura, J.I. Mengual, M. Qtaishat, Design of novel direct contact membrane distillation membranes, *Desalination* 192 (2006) 105–111. https://doi.org/10.1016/j.desal.2005.06.047.

[152] B. Li, K.K. Sirkar, Novel membrane and device for vacuum membrane distillation-based desalination process, *J. Memb. Sci.* 257 (2005) 60–75. https://doi.org/10.1016/j.memsci.2004.08.040.

[153] K.Y. Wang, S.W. Foo, T.S. Chung, Mixed matrix PVDF hollow fiberhollow-fiber membranes with nanoscale pores for desalination through direct contact membrane distillation, *Ind. Eng. Chem. Res.* 48 (2009) 4474–4483. https://doi.org/10.1021/ie8009704.

[154] Z.D. Hendren, J. Brant, M.R. Wiesner, Surface modification of nanostructured ceramic membranes for direct contact membrane distillation, *J. Memb. Sci.* 331 (2009) 1–10. https://doi.org/10.1016/j.memsci.2008.11.038.

[155] M. Zeyu, Y. Hong, M. Liyuan, M. Su, Superhydrophobic membranes with ordered arrays of nanospiked microchannels for water desalination, *Langmuir.* 25 (2009) 5446–5450. https://doi.org/10.1021/la900494u.

[156] M. Essalhi, M. Khayet, Surface segregation of fluorinated modifying macromolecule for hydrophobic/hydrophilic membrane preparation and application in air gap and direct contact membrane distillation, *J. Memb. Sci.* 417–418 (2012) 163–173. https://doi.org/10.1016/j.memsci.2012.06.028.

[157] M. Calle, Y.M. Lee, Thermally rearranged (TR) poly(ether-benzoxazole) membranes for gas separation, *Macromolecules* 44 (2011) 1156–1165. https://doi.org/10.1021/ma102878z.

[158] C.A. Scholes, B.D. Freeman, S.E. Kentish, Water vapor permeability and competitive sorption in thermally rearranged (TR) membranes, *J. Memb. Sci.* 470 (2014) 132–137. https://doi.org/10.1016/j.memsci.2014.07.024.

[159] Y. Liao, R. Wang, M. Tian, C. Qiu, A.G. Fane, Fabrication of polyvinylidene fluoride (PVDF) nanofiber membranes by electro-spinning for direct contact membrane distillation, *J. Memb. Sci.* 425–426 (2013) 30–39. https://doi.org/10.1016/j.memsci.2012.09.023.

[160] K. Gethard, O. Sae-Khow, S. Mitra, Water desalination using carbon-nanotube-enhanced membrane distillation, *ACS Appl. Mater. Interfaces* 3 (2011) 110–114. https://doi.org/10.1021/am100981s.

[161] R.R. Nair, H.A. Wu, P.N. Jayaram, I.V. Grigorieva, A.K. Geim, Unimpeded permeation of water through helium-leak-tight graphene-based membranes, *Science* 335(80) (2012) 442–444. https://doi.org/10.1126/science.1211694.

[162] D.R. Dreyer, S. Park, W. Bielawski, R.S. Ruoff, 氧化石墨烯综8FF0.Pdf, (2010). https://doi.org/10.1039/b917103g.

[163] G.R. Guillen, Y. Pan, M. Li, E.M.V. Hoek, Preparation and characterization of membranes formed by nonsolvent induced phase separation: A review, *Ind. Eng. Chem. Res.* 50 (2011) 3798–3817. https://doi.org/10.1021/ie101928r.

[164] Z. Wang, H. Yu, J. Xia, F. Zhang, F. Li, Y. Xia, Y. Li, Novel GO-blended PVDF ultrafiltration membranes, *Desalination* 299 (2012) 50–54. https://doi.org/10.1016/j.desal.2012.05.015.

[165] C. Zhao, X. Xu, J. Chen, F. Yang, Effect of graphene oxide concentration on the morphologies and antifouling properties of PVDF ultrafiltration membranes, *J. Environ. Chem. Eng.* 1 (2013) 349–354. https://doi.org/10.1016/j.jece.2013.05.014.

[166] L. Dumée, J.L. Campbell, K. Sears, J. Schütz, N. Finn, M. Duke, S. Gray, The impact of hydrophobic coating on the performance of carbon nanotube bucky-paper membranes in membrane distillation, *Desalination* 283 (2011) 64–67. https://doi.org/10.1016/j.desal.2011.02.046.

[167] L.F. Dumée, K. Sears, J. Schütz, N. Finn, C. Huynh, S. Hawkins, M. Duke, S. Gray, Characterization and evaluation of carbon nanotube Bucky-Paper membranes for direct contact membrane distillation, *J. Memb. Sci.* 351 (2010) 36–43. https://doi.org/10.1016/j.memsci.2010.01.025.

[168] E. Fontananova, V. Grosso, S.A. Aljlil, M.A. Bahattab, D. Vuono, F.P. Nicoletta, E. Curcio, E. Drioli, G. Di Profio, Effect of functional groups on the properties of multiwalled carbon nanotubes/polyvinylidenefluoride composite membranes, *J. Memb. Sci.* 541 (2017) 198–204. https://doi.org/10.1016/j.memsci.2017.07.002.

[169] S. Roy, M. Bhadra, S. Mitra, Enhanced desalination via functionalized carbon nanotube immobilized membrane in direct contact membrane distillation, *Sep. Purif. Technol.* 136 (2014) 58–65. https://doi.org/10.1016/j.seppur.2014.08.009.

[170] M.I. Siyal, C.K. Lee, C. Park, A.A. Khan, J.O. Kim, A review of membrane development in membrane distillation for emulsified industrial or shale gas wastewater treatments with feed containing hybrid impurities, *J. Environ. Manage.* 243 (2019) 45–66. https://doi.org/10.1016/j.jenvman.2019.04.105.

[171] E. Salimi, Omniphobic surfaces: State-of-the-art and future perspectives, *J. Adhes. Sci. Technol.* 33 (2019) 1369–1379. https://doi.org/10.1080/01694243.2019.1599217.

[172] K. Scott, Introduction to membrane separations, *Handb. Ind. Membr.* (1995) 3–185. https://doi.org/10.1016/b978-185617233-2/50004-0.

[173] S. Khaisri, D. deMontigny, P. Tontiwachwuthikul, R. Jiraratananon, CO2 stripping from monoethanolamine using a membrane contactor, *J. Memb. Sci.* 376 (2011) 110–118. https://doi.org/10.1016/j.memsci.2011.04.005.

[174] F.E. Ahmed, B.S. Lalia, R. Hashaikeh, A review on electrospinning for membrane fabrication: Challenges and applications, *Desalination* 356 (2015) 15–30. https://doi.org/10.1016/j.desal.2014.09.033.

[175] A. Baji, Y.W. Mai, S.C. Wong, M. Abtahi, P. Chen, Electrospinning of polymer nanofibers: Effects on oriented morphology, structures and tensile properties, *Compos. Sci. Technol.* 70 (2010) 703–718. https://doi.org/10.1016/j.compscitech.2010.01.010.

[176] Z. Peining, A.S. Nair, Y. Shengyuan, P. Shengjie, N.K. Elumalai, S. Ramakrishna, Rice grain-shaped TiO 2-CNT composite – A functional material with a novel morphology for dye-sensitized solar cells, *J. Photochem. Photobiol. A Chem.* 231 (2012) 9–18. https://doi.org/10.1016/j.jphotochem.2012.01.002.

[177] R.E. Kesting, Phase inversion membranes, *ACS Symp. Ser.* (1985) 131–164. https://doi.org/10.1021/bk-1985-0269.ch007.

[178] M.M. Teoh, S. Bonyadi, T.S. Chung, Investigation of different hollow fiber module designs for flux enhancement in the membrane distillation process, *J. Memb. Sci.* 311 (2008) 371–379. https://doi.org/10.1016/j.memsci.2007.12.054.

[179] M. Su, M.M. Teoh, K.Y. Wang, J. Su, T.S. Chung, Effect of inner-layer thermal conductivity on flux enhancement of dual-layer hollow fiberhollow-fiber membranes in direct contact membrane distillation, *J. Memb. Sci.* 364 (2010) 278–289. https://doi.org/10.1016/j.memsci.2010.08.028.

[180] N. Tang, C. Feng, H. Han, X. Hua, L. Zhang, J. Xiang, P. Cheng, W. Du, X. Wang, High permeation flux polypropylene/ethylene vinyl acetate co-blending membranes via thermally induced phase separation for vacuum membrane

distillation desalination, *Desalination* 394 (2016) 44–55. https://doi.org/10.1016/j.desal.2016.04.024.

[181] Z. Song, M. Xing, J. Zhang, B. Li, S. Wang, Determination of phase diagram of a ternary PVDF/γ-BL/DOP system in TIPS process and its application in preparing hollow fiberhollow-fiber membranes for membrane distillation, *Sep. Purif. Technol.* 90 (2012) 221–230. https://doi.org/10.1016/j.seppur.2012.02.043.

[182] M. Ulbricht, Advanced functional polymer membranes, *Polymer (Guildf)* 47 (2006) 2217–2262. https://doi.org/10.1016/j.polymer.2006.01.084.

[183] Z.Q. Dong, X. hua Ma, Z.L. Xu, W.T. You, F. bing Li, Superhydrophobic PVDF-PTFE electrospun nanofibrous membranes for desalination by vacuum membrane distillation, *Desalination* 347 (2014) 175–183. https://doi.org/10.1016/j.desal.2014.05.015.

[184] Z.X. Low, Y.T. Chua, B.M. Ray, D. Mattia, I.S. Metcalfe, D.A. Patterson, Perspective on 3D printing of separation membranes and comparison to related unconventional fabrication techniques, *J. Memb. Sci.* 523 (2017) 596–613. https://doi.org/10.1016/j.memsci.2016.10.006.

[185] M. Yao, Y.C. Woo, L.D. Tijing, W.G. Shim, J.S. Choi, S.H. Kim, H.K. Shon, Effect of heat-press conditions on electrospun membranes for desalination by direct contact membrane distillation, *Desalination* 378 (2016) 80–91. https://doi.org/10.1016/j.desal.2015.09.025.

[186] J.Y. Lee, W.S. Tan, J. An, C.K. Chua, C.Y. Tang, A.G. Fane, T.H. Chong, The potential to enhance membrane module design with 3D printing technology, *J. Memb. Sci.* 499 (2016) 480–490. https://doi.org/10.1016/j.memsci.2015.11.008.

[187] Y. Liao, C.H. Loh, R. Wang, A.G. Fane, Electrospun superhydrophobic membranes with unique structures for membrane distillation, *ACS Appl. Mater. Interfaces* 6 (2014) 16035–16048. https://doi.org/10.1021/am503968n.

[188] D.E. Suk, T. Matsuura, H.B. Park, Y.M. Lee, Development of novel surface modified phase inversion membranes having hydrophobic surface-modifying macromolecule (nSMM) for vacuum membrane distillations, *Desalination* 261 (2010) 300–312. https://doi.org/10.1016/j.desal.2010.06.058.

[189] D. Hou, J. Wang, X. Sun, Z. Ji, Z. Luan, Preparation and properties of PVDF composite hollow fiber membranes for desalination through direct contact membrane distillation, *J. Memb. Sci.* 405–406 (2012) 185–200. https://doi.org/10.1016/j.memsci.2012.03.008.

[190] J.A. Prince, V. Anbharasi, T.S. Shanmugasundaram, G. Singh, Preparation and characterization of novel triple layer hydrophilic- hydrophobic composite membrane for desalination using air gap membrane distillation, *Sep. Purif. Technol.* 118 (2013) 598–603. https://doi.org/10.1016/j.seppur.2013.08.006.

[191] F. Edwie, M.M. Teoh, T.S. Chung, Effects of additives on dual-layer hydrophobic-hydrophilic PVDF hollow fiber membranes for membrane distillation and continuous performance, *Chem. Eng. Sci.* 68 (2012) 567–578. https://doi.org/10.1016/j.ces.2011.10.024.

[192] M.M. Teoh, T.S. Chung, Y.S. Yeo, Dual-layer PVDF/PTFE composite hollow fibers with a thin macrovoid-free selective layer for water production via membrane distillation, *Chem. Eng. J.* 171 (2011) 684–691. https://doi.org/10.1016/j.cej.2011.05.020.

[193] Z.M. Huang, Y.Z. Zhang, M. Kotaki, S. Ramakrishna, A review on polymer nanofibers by electrospinning and their applications in nanocomposites, *Compos. Sci. Technol.* 63 (2003) 2223–2253. https://doi.org/10.1016/S0266-3538(03)00178-7.

[194] M. Yao, Y.C. Woo, L.D. Tijing, C. Cesarini, H.K. Shon, Improving nanofiber membrane characteristics and membrane distillation performance of heat-pressed

Membranes for MD/MCr **203**

membranes via annealing post-treatment, *Appl. Sci.* 7 (2017). https://doi.org/10.3390/app7010078.

[195] A. Razmjou, E. Arifin, G. Dong, J. Mansouri, V. Chen, Superhydrophobic modification of TiO 2 nanocomposite PVDF membranes for applications in membrane distillation, *J. Memb. Sci.* 415–416 (2012) 850–863. https://doi.org/10.1016/j.memsci.2012.06.004.

[196] Y. Liao, R. Wang, A.G. Fane, Engineering superhydrophobic surface on poly (vinylidene fluoride) nanofiber membranes for direct contact membrane distillation, *J. Memb. Sci.* 440 (2013) 77–87. https://doi.org/10.1016/j.memsci.2013.04.006.

[197] S. Meng, J. Mansouri, Y. Ye, V. Chen, Effect of templating agents on the properties and membrane distillation performance of TiO2-coated PVDF membranes, *J. Memb. Sci.* 450 (2014) 48–59. https://doi.org/10.1016/j.memsci.2013.08.036.

[198] J. Zhang, Z. Song, B. Li, Q. Wang, S. Wang, Fabrication and characterization of superhydrophobic poly (vinylidene fluoride) membrane for direct contact membrane distillation, *Desalination* 324 (2013) 1–9. https://doi.org/10.1016/j.desal.2013.05.018.

[199] X. Yang, R. Wang, L. Shi, A.G. Fane, M. Debowski, Performance improvement of PVDF hollow fiber-based membrane distillation process, *J. Memb. Sci.* 369 (2011) 437–447. https://doi.org/10.1016/j.memsci.2010.12.020.

[200] X. Wei, B. Zhao, X.M. Li, Z. Wang, B.Q. He, T. He, B. Jiang, CF 4 plasma surface modification of asymmetric hydrophilic polyethersulfone membranes for direct contact membrane distillation, *J. Memb. Sci.* 407–408 (2012) 164–175. https://doi.org/10.1016/j.memsci.2012.03.031.

[201] C. Su, J. Chang, K. Tang, F. Gao, Y. Li, H. Cao, Novel three-dimensional superhydrophobic and strength-enhanced electrospun membranes for long-term membrane distillation, *Sep. Purif. Technol.* 178 (2017) 279–287. https://doi.org/10.1016/j.seppur.2017.01.050.

[202] K. Sears, L. Dumée, J. Schütz, M. She, C. Huynh, S. Hawkins, M. Duke, S. Gray, Recent developments in carbon nanotube membranes for water purification and gas separation, *Materials (Basel)* 3 (2010) 127–149. https://doi.org/10.3390/ma3010127.

[203] P.S. Goh, A.F. Ismail, B.C. Ng, Carbon nanotubes for desalination: Performance evaluation and current hurdles, *Desalination* 308 (2013) 2–14. https://doi.org/10.1016/j.desal.2012.07.040.

[204] M. Bhadra, S. Mitra, Nanostructured membranes in analytical chemistry, *TrAC – Trends Anal. Chem.* 45 (2013) 248–263. https://doi.org/10.1016/j.trac.2012.12.010.

[205] W.C. Lai, L.T. Cheng, Preparation and characterization of novel poly(vinylidene fluoride) membranes using self-assembled dibenzylidene sorbitol for membrane distillation, *Desalination* 332 (2014) 7–17. https://doi.org/10.1016/j.desal.2013.10.020.

[206] M. Qtaishat, D. Rana, M. Khayet, T. Matsuura, Preparation and characterization of novel hydrophobic/hydrophilic polyetherimide composite membranes for desalination by direct contact membrane distillation, *J. Memb. Sci.* 327 (2009) 264–273. https://doi.org/10.1016/j.memsci.2008.11.040.

[207] M. Qtaishat, M. Khayet, T. Matsuura, Novel porous composite hydrophobic/hydrophilic polysulfone membranes for desalination by direct contact membrane distillation, *J. Memb. Sci.* 341 (2009) 139–148. https://doi.org/10.1016/j.memsci.2009.05.053.

[208] A. Figoli, C. Ursino, F. Galiano, E. Di Nicolò, P. Campanelli, M.C. Carnevale, A. Criscuoli, Innovative hydrophobic coating of perfluoropolyether (PFPE) on commercial hydrophilic membranes for DCMD application, *J. Memb. Sci.* 522 (2017) 192–201. https://doi.org/10.1016/j.memsci.2016.08.066.

[209] C. Capello, U. Fischer, K. Hungerbühler, What is a green solvent? A comprehensive framework for the environmental assessment of solvents, *Green Chem.* 9 (2007) 927–934. https://doi.org/10.1039/b617536h.

[210] M. Razali, J.F. Kim, M. Attfield, P.M. Budd, E. Drioli, Y.M. Lee, G. Szekely, Sustainable wastewater treatment and recycling in membrane manufacturing, *Green Chem.* 17 (2015) 5196–5205. https://doi.org/10.1039/c5gc01937k.

[211] A. Figoli, A. Creiscuoli, *Green chemistry and sustainable technology sustainable membrane technology for water and wastewater treatment*, Springer Nature, Singapore, 2017.

[212] Y.K. Ong, N. Widjojo, T.S. Chung, Fundamentals of semi-crystalline poly(vinylidene fluoride) membrane formation and its prospects for biofuel (ethanol and acetone) separation via pervaporation, *J. Memb. Sci.* 378 (2011) 149–162. https://doi.org/10.1016/j.memsci.2011.04.037.

[213] D. Prat, A. Wells, J. Hayler, H. Sneddon, C.R. McElroy, S. Abou-Shehada, P. J. Dunn, CHEM21 selection guide of classical- and less classical-solvents, *Green Chem.* 18 (2015) 288–296. https://doi.org/10.1039/c5gc01008j.

[214] R.L. Maynard, Late lessons from early warnings: The precautionary principle 1896–2000: European Environment Agency. Editorial team chaired by Poul Harromoes. Environmental Issue Report No. 22 (p. 212; free of charge) 2001. ISBN 92 9167 323 4. Catalogue no. TH-39-01-82, 2002. https://doi.org/10.1136/oem.59.11.789-a.

[215] K. Alfonsi, J. Colberg, P.J. Dunn, T. Fevig, S. Jennings, T.A. Johnson, H.P. Kleine, C. Knight, M.A. Nagy, D.A. Perry, M. Stefaniak, Green chemistry tools to influence a medicinal chemistry and research chemistry based organisation, *Green Chem.* 10 (2008) 31–36. https://doi.org/10.1039/b711717e.

[216] G. Szekely, M.F. Jimenez-Solomon, P. Marchetti, J.F. Kim, A.G. Livingston, Sustainability assessment of organic solvent nanofiltration: From fabrication to application, *Green Chem.* 16 (2014) 4440–4473. https://doi.org/10.1039/c4gc00701h.

[217] V. Faggian, P. Scanferla, S. Paulussen, S. Zuin, Combining the European chemicals regulation and an (eco)toxicological screening for a safer membrane development, *J. Clean. Prod.* 83 (2014) 404–412. https://doi.org/10.1016/j.jclepro.2014.07.017.

[218] J. Sherwood, M. De Bruyn, A. Constantinou, L. Moity, C.R. McElroy, T. J. Farmer, T. Duncan, W. Raverty, A.J. Hunt, J.H. Clark, Dihydrolevoglucosenone (Cyrene) as a bio-based alternative for dipolar aprotic solvents, *Chem. Commun.* 50 (2014) 9650–9652. https://doi.org/10.1039/c4cc04133j.

[219] T. Marino, F. Galiano, A. Molino, A. Figoli, New frontiers in sustainable membrane preparation: Cyrene^{TM} as green bioderived solvent, *J. Memb. Sci.* 580 (2019) 224–234. https://doi.org/10.1016/j.memsci.2019.03.034.

[220] J.E. Camp, Bio-available solvent cyrene: Synthesis, derivatization, and applications, *ChemSusChem.* 11 (2018) 3048–3055. https://doi.org/10.1002/cssc.201801420.

[221] J.F. Kim, J.T. Jung, H.H. Wang, S.Y. Lee, T. Moore, A. Sanguineti, E. Drioli, Y. M. Lee, Microporous PVDF membranes via thermally induced phase separation (TIPS) and stretching methods, *J. Memb. Sci.* 509 (2016) 94–104. https://doi.org/10.1016/j.memsci.2016.02.050.

[222] S. Ichi Sawada, C. Ursino, F. Galiano, S. Simone, E. Drioli, A. Figoli, Effect of citrate-based non-toxic solvents on poly(vinylidene fluoride) membrane preparation via thermally induced phase separation, *J. Memb. Sci.* 493 (2015) 232–242. https://doi.org/10.1016/j.memsci.2015.07.003.

[223] Y. Su, C. Chen, Y. Li, J. Li, PVDF membrane formation via thermally induced phase separation, *J. Macromol. Sci. Part A Pure Appl. Chem.* 44 (2007) 99–104. https://doi.org/10.1080/10601320601044575.

[224] B.J. Cha, J.M. Yang, Preparation of poly(vinylidene fluoride) hollow fiber membranes for microfiltration using modified TIPS process, *J. Memb. Sci.* 291 (2007) 191–198. https://doi.org/10.1016/j.memsci.2007.01.008.

[225] S. Rajabzadeh, T. Maruyama, T. Sotani, H. Matsuyama, Preparation of PVDF hollow fiber membrane from a ternary polymer/solvent/nonsolvent system via thermally induced phase separation (TIPS) method, *Sep. Purif. Technol.* 63 (2008) 415–423. https://doi.org/10.1016/j.seppur.2008.05.027.

[226] N.T. Hassankiadeh, Z. Cui, J.H. Kim, D.W. Shin, S.Y. Lee, A. Sanguineti, V. Arcella, Y.M. Lee, E. Drioli, Microporous poly(vinylidene fluoride) hollow fiber membranes fabricated with PolarClean as water-soluble green diluent and additives, *J. Memb. Sci.* 479 (2015) 204–212. https://doi.org/10.1016/j.memsci.2015.01.031.

[227] J.T. Jung, H.H. Wang, J.F. Kim, J. Lee, J.S. Kim, E. Drioli, Y.M. Lee, Tailoring nonsolvent-thermally induced phase separation (N-TIPS) effect using triple spinneret to fabricate high performance PVDF hollow fiber membranes, *J. Memb. Sci.* 559 (2018) 117–126. https://doi.org/10.1016/j.memsci.2018.04.054.

[228] C. Meringolo, T.F. Mastropietro, T. Poerio, E. Fontananova, G. De Filpo, E. Curcio, G. Di Profio, Tailoring PVDF membranes surface topography and hydrophobicity by a sustainable two-steps phase separation process, *ACS Sustain. Chem. Eng.* 6 (2018) 10069–10077. https://doi.org/10.1021/acssuschemeng.8b01407.

[229] Q. Wang, Z. Wang, Z. Wu, Effects of solvent compositions on physicochemical properties and anti-fouling ability of PVDF microfiltration membranes for wastewater treatment, *Desalination* 297 (2012) 79–86. https://doi.org/10.1016/j.desal.2012.04.020.

[230] S. Fadhil, T. Marino, H.F. Makki, Q.F. Alsalhy, S. Blefari, F. Macedonio, E. Di Nicolò, L. Giorno, E. Drioli, A. Figoli, Novel PVDF-HFP flat sheet membranes prepared by triethyl phosphate (TEP) solvent for direct contact membrane distillation, *Chem. Eng. Process. Process Intensif.* 102 (2016) 16–26. https://doi.org/10.1016/j.cep.2016.01.007.

[231] F. Russo, F. Galiano, F. Pedace, F. Arico, A. Figoli, Dimethyl isosorbide as a green solvent for sustainable ultrafiltration and microfiltration membrane preparation (2019). https://doi.org/10.1021/acssuschemeng.9b06496.

[232] A. Figoli, T. Marino, S. Simone, E. Di Nicolò, X.M. Li, T. He, S. Tornaghi, E. Drioli, Towards non-toxic solvents for membrane preparation: A review, *Green Chem.* 16 (2014) 4034–4059. https://doi.org/10.1039/c4gc00613e.

[233] M.A. Rasool, P.P. Pescarmona, I.F.J. Vankelecom, Applicability of organic carbonates as green solvents for membrane preparation, *ACS Sustain. Chem. Eng.* 7 (2019) 13774–13785. https://doi.org/10.1021/acssuschemeng.9b01507.

[234] T. Marino, F. Russo, A. Criscuoli, A. Figoli, TamiSolve® NxG as novel solvent for polymeric membrane preparation, *J. Memb. Sci.* 542 (2017) 418–429. https://doi.org/10.1016/j.memsci.2017.08.038.

[235] H.H. Chang, L.K. Chang, C.D. Yang, D.J. Lin, L.P. Cheng, Effect of solvent on the dipole rotation of poly(vinylidene fluoride) during porous membrane formation by precipitation in alcohol baths, *Polymer (Guildf)* 115 (2017) 164–175. https://doi.org/10.1016/j.polymer.2017.03.044.

[236] L. Wu, J. Sun, An improved process for polyvinylidene fluoride membrane preparation by using a water soluble diluent via thermally induced phase separation technique, *Mater. Des.* 86 (2015) 204–214. https://doi.org/10.1016/j.matdes.2015.07.053.

[237] Z. Cui, N.T. Hassankiadeh, S.Y. Lee, K.T. Woo, J.M. Lee, A. Sanguineti, V. Arcella, Y.M. Lee, E. Drioli, Tailoring novel fibrillar morphologies in poly (vinylidene fluoride) membranes using a low toxic triethylene glycol diacetate

(TEGDA) diluent, *J. Memb. Sci.* 473 (2015) 128–136. https://doi.org/10.1016/j.memsci.2014.09.019.

[238] M.L. Yeow, Y.T. Liu, K. Li, Morphological study of poly(vinylidene fluoride) asymmetric membranes: Effects of the solvent, additive, and dope temperature, *J. Appl. Polym. Sci.* 92 (2004) 1782–1789. https://doi.org/10.1002/app.20141.

[239] M.mi Tao, F. Liu, B.rong Ma, L.xin Xue, Effect of solvent power on PVDF membrane polymorphism during phase inversion, *Desalination* 316 (2013) 137–145. https://doi.org/10.1016/j.desal.2013.02.005.

[240] S. Nejati, C. Boo, C.O. Osuji, M. Elimelech, Engineering flat sheet microporous PVDF films for membrane distillation, *J. Memb. Sci.* 492 (2015) 355–363. https://doi.org/10.1016/j.memsci.2015.05.033.

[241] T. Marino, S. Blefari, E. Di Nicolò, A. Figoli, A more sustainable membrane preparation using triethyl phosphate as solvent, *Green Process. Synth.* 6 (2017) 295–300. https://doi.org/10.1515/gps-2016-0165.

[242] Z. Cui, Y. Cheng, K. Xu, J. Yue, Y. Zhou, X. Li, Q. Wang, S.P. Sun, Y. Wang, X. Wang, Z. Wang, Wide liquid-liquid phase separation region enhancing tensile strength of poly(vinylidene fluoride) membranes via TIPS method with a new diluent, *Polymer (Guildf)* 141 (2018) 46–53. https://doi.org/10.1016/j.polymer.2018.02.054.

[243] M. Khayet, A.O. Imdakm, T. Matsuura, Monte Carlo simulation and experimental heat and mass transfer in direct contact membrane distillation, *Int. J. Heat Mass Transf.* 53 (2010) 1249–1259. https://doi.org/10.1016/j.ijheatmasstransfer.2009.12.043.

[244] R.W. Schofield, A.G. Fane, C.J.D. Fell, R. Macoun, Factors affecting flux in membrane distillation, *Desalination* 77 (1990) 279–294. https://doi.org/10.1016/0011-9164(90)85030-E.

[245] M.A. Izquierdo-Gil, M.C. García-Payo, C. Fernández-Pineda, Air gap membrane distillation of sucrose aqueous solutions, *J. Memb. Sci.* 155 (1999) 291–307. https://doi.org/10.1016/S0376-7388(98)00323-8.

[246] A. Criscuoli, M.C. Carnevale, E. Drioli, Evaluation of energy requirements in membrane distillation, *Chem. Eng. Process. Process Intensif.* 47 (2008) 1098–1105. https://doi.org/10.1016/j.cep.2007.03.006.

[247] V.V.U.I.B.V. and V.N.N. P.P. Zolotarev, Karpov Physico-Chemical Research Institute, ul. Obukha 10. Moscow 103064. Russia, *J. Hazard. Mater.* 37 (1994) 77–82.

[248] K.W. Lawson, D.R. Lloyd, Membrane distillation. I. Module design and performance evaluation using vacuum membrane distillation, *J. Memb. Sci.* 120 (1996) 111–121. https://doi.org/10.1016/0376-7388(96)00140-8.

[249] K.W. Lawson, D.R. Lloyd, Membrane distillation. II. Direct contact MD, *J. Memb. Sci.* 120 (1996) 123–133. https://doi.org/10.1016/0376-7388(96)00141-X.

[250] M.I. VáZquez-GonzáLez, L. MartíNez, Nonisothermal water transport through hydrophobic membranes in a stirred cell, *Sep. Sci. Technol.* 29 (1994) 1957–1966. https://doi.org/10.1080/01496399408002183.

[251] M. Gryta, Influence of polypropylene membrane surface porosity on the performance of membrane distillation process, *J. Memb. Sci.* 287 (2007) 67–78. https://doi.org/10.1016/j.memsci.2006.10.011.

[252] S. Al-Obaidani, E. Curcio, F. Macedonio, G. Di Profio, H. Al-Hinai, E. Drioli, Potential of membrane distillation in seawater desalination: Thermal efficiency, sensitivity study and cost estimation, *J. Memb. Sci.* 323 (2008) 85–98. https://doi.org/10.1016/j.memsci.2008.06.006.

[253] E. Drioli, Y. Wu, V. Calabro, Membrane distillataion in the treatment of aqueous solutions, *J. Memb. Sci.* 33 (1987) 277–284. https://doi.org/10.1016/S0376-7388(00)80285-9.

[254] C.M. Guijt, G.W. Meindersma, T. Reith, A.B. De Haan, Method for experimental determination of the gas transport properties of highly porous fibre membranes: A first step before predictive modelling of a membrane distillation process, *Desalination* 147 (2002) 127–132. https://doi.org/10.1016/S0011-9164(02)00598-2.

[255] C.H. Lee, W.H. Hong, Effect of operating variables on the flux and selectivity in sweep gas membrane distillation for dilute aqueous isopropanol, *J. Memb. Sci.* 188 (2001) 79–86. https://doi.org/10.1016/S0376-7388(01)00373-8.

[256] E.O.N. Dow, J. Zhang, M. Duke, J. Li, S.R. Gray, *Membrane distillation of brine wastes*, CRC for Water Quality and Treatment, 2008. www.xzero.se/doc/MembraneDistillationofBrineWastesWQRA.pdf.

[257] E. Guillén-Burrieza, G. Zaragoza, S. Miralles-Cuevas, J. Blanco, Experimental evaluation of two pilot-scale membrane distillation modules used for solar desalination, *J. Memb. Sci.* 409–410 (2012) 264–275. https://doi.org/10.1016/j.memsci.2012.03.063.

[258] E. Guillén-Burrieza, J. Blanco, G. Zaragoza, D.C. Alarcón, P. Palenzuela, M. Ibarra, W. Gernjak, Experimental analysis of an air gap membrane distillation solar desalination pilot system, *J. Memb. Sci.* 379 (2011) 386–396. https://doi.org/10.1016/j.memsci.2011.06.009.

[259] http://aquastill.nl/technology/modules/ (accessed on May 2020).

[260] D. Winter, J. Koschikowski, S. Ripperger, Desalination using membrane distillation: Flux enhancement by feed water deaeration on spiral-wound modules, *J. Memb. Sci.* 423–424 (2012) 215–224. https://doi.org/10.1016/j.memsci.2012.08.018.

[261] G. Zakrzewska-Trznadel, M. Harasimowicz, A.G. Chmielewski, Concentration of radioactive components in liquid low-level radioactive waste by membrane distillation, *J. Memb. Sci.* 163 (1999) 257–264. https://doi.org/10.1016/S0376-7388(99)00171-4.

[262] W.L. Gore, R.W. Gore, D.W. Gore, Desalination device and process, US patent 4,545,862 (1985).

[263] F. Laganà, G. Barbieri, E. Drioli, Direct contact membrane distillation: Modelling and concentration experiments, *J. Macromol. Sci. Part A Pure Appl. Chem.* 166 (2000) 1–11. https://doi.org/10.1080/01933922.2013.747351.

[264] D. Zhao, J. Zuo, K.J. Lu, T.S. Chung, Fluorographite modified PVDF membranes for seawater desalination via direct contact membrane distillation, *Desalination* 413 (2017) 119–126. https://doi.org/10.1016/j.desal.2017.03.012.

[265] H. Yan, X. Lu, C. Wu, X. Sun, W. Tang, Fabrication of a super-hydrophobic polyvinylidene fluoride hollow fiber membrane using a particle coating process, *J. Memb. Sci.* 533 (2017) 130–140. https://doi.org/10.1016/j.memsci.2017.03.033.

[266] H. Fang, J.F. Gao, H.T. Wang, C.S. Chen, Hydrophobic porous alumina hollow fiber for water desalination via membrane distillation process, *J. Memb. Sci.* 403–404 (2012) 41–46. https://doi.org/10.1016/j.memsci.2012.02.011.

[267] Z.Q. Dong, B.J. Wang, X.H. Ma, Y.M. Wei, Z.L. Xu, FAS grafted electrospun poly(vinyl alcohol) nanofiber membranes with robust superhydrophobicity for membrane distillation, *ACS Appl. Mater. Interfaces.* 7 (2015) 22652–22659. https://doi.org/10.1021/acsami.5b07454.

[268] Z.Q. Dong, X.H. Ma, Z.L. Xu, Z.Y. Gu, Superhydrophobic modification of PVDF-SiO2 electrospun nanofiber membranes for vacuum membrane distillation, *RSC Adv.* 5 (2015) 67962–67970. https://doi.org/10.1039/c5ra10575g.

[269] F. Guo, A. Servi, A. Liu, K.K. Gleason, G.C. Rutledge, Desalination by membrane distillation using electrospun polyamide fiber membranes with surface fluorination by chemical vapor deposition, *ACS Appl. Mater. Interfaces* 7 (2015) 8225–8232. https://doi.org/10.1021/acsami.5b01197.

[270] K.J. Lu, J. Zuo, T.S. Chung, Tri-bore PVDF hollow fibers with a super-hydrophobic coating for membrane distillation, *J. Memb. Sci.* 514 (2016) 165–175. https://doi.org/10.1016/j.memsci.2016.04.058.

[271] J. Lee, C. Boo, W.H. Ryu, A.D. Taylor, M. Elimelech, Development of omniphobic desalination membranes using a charged electrospun nanofiber scaffold, *ACS Appl. Mater. Interfaces* 8 (2016) 11154–11161. https://doi.org/10.1021/acsami. 6b02419.

[272] B.S. Lalia, E. Guillen-Burrieza, H.A. Arafat, R. Hashaikeh, Fabrication and characterization of polyvinylidenefluoride-co-hexafluoropropylene (PVDF-HFP) electrospun membranes for direct contact membrane distillation, *J. Memb. Sci.* 428 (2013) 104–115. https://doi.org/10.1016/j.memsci.2012.10.061.

[273] Z. Li, Y. Peng, Y. Dong, H. Fan, P. Chen, L. Qiu, Q. Jiang, Effects of thermal efficiency in DCMD and the preparation of membranes with low thermal conductivity, *Appl. Surf. Sci.* 317 (2014) 338–349. https://doi.org/10.1016/j. apsusc.2014.07.080.

[274] Y. Peng, Y. Dong, H. Fan, P. Chen, Z. Li, Q. Jiang, Preparation of polysulfone membranes via vapor-induced phase separation and simulation of direct-contact membrane distillation by measuring hydrophobic layer thickness, *Desalination* 316 (2013) 53–66. https://doi.org/10.1016/j.desal.2013.01.021.

[275] S. Munirasu, F. Banat, A.A. Durrani, M.A. Haija, Intrinsically superhydrophobic PVDF membrane by phase inversion for membrane distillation, *Desalination* 417 (2017) 77–86. https://doi.org/10.1016/j.desal.2017.05.019.

[276] T.L.S. Silva, S. Morales-Torres, J.L. Figueiredo, A.M.T. Silva, Multi-walled carbon nanotube/PVDF blended membranes with sponge- and finger-like pores for direct contact membrane distillation, *Desalination* 357 (2015) 233–245. https:// doi.org/10.1016/j.desal.2014.11.025.

[277] M. Baghbanzadeh, D. Rana, T. Matsuura, C.Q. Lan, Effects of hydrophilic CuO nanoparticles on properties and performance of PVDF VMD membranes, *Desalination* 369 (2015) 75–84. https://doi.org/10.1016/j.desal.2015.04.032.

[278] T. Xiao, P. Wang, X. Yang, X. Cai, J. Lu, Fabrication and characterization of novel asymmetric polyvinylidene fluoride (PVDF) membranes by the nonsolvent thermally induced phase separation (NTIPS) method for membrane distillation applications, *J. Memb. Sci.* 489 (2015) 160–174. https://doi.org/10.1016/j. memsci.2015.03.081.

[279] M.A. Tooma, T.S. Najim, Q.F. Alsalhy, T. Marino, A. Criscuoli, L. Giorno, A. Figoli, Modification of polyvinyl chloride (PVC) membrane for vacuum membrane distillation (VMD) application, *Desalination* 373 (2015) 58–70. https://doi. org/10.1016/j.desal.2015.07.008.

[280] H. Abdallah, A.F. Moustafa, A.A.H. AlAnezi, H.E.M. El-Sayed, Performance of a newly developed titanium oxide nanotubes/polyethersulfone blend membrane for water desalination using vacuum membrane distillation, *Desalination* 346 (2014) 30–36. https://doi.org/10.1016/j.desal.2014.05.003.

[281] R. Thomas, E. Guillen-Burrieza, H.A. Arafat, Pore structure control of PVDF membranes using a 2-stage coagulation bath phase inversion process for application in membrane distillation (MD), *J. Memb. Sci.* 452 (2014) 470–480. https:// doi.org/10.1016/j.memsci.2013.11.036.

[282] S. Devi, P. Ray, K. Singh, P.S. Singh, Preparation and characterization of highly micro-porous PVDF membranes for desalination of saline water through vacuum membrane distillation, *Desalination* 346 (2014) 9–18. https://doi.org/10.1016/j. desal.2014.05.004.

[283] D. Sun, M.Q. Liu, J.H. Guo, J.Y. Zhang, B.B. Li, D.Y. Li, Preparation and characterization of PDMS-PVDF hydrophobic microporous membrane for membrane

distillation, *Desalination* 370 (2015) 63–71. https://doi.org/10.1016/j.desal.2015.05.017.

[284] D. Hou, H. Fan, Q. Jiang, J. Wang, X. Zhang, Preparation and characterization of PVDF flat-sheet membranes for direct contact membrane distillation, *Sep. Purif. Technol.* 135 (2014) 211–222. https://doi.org/10.1016/j.seppur.2014.08.023.

[285] D. Hou, G. Dai, J. Wang, H. Fan, L. Zhang, Z. Luan, Preparation and characterization of PVDF/nonwoven fabric flat-sheet composite membranes for desalination through direct contact membrane distillation, *Sep. Purif. Technol.* 101 (2012) 1–10. https://doi.org/10.1016/j.seppur.2012.08.031.

[286] Y.C. Woo, Y. Kim, W.G. Shim, L.D. Tijing, M. Yao, L.D. Nghiem, J.S. Choi, S.H. Kim, H.K. Shon, Graphene/PVDF flat-sheet membrane for the treatment of RO brine from coal seam gas produced water by air gap membrane distillation, *J. Memb. Sci.* 513 (2016) 74–84. https://doi.org/10.1016/j.memsci.2016.04.014.

[287] X. Feng, L.Y. Jiang, T. Matsuura, P. Wu, Fabrication of hydrophobic/hydrophilic composite hollow fibers for DCMD: Influence of dope formulation and external coagulant, *Desalination* 401 (2017) 53–63. https://doi.org/10.1016/j.desal.2016.07.026.

[288] J.A. Kharraz, M.R. Bilad, H.A. Arafat, Flux stabilization in membrane distillation desalination of seawater and brine using corrugated PVDF membranes, *J. Memb. Sci.* 495 (2015) 404–414. https://doi.org/10.1016/j.memsci.2015.08.039.

[289] P. Wang, T.S. Chung, Design and fabrication of lotus-root-like multi-bore hollow fiber membrane for direct contact membrane distillation, *J. Memb. Sci.* 421–422 (2012) 361–374. https://doi.org/10.1016/j.memsci.2012.08.003.

[290] Z. Wang, D. Hou, S. Lin, Composite membrane with underwater-oleophobic surface for anti-oil-fouling membrane distillation, *Environ. Sci. Technol.* 50 (2016) 3866–3874. https://doi.org/10.1021/acs.est.5b05976.

[291] N. Hamzah, C.P. Leo, Membrane distillation of saline with phenolic compound using superhydrophobic PVDF membrane incorporated with TiO2 nanoparticles: Separation, fouling and self-cleaning evaluation, *Desalination* 418 (2017) 79–88. https://doi.org/10.1016/j.desal.2017.05.029.

[292] S. Simone, A. Figoli, A. Criscuoli, M.C. Carnevale, S.M. Alfadul, H.S. Al-Romaih, F.S. Al Shabouna, O.A. Al-Harbi, E. Drioli, Effect of selected spinning parameters on PVDF hollow fiber morphology for potential application in desalination by VMD, *Desalination* 344 (2014) 28–35. https://doi.org/10.1016/j.desal.2014.03.004.

[293] Y. Tang, N. Li, A. Liu, S. Ding, C. Yi, H. Liu, Effect of spinning conditions on the structure and performance of hydrophobic PVDF hollow fiber membranes for membrane distillation, *Desalination* 287 (2012) 326–339. https://doi.org/10.1016/j.desal.2011.11.045.

[294] K.J. Lu, J. Zuo, T.S. Chung, Novel PVDF membranes comprising n-butylamine functionalized graphene oxide for direct contact membrane distillation, *J. Memb. Sci.* 539 (2017) 34–42. https://doi.org/10.1016/j.memsci.2017.05.064.

[295] L. Zhao, C. Wu, Z. Liu, Q. Zhang, X. Lu, Highly porous PVDF hollow fiber membranes for VMD application by applying a simultaneous co-extrusion spinning process, *J. Memb. Sci.* 505 (2016) 82–91. https://doi.org/10.1016/j.memsci.2016.01.014.

[296] J. Zhu, L. Jiang, T. Matsuura, New insights into fabrication of hydrophobic/hydrophilic composite hollow fibers for direct contact membrane distillation, *Chem. Eng. Sci.* 137 (2015) 79–90. https://doi.org/10.1016/j.ces.2015.05.064.

[297] E. Drioli, A. Ali, S. Simone, F. MacEdonio, S.A. Al-Jlil, F.S. Al Shabonah, H.S. Al-Romaih, O. Al-Harbi, A. Figoli, A. Criscuoli, Novel PVDF hollow fiber

membranes for vacuum and direct contact membrane distillation applications, *Sep. Purif. Technol.* 115 (2013) 27–38. https://doi.org/10.1016/j.seppur.2013.04.040.

[298] C. Feng, K.C. Khulbe, T. Matsuura, R. Gopal, S. Kaur, S. Ramakrishna, M. Khayet, Production of drinking water from saline water by air-gap membrane distillation using polyvinylidene fluoride nanofiber membrane, *J. Memb. Sci.* 311 (2008) 1–6. https://doi.org/10.1016/j.memsci.2007.12.026.

[299] L. Francis, H. Maab, A. Alsaadi, S. Nunes, N. Ghaffour, G.L. Amy, Fabrication of electrospun nanofibrous membranes for membrane distillation application, Desalin. *Water Treat.* 51 (2013) 1337–1343. https://doi.org/10.1080/19443994.2012.700037.

[300] T. Zhou, Y. Yao, R. Xiang, Y. Wu, Formation and characterization of polytetra-fluoroethylene nanofiber membranes for vacuum membrane distillation, *J. Memb. Sci.* 453 (2014) 402–408. https://doi.org/10.1016/j.memsci.2013.11.027.

[301] B.S. Lalia, E. Guillen, H.A. Arafat, R. Hashaikeh, Nanocrystalline cellulose reinforced PVDF-HFP membranes for membrane distillation application, *Desalination* 332 (2014) 134–141. https://doi.org/10.1016/j.desal.2013.10.030.

[302] J.A. Prince, G. Singh, D. Rana, T. Matsuura, V. Anbharasi, T. S. Shanmugasundaram, Preparation and characterization of highly hydrophobic poly(vinylidene fluoride) – Clay nanocomposite nanofiber membranes (PVDF-clay NNMs) for desalination using direct contact membrane distillation, *J. Memb. Sci.* 397–398 (2012) 80–86. https://doi.org/10.1016/j.memsci.2012.01.012.

[303] Y. Huang, Q.L. Huang, H. Liu, C.X. Zhang, Y.W. You, N.N. Li, C.F. Xiao, Preparation, characterization, and applications of electrospun ultrafine fibrous PTFE porous membranes, *J. Memb. Sci.* 523 (2017) 317–326. https://doi.org/10.1016/j.memsci.2016.10.019.

[304] E.J. Lee, A.K. An, P. Hadi, S. Lee, Y.C. Woo, H.K. Shon, Advanced multi-nozzle electrospun functionalized titanium dioxide/polyvinylidene fluoride-co-hexafluoropropylene (TiO2/PVDF-HFP) composite membranes for direct contact membrane distillation, *J. Memb. Sci.* 524 (2017) 712–720. https://doi.org/10.1016/j.memsci.2016.11.069.

[305] Y. Liao, R. Wang, A.G. Fane, Fabrication of bioinspired composite nanofiber membranes with robust superhydrophobicity for direct contact membrane distillation, *Environ. Sci. Technol.* 48 (2014) 6335–6341. https://doi.org/10.1021/es405795s.

[306] X. Li, C. Wang, Y. Yang, X. Wang, M. Zhu, B.S. Hsiao, Dual-biomimetic super-hydrophobic electrospun polystyrene nanofibrous membranes for membrane distillation, *ACS Appl. Mater. Interfaces* 6 (2014) 2423–2430. https://doi.org/10.1021/am4048128.

[307] J.A. Prince, D. Rana, G. Singh, T. Matsuura, T. Jun Kai, T.S. Shanmugasundaram, Effect of hydrophobic surface modifying macromolecules on differently produced PVDF membranes for direct contact membrane distillation, *Chem. Eng. J.* 242 (2014) 387–396. https://doi.org/10.1016/j.cej.2013.11.039.

[308] L.D. Tijing, Y.C. Woo, M.A.H. Johir, J.S. Choi, H.K. Shon, A novel dual-layer bicomponent electrospun nanofibrous membrane for desalination by direct contact membrane distillation, *Chem. Eng. J.* 256 (2014) 155–159. https://doi.org/10.1016/j.cej.2014.06.076.

6 Artificial Water Channels

Toward Next-Generation Reverse Osmosis Membranes

Maria Di Vincenzo, Sophie Cerneaux, Istvan Kocsis, and Mihail Barboiu
Institut Européen des Membranes
University of Montpellier
ENSCM-CNRS
Montpellier (France)

6.1 INTRODUCTION

Universal access to safe drinking water is a fundamental need and human right [1]. Competition over increasingly contaminated freshwater in recent decades has made the availability of this resource a serious global challenge for the 21st century. Water scarcity and the problems associated with the lack of the clean water are well known. Because oceans hold 97.5% of the total water in our planet [2], desalination is considered, along with environmental remediation technologies and conservation policies, as an important solution to the problem of freshwater scarcity. Currently, about 100 million m^3 of desalinated water is produced per day, with an annual increase of more than 10% [3].

Treated as a lack of essential resource; it is becoming increasingly urgent in the context of simultaneous growth of population and economic activities. To address the increasing demand for water supply, several cutting-edge technologies have been explored in recent decades [4]. Current technologies are based on thermal processes [Multieffect Desalination (MED) or Multistage Flash (MSF)] and membrane processes [pressure driven Reverse Osmosis (RO), voltage driven Electrodialysis (ED), Membrane Distillation (MD) and Forward Osmosis (FO)], with RO being the dominant process used for seawater or brackish water desalination [5]. Most of the reverse osmosis membranes rely on an anisotropic polymerized thin-film composite (TFC) polyamide-PA layer produced by interfacial polymerization on an underlying porous support layer, commercially available polysulfone, all integrated into spiral wound modules for fitting in pressure vessels [6]. Current reverse osmosis-RO desalination can produce water at < \$ 1/m^3, which is about five times the cost of treated surface water. Energy use has dropped but is still about twice the thermodynamic minimum required when operating at 60% recovery, around 1.2 kWh/m^3. and this value has increased above 1.5 kWh/m^3 at 80% recovery [6]. Closed-circuit RO will soon allow operating at these high recovery regimes but with higher minimum energy requirements.

There is need for membrane innovation. The target of the future membrane is to increase both the water permeability while achieving high salt rejection. As pointed out recently, in many cases selectivity matters as much if not more than permeability. Efficiency may increase in the coming years by changing the process conditions, or by changing the membranes [7].

The RO desalination, using thin-film composite-TFC membranes, has evolved over the years bringing down the costs significantly [8]. However, the desalination is still relatively expensive. More than half of a century has passed since the first functional RO membrane was designed [9]. Despite fundamental knowledge, the material structure understanding is highly important for the up-scale from the laboratory films toward membranes, modules, and further process designs. New materials bearing promise for higher productivity in desalination have been developed in recent years, consequentially leading to the emerging research fields of thin-film nanocomposite-TFN [10] and biomimetic membranes [11]. The prime objective nowadays is to explore the naturally evolved desalination pathways and assess the possibilities of using them as the basis for *engineered* desalination processes of enhanced performance. Recent research projects propose to find an easily scalable Artificial Water Channels AWC approach that could be immediately applied to desalination systems to increase their energy efficiency. This chapter describes the incipient context and evolution of the artificial water channels field.

6.2 NATURAL WATER CHANNELS -THE AQUAPORINS

Natural systems have evolved for millions of years to accept functional biostructures to transport metabolites across cell membranes [12, 13]. Most natural physiological processes depend on selective exchanges between a cell and its environment, and water plays an important role [14]. The natural water channels, the Aquaporins (AQP) are known for their fast transport rates ($\sim 10^8$-10^9 water molecules/s/channel) and the perfect rejection of ions [15–18]. There are several features that are important for the AQPs efficient transport:

A) *Selectivity filter*-SF: AQPs present a hourglass SF structure offering size restriction of ~ 3 Å and selectivity against cations, reinforced through electrostatic repulsion in the region known as the aromatic arginine (ar/R) constriction. Water in the SF is H-bonded to the protein and to other adjacent water molecules, allowing the passage of only a single file of molecules. The narrowest diameter of the pore itself is large enough for the water molecules to pass, but restrictive enough to block the passage of hydrated Na^+ or Mg^{2+} cations, with hydration shells > 3 Å.

B) *Dipolar orientation of water wires*: An interesting collective structuration of water takes place in the SF, the interconnected water molecules adopting a dipolar orientation outward from the center of the pore. Although protons can pass from one water molecule to another through a Grotthuss mechanism, the inversion center breaks the water wire dipole and prohibits the proton translocation. More recent studies confirm that precise dynamic clustering of water molecules in SF is strongly favored via donor – acceptor H-bonding, excluding other small molecules, such as the hydroxide anions, because they are not able to form a stable H-bonding within the SF [18].

6.3 BIOASSISTED AQP-EMBEDDED MEMBRANES

Their high permeability/selectivity inspired the incorporation of AQPs into membranes for desalination [19]. To tackle this problem, an innovative approach was developed by using an amphiphilic artificial matrix hosting the AQPs. In terms of flexibility and stability, the block copolymers offered a good solution. The first work successfully incorporated AQPs into polymer-based vesicles, called *polymersomes* and proved the functionality of the protein [20]. The vesicles were made of a block copolymer with poly-(2-methyloxazoline)-poly-(dimethyl- siloxane)-poly-(2-methyloxazoline) (PMOXA-PDMS-PMOXA) units and were used to host AQPZ, a bacterial type protein from the Aquaporin family. Using the stopped flow setup, it has been shown that the otherwise impermeable polymersomes undergo an increase of up to 800 times in water permeability when the AQPs are incorporated (Figure 6.1).

The activation energy for water to pass through the polymersomes was comparable to that obtained in the case of native protein. The typical RO membrane shows a productivity of 1 µm/s/bar, while the modified AQPZ-ABA polymer not as a membrane is estimated for a productivity of 167 µm/s/bar. Indeed, an improvement in permeability of ~2 orders of magnitude does not yield to a decrease in energy consumption by the same factor.

Due to its energy efficiency, membrane technologies start to adapt the AQP water purification technologies [21, 22]. The high permeability/selectivity of AQPs inspired their incorporation into membranes for water purification applications. Thus, a new branch of membrane technologies for water purification, called AQP-bioassisted membranes, was developed. As a result, it was an attractive option to combine the potential of the AQPs with the current existing technology in membrane-based water filtration. As soon as the functions of the AQPs became known, it seems logical that in order to make use of their high selectivity and water transport, the proteins need to be used in conjunction with a membrane barrier. Their large-scale applications are of interest, owing to the high costs of the AQPs production, low

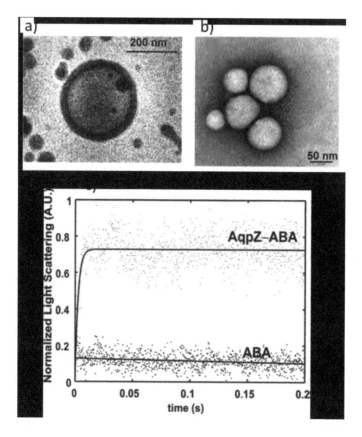

FIGURE 6.1 a) Cryogenic transmission electron micrograph of a polymer vesicle, b) electron micrograph of a cluster of vesicles and c) Stopped-flow light-scattering experiments; increase in relative light scattering with and without reconstituted AQPZ into the polymer at 5.5°C (image adapted with permission from reference [20]).

stability, and practical synthetic constraints of membrane fabrication processes [21–23].

In the simplest fashion, these membranes consisted of three main components: 1) AQP proteins, 2) amphiphilic molecules as building blocks for the construction of protecting liposomes in which the AQPs are embedded, and 3) a polymeric support for mechanical resistance. The choice for the amphiphilic molecules is supported by the fact that the AQPs are transmembrane proteins and their native environment is partly the hydrophobic region of the lipids found in cellular membranes, which meanwhile keeps the *mouths* of the protein exposed to the water from the cytosol. As such, a first reasonable strategy was to fabricate membranes having an active barrier, a lipid bilayer containing AQP for the enhanced water permeation. It was shown that the supported lipid bilayer (SLBs) membranes are suitable impermeable platforms for incorporation

Artificial Water Channels

of AQPs for water filtration. However, no reliable results have been obtained with such AQPs/bilayers (Figure 6.2a) [24].

It was also demonstrated that the use of covalently fixed SLBs on a Psf support can yield functional membranes with higher stability [25]. This was achieved through the covalent attachment of SLBs to a polydopamine coated polyehtersulfone. Using an amine functionalized lipid, the bottom half of the lipid bilayer is attached to the free carboxylic groups that can be found on the polydopamine. The covalent bonding between the bilayer and the support prevents the easy desorption of the lipid matrix, offering stability to the active layer of the membrane (Figure 6.2b). Nevertheless, the SLB is still prone to degradation when in contact with solutions containing detergents or high salinity, which can readily disrupt the lipid bilayer. The performances of these SLB membranes are remarkable in water filtration, although the reliability of these remains an issue for long term applications and upscaling.

In a compromise to obtain high stability and relatively good performance, supprimer bioassited membranes a different synthetic strategy was applied by using a mixed-matrix approach, when bilayer vesicles incorporating AQPs have been incorporated into traditional polyamide thin films [26]. This way, the AQPs can maintain their efficiency, having a basically native lipid membrane environment and at the same time, having minimal defect influences on the composite membrane in interaction with the polyamide matrix (Figure 6.3). The hybrid AQP-PA based membranes proved to be 40% more permeable than their simple thin film composites (TFC) counterparts, meanwhile being stable for periods relevant for industrial applications at lower energy consumption and operating costs [27].

Most of the current designs for AQP membranes involve polymeric host materials, which can make it viable for upscaling and use in industry [28]. Active high-flux AQP membranes have become the commercially available products. Production of AQP proteins is a feasible task since biological protocols and

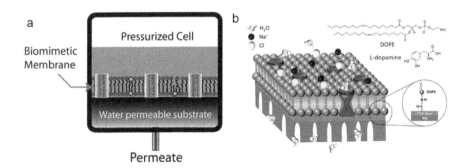

FIGURE 6.2 Schematic presentation of the structure of the AQPZ-incorporated SLB: a) on charged NF membrane supports [24]; and b) on a covalent amide bond 1,2-Dioleoyl-sn-glycero-3-phosphoethanolamine (DOPE) on top of a polydopamine (PDA) layer (orange), coated on a polyethersulfone (PSf) (gray) porous support [25]. (image adapted with permission from references [24, 25]).

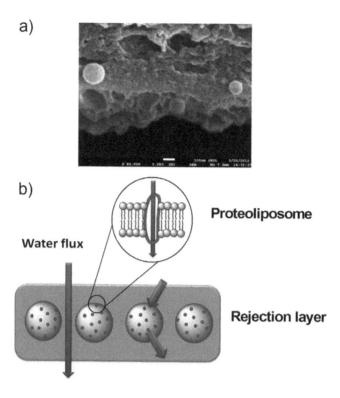

FIGURE 6.3 a) Scanning electron microscopy (SEM) image of the cross section of an AQPZ-thin film composite membrane; and b) Conceptual model of AQP proteoliposomes integrated into a polyamide matrix (image adapted with permissions from reference [26]).

reagents are in use, but the purification setup still represents an expensive and time-consuming obstacle. The question "how the natural proteins/artificial host matrix systems can be efficiently used together" has been thus partially answered [26–28].

However, no clear strategies converge to the easy synthetic development of robust AQP *bioassisted reverse osmosis (RO) or forward osmosis (FO) membranes,* working at high-pressure and high-salt concentrations, incompatible with biological AQPs. Most importantly, today, no readily usable AQP-based membrane exists for desalination.

They are many challenges to use the AQP as selective transporters related to their:

a) Long term stability when embedded in artificial membranes;
b) Bio-stability and unconventional processing requirements: high pressure and salinity;

Artificial Water Channels

c) Production – membrane proteins are challenging to mass produce; and

d) AQPs present an important flux, but not an acceptable permeability per active area, the selectivity is not reaching more than 97%, a critical problem for desalination.

A possible future approach is to replace AQPs with synthetic channels. Recent advancements have made it possible to synthesize artificial water channels (AWCs) that feature efficient, yet selective, transport of water based on size exclusion, as well as H-bond clustering at the molecular level [29]. Another current disadvantage of biomimetic AQPs membranes is the low density of water channels per unit area of roughly 40 channels/μm^2 [23].

6.4 WHY ARE ARTIFICIAL WATER CHANNELS IMPORTANT?

Biomimetics or biomimicry is the science behind mimicking the well-tuned natural processes. Looking to nature as a model system paves the way for new ideas in exploring sustainable water solutions. Nature should provide answers for efficient desalination. Despite the tremendous imagination of synthetic chemists to produce sophisticated architectures confining water clusters, most strategies to selectively transport water have been related to the use of natural Aquaporin as the selectivity components. A possible future approach is to replace AQPs with synthetic AWCs. Ion-rejecting, highly permeable artificial *biomimetic* channels may provide a route to producing highly efficient membranes for separations, particularly desalination. Solute-rejecting thin-film composite reverse osmosis is currently the best available technology for desalination, but it suffers from high energy consumption. In practice, there is currently no economically viable RO membrane on the market. Therefore, there are incentives to further reduce cost and energy use for desalination as it becomes more widespread. Recent advancements have made it possible to synthesize AWCs that feature efficient transport of water. AWCs exhibit the following potential advantages:

a) They are easily tunable and may be chemically optimized for desirable properties.

b) They can be prepared on a large scale at low cost.

c) As engineering processing without biorelated steps becomes simpler and more reproducible, they should be easily immobilized in membrane in a scalable manner.

d) They should also be robust with a longer lifetime after incorporation into membranes.

Within this context, the next impact may be related to increases in the incipient potential, imagining that high-water conduction activity obtained *with* natural AQPs can be bio-mimicked using simpler compounds that can display functions *like* the natural ones, the *AWCs*, using synthetic biomimetic approaches. AWCs have been recently proposed as a biomimetic AQPs [30, 31]. They are constructed from artificial molecules defining a water-pore superstructure, surrounded by a hydrophobic exterior accommodating the membrane environment [32–36]. Despite the incipient efforts, only a few AWCs have been produced in attempts to attain the high-water permeability and significant ion rejection of AQPs [34]. Within this context, mimicking AQPs by using simple AWCs is an important endeavor from both a) the fundamental science perspective related to deeper understanding of water structure and its transport function and b) the applied science perspective because WCs might have a timely impact on increasing water filtration efficiency by using natural principles to change the desalination paradigms.

Artificial ion-channels [37–40] have been extensively developed for decades (thousands of papers in the literature) with the hope of mimicking the ionic conduction of natural proteins [41, 42]. Amazingly, there was no systematic and little progress in the area of *Artificial water channels* before 2011 [29]. It was shown that structuration of water clusters within the pores, effects on their stability, dynamic behaviors or selective translocation, once inserted in the bilayer membranes. These studies provide detailed experimental data correlated with extensive theoretical simulations on water transport with efficiency similar to that of natural channels.

6.5 WHERE ARE WE NOW? WHERE ARE WE GOING IN THIS NEWBORN FIELD?

The target of the future membrane is to increase the water permeability by 3 to 5 folds of the currently best TFC membranes while achieving/maintaining 99.5% salt rejection. As pointed out recently, in many cases, selectivity matters as much, if not more than permeability. Forthcoming research might provide answers to essential questions related to:

a) complicated interactional scenario related to water transport through biomimetic pores; and
b) bioinspired desalination with membranes that mimic mechanisms of natural desalinators.

We know that the AQPs are the representative proteins for the water transport across cell membrane, but their structures are tremendously complicated. Simpler architectures encapsulating water may provide excellent and valuable models for understanding/designing novel biomimetic water channel systems. Although there is a rich inventory of molecular encapsulation of water, in most cases, the host systems are highly polar, which prohibits them to be used in conditions similar to the one offered by the membrane to AQPs. As a result, the development of the AWCs is highly important. The first synthetic water transporting system was reported in 2011 [30] and the term *artificial water channels* was coined practically one year later [31]. The common features of AWCs are the presence of a central pore, which is a confining hydrophilic or hydrophobic region that allows the

Artificial Water Channels

passage of water, and a hydrophobic shell that permits the insertion into the lipid membranes and subsequent analysis for water transport. Following are the main approaches to designing these systems, when considering the structure of channels:

- *Single molecular channels,* as one molecular entity that can span the length of the bilayer;
- *Self-assembled supramolecular channels* that can self-assemble from multiple molecular components into architectures spanning the bilayer; and
- *Nanochannels* related to carbon nanotubes as well as Graphene Oxyde (GO) materials and membranes.

6.5.1 SINGLE MOLECULAR CHANNELS

The pillar[5]arenes (PAP1) are the first reported single molecular channels (Figure 6.4a) [43]. Intramolecular H-bonds between lateral hydrazide arms confer robustness, keeping the membrane spanning channels intact, when inserted into a lipid bilayer. PAP1 have a relatively low water transport rate: up to 40 molecules of water/s/channel. As in AQPs, PAP1 display water selectivity against H^+ and OH^- ions. The explanation for this is the impeded proton hoping through a broken, discontinuous water wire, while the cation transport can be attributed to the carrier transport mechanism. Following up on these results, a second generation of more hydrophobic pillar [5] arenes, PAP2, containing poly-Phe arms, showed an osmotic water permeability of 3.5×10^8 water/s/channel [44], in the range of AQPs and carbon nanotubes (CNTs). Nevertheless, PAP2 channels present a drawback when it comes to ionic selectivity.

Having an inner pore size of ~5 Å, they allow the passage of cations according to their hydration energy. More interestingly, PAP2 channels may be packed in a hexagonal arrangement in 2D membranes, thus the fabrication of active thin layers is possible [47–49].

6.5.2 SELF-ASSEMBLED SUPRAMOLECULAR CHANNELS

The second category of AWCs uses a noncovalent self-assembly strategy, that often confers the supramolecular channels properties that are not predictable and differ completely from that of their building blocks. Based on this concept, it has been reported that AWCs based on alkylureido-imidazoles, can self-assembly in tubular imidazole I-quartet channels, mutually stabilized by water wires (Figure 6.4b) [34]. They show similar structural features as those found in the AQP pores, with a central pore size of ~2.6 Å. The water molecules form a single molecular wire adopting a unique dipolar orientation inside the chiral I-quartets. The I-quartet channels are able to transport ~1.5×10^6 water/s/channel, which is within two orders of magnitude of AQPs' rates, but most importantly, rejecting all ions except protons [34]. This ion-exclusion selectivity is important and unique for the AWCs. It is based on dimensional steric reasons, whereas hydrophobic and hydrodynamic effects appear to be less

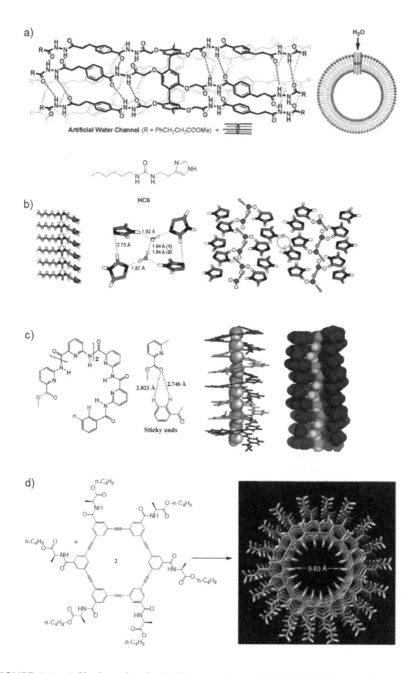

FIGURE 6.4 a) Single molecular PAP1 water channels [43]; b) Single crystal structure of hexyl-ureidoethylimidazole and its related urea H-bonding ribbons and of I-quartets that present d) dipolar orientation of water [34]; c) X-ray single crystal structure and d) representation of a single water file aquafoldamers [45]; and d) Simulated structure of macrocyclic channels [46]. (image adapted with permissions from reference [34, 43, 45, 46]).

Artificial Water Channels

important. The I-quartets have been found to efficiently transport protons when an osmotic pressure is applied, in the absence of any pH gradient. This highlights the importance of oriented dipolar water wires inside the channels, allowing polarization and synergetic *antiport* proton translocation through the bilayer membrane.

Small olygopyridine molecules that are able to self-assemble into helical *aquafoldamers* with a pore of ~2.8 Å, close to the size of narrowest region in AQPs (Figure 6.4c) [45]. The supramolecular nanopore contains a single file of water molecules. Interestingly, when water permeability tests were performed, the channels showed no osmotic pressure induced water transport, but when proton gradient was applied, they presented observable – water transport. This highlights the dependence between the chainlike confinement of the water inside the pore and translocation capabilities of the channel; however, a standardized value for water channel permeability needs to be determined. The channels showed no transport of Na^+ and K^+ cations, making them a viable option for desalination applications.

With a similar approach, giant macrocycles self-assemblies into tubular channels *via* stacking and H-bonding interactions have been recently synthesized (Figure 6.4d) [46]. Although the size of the pore of 4–5 Å would normally offer little selectivity, the channels can remarkably transport K^+ selectively against both Li^+ and Na^+ cations. The transport of water occurs in a similar fashion as the one observed in the case of CNTs, with no specific binding to the walls of the pore. The determined osmotic water permeability of these hydrophobic channels is ~4.9×10^7 water/s/channel.

Recently, the research on water channels advanced further, showing that a self-assembled *T-channel* is formed with morphological characteristics reminiscent of natural Gramicidin-A [32]. The formation of an aqua-lubricated channel is generated via H-bonding between water molecules and carbonyl groups identical to the carbonyl strings in AQPs. Vesicles incorporating *T-channels* show 50 times higher water permeability than control vesicles, but this system does not show water/ion selectivity.

6.5.3 NANOCHANNELS

At the nanolevel, carbon nanotubes (CNTs) offer an exciting source of inspiration to mimic natural channels [47–49]. The CNTs can be considered as artificial systems where the selectivity can be modulated at the entrance, and the net dipolar orientation of water molecules and the electroosmotic flow of actively pumped chemicals can be controlled through the inner CNTs cores at important flow rates. The frictionless water flow is 4–5 orders of magnitude increasing over what would be seen in other conventional nanoporous or AQP structures. Important studies show ion rejection at low salt concentrations, which is promising for applications. In the past few years, membranes based on graphene and graphene oxide (GO) have gained considerable interest for their potential for desalination [50]. The swollen membrane allows selectivity as water molecules are transported with ions. However, the spacing in a GO laminar membrane can be controlled by physically restraining the membrane from swelling. When inter-layer spacing is reduced to 7.0 Å, the ion permeation rate was reduced by two orders of magnitude while the water permeation rate was only halved.

6.6 OUTLINE BEYOND STATE OF ART

All of the examples studied must restore some balance to our understanding of the biomimicking of natural AQPs, which remains an important exploring challenge. These findings show that natural channels can be biomimicked using simpler artificial compounds displaying constitutional functions like natural compounds (Table 6.1). Biomimicking the complex superstructures of proteins at the molecular level is an important challenge. The artificial water channels with a simplified structure allow the dynamic structural behaviours of confined superstructures of water to be elucidated. The current research on AWCs is focused on their synthesis, molecular simulations, and transport properties in lipid bilayer membranes. The osmotic water permeability of artificial systems ranges from several orders of magnitude lower than that of AQP1, with similar or even better values for Carbon Nanotubes (CNTs) (Table 6.1).

The large cross-sectional area of AQPs (9.0 nm^2) is somewhat disproportionate with regard to the size of its active water pore. It is important to note that the effective cross-sectional area of the AWCs channels in 2D layers takes up significantly less space than proteins. From an engineering point of view, this could potentially increase their permeability per surface area by orders of magnitude, when compared with AQPs.

The two main challenges for AWCs are: i) to design tailored channels that reject salt but allow the easy passage of water molecules, and ii) the transfer of the AWCs from the nanomolecular level to that of large-scale membrane materials. Although this has been partly achieved using AQPs, no study has yet been reported on integrating the AWCs within active layers casted onto polymeric membrane supports.

Designing stable polymer matrices capable of keeping AWCs in their active state will prove crucial for future hybrid membranes. The use of lipid bilayers or hydrophobic polymers as host matrices, which certainly can be affected by the presence of organic solutes or surfactants in water, is not a necessary option. These polymeric matrices have been used until now to compare AWCs with AQPs. One strategy to obtain active layers is to incorporate the AWC networks while forming the Polyamide PA layer of RO membrane or to imagine new cross-linked polymers in order to obtain networks of AWC within the thin layers. If AWCs are covalently attached to TFC, there is no degradation of layer — only temporary channel destabilization. Thus, the AWCs are dense and highly robust, and would be encapsulated in the dense part of the membrane while the PA matrix would be responsible for the rejections of organics.

To compete with polymeric reverse osmosis membranes, the current gold standard for desalinating seawater, water flux, and rejection rates for artificial water channels need to be very close to that of AQPs. While having even higher flux values will not have a substantial impact on membrane filtration performances,

Artificial Water Channels

223

TABLE 6.1

Performance overview of artificial water channels and pores (adapted with permission from reference [3])

Water Channels	Net Permeability for water/ Single channel permeability	Selectivity	Ref.
Aquaporins – AQPs.	167 μm/s/bar 4×10^9 water molecules/s/ channel	high selectivity for water, perfect rejection of ions/H^+	[18, 20]
Carbon Nanotubes- CTNs Estimated diameter = 15 Å	300 μm/s/bar, 1.9×10^9 water molecules/s/channel	—	[45]
Carbon Nanotubes- CTNs Estimated diameter = 8 Å	520 μm/s/bar, 2.3×10^{10} water molecules/s/channel	ion rejection in dilute solutions	[45]
Hydrophilic I-quartets, channel (2.6 Å)	3-4 μm/s/bar 1.5×10^6 water molecules/s/ channel	high selectivity for water, reject all ions except protons	[34]
Aquafoldamers, Hydrophilic channel (2.8 Å)	No permeability reported	high selectivity for water, reject all ions except protons	[48]
Hydrophobic peptide tubular macrocyclic pores (6.4 Å)	51 μm/s/bar 4.9×10^7 water molecules/s/ channel	no selectivity for water, high conduction for K^+ and H^+	[49]
Hydrazide appended Pillar [5] arene, PAP1 unimolecular channel (6.5 Å)	8.6×10^{-6} μm/s/bar 40 molecules/s/channel	no selectivity for water conduction for alkali cations no conduction for H^+	[43]
Peptide appended Pillar [5] arenes, PAP2 unimolecular channel (5 Å)	swelling: 30 μm/s/bar or 3.5 10^8 shrinking: 1 μm/s/bar or 3.7×10^6 molecules/s/channel	no selectivity for water, good conduction for alkali cations	[44]
Double helical water T-channels (~2.5–4 Å)	No permeability reported	Enhanced conduction states for cations and for H^+	[32]
RO membrane	2 μm/s/bar	99,8% rejection of salts versus water	[20]
FO membrane RO membrane	2,5 μm/s 5,3 μm/s	—	[5, 6]

highly water specific membranes will improve the quality of water produced, as well as remove unnecessary pretreatment processes.

It was discovered that imidazole I-quartet, i.e., stacks of four imidazoles and two water molecules, that can mutually stabilize oriented dipolar water-wires within 2.6-Å pores [34]. We know that a pore with a diameter of ~ 3 Å very close to the narrowest constriction (2.8 Å) observed in AQP, is a critical prerequisite for ion-exclusion behaviors. Bigger pore diameters (> 3 Å) are finally beginning to meet encouraging signs for high water permeability (10^6–10^8 water molecules/s/channel), while most of these channels present ionic transport activity, thus, no selectivity yet. The total ionic-exclusion by I – quartet channels suggests that they hold significant promise for the incipient development of the first effective artificial AQPs — high water permeability (10^6 water molecules/s/channel) and total ionic rejection. These key features must be combined to exploit both the selectivity of hydrophilic H-bonding channels such as I-quartets [30] and the high permeability of hydrophobic carbon nanotubes. Shorter channels lengths, and more dense channels per unit area of surface will play an important role in developing efficient water filtration systems based on AWC technology. Given the importance of developing even more efficient membranes for desalination, this will be an exciting and challenging project, potentially redefining the paradigms of desalination (Figure 6.5).

6.7 CONCLUSIONS

The *biomimetic water transport* can solve/improve the performance and lifetime of bioassisted membrane systems in which highly selective but labile AQPs are used. In terms of *economic gain* and the *economy of matter* the artificial systems might represent an interesting alternative. This project, scientifically *challenging and of great fundamental interest*, is leading to pioneering scientific or applicative discoveries, as literature can attest. Biomimetic water transport, fundamental in nature, should allow not only development of new knowledge and a greater understanding on bottom-up design of multifunctional systems, but also addressing different societal challenges. The thin-layer polyamide membranes, used for water desalination since 1972, has improved dramatically over the last 50 years [51]; however, their low-resolution structure and the transport mechanism are still poorly understood. The ultimate performance of a membrane is usually

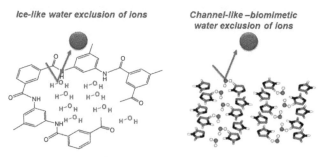

FIGURE 6.5 RO versus biomimetic artificial water channels for desalination strategies (image adapted with permissions from reference [3]).

Artificial Water Channels　225

determined by an increased flux (or permeance) due to the introduction of defects in the membrane structure lowering selectivity, which is always decreasing. The first identification of AWCs opened the door to new applicative desalination processes similar to natural ones. We speculate at this time that biomimetic AWCs might have timely important influences to increase this driving force for transport by using natural principles and to change the water desalination paradigms.

REFERENCES

[1] Eliasson, J. 2015. The rising pressure of global water shortages. *Nature* 517: 6–7.

[2] Schewe, J., Heinke, J., Gerten, D., Haddeland, I., Arnell, N. W., Clark, D. B., Dankers, R., Eisner, S., Fekete, B. M., Colón-González, F. J., Gosling, S. N., Kim, H., Liu, X., Masaki, Y., Portmann, F. T., Satoh, Y., Stacke, T., Tang, Q., Wada, Y., Wisser, D., Albrecht, T., Frieler, K., Piontek, F., Warszawski, L., Kabat, P. 2014. Multimodel assessment of water scarcity under climate change. *Proc. Natl Acad Sci.* 111: 3245–3250.

[3] Kocsis, I., Sun, Z., Legrand, Y. M., Barboiu, M. 2018. Artificial water channels – Deconvolution of natural aquaporins through synthetic design. *npj Clean Water* 1: 1–11.

[4] Mekonnen, M. M., Hoekstra, A. Y. 2016. Four billion people facing severe water scarcity. *Sci. Adv.* 2: e1500323–e1500323.

[5] Imbrogno, J., Belfort, G. 2016. Membrane desalination: Where are we, and what can we learn from fundamentals? *Annu. Rev. Chem. Biomol. Eng.* 7: 1.1–1.36.

[6] Park, H. B., Kamcev, J., Robeson, L. M., Elimelech, M., Freeman, B. D. 2017. Maximizing the right stuff: The trade-off between membrane permeability and selectivity. *Science* 356: eaab0530.

[7] Werber, J. R., Osuji, C. O., Elimelech, M. 2016. Materials for next-generation desalination and water purification membranes. *Nature Rev. Mat.* 1: 16018.

[8] Kumar, M., Culp, Y., Shen, Y.-X. 2016. Water desalination: History, advances, and challenges. *Bridge* 46(4): 22–29.

[9] Cadotte, J. E. 1981. Interfacially synthesized reverse osmosis membrane. US Patent 4,277,344 A. C.

[10] Pendergast, M. T. M., Hoek, E. M. V. 2011. A review of water treatment membrane nanotechnologies. *Energy Environ. Sci.* 4: 1946–1971.

[11] Barboiu, M., Gilles, A. 2013. From natural to bio-assisted and biomimetic artificial water channel systems. *Acc. Chem. Res.* 46: 2814–2823.

[12] Ball, P. 2008. Water as an active constituent in cell biology. *Chem. Rev.* 108: 78–108.

[13] Ball, P. 2017. Water is an active matrix of life for cell and molecular biology. *Proc. Natl. Acad. Sci.* 51: 13327–13335.

[14] Zhong, D., Pal, S. K., Zewail, A. H. 2011. Biological water: A critique. *Chem. Phys. Lett.* 503: 1–11.

[15] Agre, P. 2004. Aquaporin water channels (nobel lecture). *Angew. Chem. Int. Ed.* 43: 4278–4290.

[16] Tajkhorshid, E., Nollert, P., Jensen, M. Ø., Miercke, L. J., O'Connell, J., Stroud, R. M., Schulten, K. 2002. Control of the selectivity of the aquaporin water channel family by global orientational tuning. *Science* 296: 525–530.

[17] Zeidel, M. L., Ambudkar, S. V., Smith, B. L., Agre, P. 1992. Reconstitution of functional water channels in liposomes containing purified red cell CHIP28 protein. *Biochemistry* 31: 7436–7440.

[18] Kosinska Eriksson, U., Fischer, G., Friemann, R., Enkavi, G., Tajkhorshid, E., Neutze, R. 2013. Subangstrom resolution x-ray structure details aquaporin-water interactions. *Science* 340: 1346–1349.

[19] Shen, Y.-X., Saboe, P. O., Sines, I. T., Erbakan, M., Kumar, M. 2014. Biomimetic membranes: A review. *J. Membr. Sci.* 454: 359–381.

[20] Kumar, M., Grzelakowski, M., Zilles, J., Clark, M., Meier, W. 2007. Highly permeable polymeric membranes based on the incorporation of the functional water channel protein Aquaporin Z. *Proc. Natl. Acad. Sci.* 104: 20719–20724.

[21] Tang, C. Y., Zhao, Y., Wang, R., Hélix-Nielsen, C., Fane, A. G. 2013. Desalination by biomimetic aquaporin membranes: Review of status and prospects. *Desalination* 308: 34–40.

[22] Shannon, M. A., Bohn, P. W., Elimelech, M., Georgiadis, J. G., Marinas, B. J., Mayes, A. M. 2008. Science and technology for water purification in the coming decades. *Nature* 452: 301–310.

[23] Song, W., Lang, C., Shen, Y. X., Kumar, M. 2018. Design considerations for artificial water channel–based membranes. *Annu. Rev. Mat. Res.* 48: 57–82.

[24] Kaufman, Y., Berman, A., Freger, V. 2010. Supported lipid bilayer membranes for water purification by reverse osmosis. *Langmuir* 26: 7388–7395.

[25] Ding, W., Cai, J., Yu, Z., Wang, Q., Xu, Z., Wang, Z., Gao, C. 2015. Fabrication of an aquaporin-based forward osmosis membrane through covalent bonding of a lipid bilayer to a microporous support. *J Mater Chem A* 3: 20118–20126.

[26] Zhao, Y., Qiu, C., Li, X., Vararattanavech, A., Shen, W., Torres, J., Nielsen, C. H., Wang, R., Hu, X., Fane, A. G., Tang, C. Y. 2012. Synthesis of robust and high-performance aquaporin-based biomimetic membranes by interfacial polymerization-membrane preparation and RO performance characterization. *J. Membr. Sci.* 423–424: 422–428.

[27] Qi, S., Wang, R., Chaitra, G. K. M., Torres, J., Hu, X., Fane, A. G. 2016. Aquaporin-based biomimetic reverse osmosis membranes: Stability and long-term performance. *J. Membr. Sci.* 508: 94–103.

[28] See https://aquaporin.com

[29] Barboiu, M. 2016. Artificial water channels – Incipient innovative developments. *Chem. Commun.* 52: 5657–5665.

[30] Le Duc, Y., Michau, M., Gilles, A., Gence, V., Legrand, Y.-M., van der Lee, A., Tingry, S., Barboiu, M. 2011. Imidazole-quartet water and proton dipolar channels. *Angew. Chem. Int. Ed.* 50: 11366–11372.

[31] Barboiu, M. 2012. Artificial water channels. *Angew. Chem. Int. Ed.* 51: 11674–11676.

[32] Barboiu, M., Le Duc, Y., Gilles, A., Cazade, P.-A., Michau, M., Legrand, Y.-M., van der Lee, A., Coasne, B., Parvizi, P., Post, J., Fyles, T. 2014. An artificial primitive mimic of the Gramicidin – A channel. *Nat. Commun.* 5: 4142.

[33] Barboiu, M., Cazade, P. A., Le Duc, Y., Legrand, Y. M., van der Lee, A., Coasne, B. 2015. Polarized water wires under confinement in chiral channels. *J. Phys. Chem. B* 119: 8707–8717.

[34] Licsandru, E., Kocsis, I., Shen, Y. X., Murail, S., Legrand, Y. M., Van Der Lee, A., Barboiu, M. 2016. Salt-excluding artificial water channels exhibiting enhanced dipolar water and proton translocation. *J. Am. Chem. Soc.* 138: 5403–5409.

[35] Kocsis, I., Sorci, M., Vanselous, H., Murail, S., Sanders, S. E., Licsandru, E., Legrand, Y.-M., van der Lee, A., Baaden, M., Petersen, P. B., Belfort, G., Barboiu, M. 2018. Oriented chiral water wires in artificial transmembrane channels. *Science Adv.* 4: eaao5603.

[36] Murail, S., Vasiliu, T., Neamtu, A., Barboiu, M., Sterpone, F., Baaden, M. 2018. Water permeation across artificial I-quartet membrane channels: From structure to chaos. *Faraday Discuss* 209: 125–148.

Artificial Water Channels

[37] Gokel, G. W., Mukhopadhyay, A. 2001. Synthetic models of cation-conducting channels. *Chem. Soc Rev.* 30: 274–286.

[38] Matile, S. 2001. En route to supramolecular functional plasticity: Synthetic β-barrels, the barrel-stave motif, and related approaches. *Chem. Soc. Rev.* 30: 158–167.

[39] Cazacu, A., Tong, C., van der Lee, A., Fyles, T. M., Barboiu, M. 2006. Columnar self-assembled ureidocrown-ethers – An example of ion-channel organization in lipid bilayers. *J. Am. Chem. Soc.* 128: 9541–9548.

[40] Fyles, T. M. 2007. Synthetic ion channels in bilayer membranes. *J. Org. Chem.* 36: 335–347.

[41] Hille, B. *Ion Channels of Excitable Membranes*, Sinauer Associates, Sunderland, MA, 3rd Ed., 2001.

[42] Hurtley, S. M. 2005. Crossing the bilayer. *Science* 310: 1451.

[43] Hu, X.-B., Chen, Z., Tang, G., Hou, J.-L., Li, Z.-T. 2012. Single-molecular artificial transmembrane water channels. *J. Am. Chem. Soc.* 134: 8384–8387.

[44] Shen, Y., Si, W., Erbakan, M., Decker, K., De Zorzi, R., Saboe, P. O., Kang, Y. J., Majd, S., Butler, P. J., Walz, T., Aksimentiev, A., Hou, J. L., Kumar, M. 2015. Highly permeable artificial water channels that can self-assemble into two-dimensional arrays. *Proc. Natl. Acad. Sci.* 112: 9810–9815.

[45] Huo, Y., Zeng, H. 2016. "Sticky"-ends-guided creation of functional hollow nano-pores for guest encapsulation and water transport. *Acc. Chem. Res.* 49: 922–930.

[46] Zhou, X.,, Liu, G. D., Yamato, K., Shen, Y., Cheng, R. X., Wei, X. X., Bai, W. L., Gao, Y., Li, H., Liu, Y., Liu, F. T., Czajkowsky, D. M., Wang, J. F., Dabney, M. J., Cai, Z. H., Hu, J., Bright, F. V., He, L., Zeng, X. C., Shao, Z. F., Gong, B. 2012. Self-assembling subnanometer pores with unusual mass-transport properties. *Nature Commun.* 3: 949.

[47] Hinds, B. J., Chopra, N., Rantell, T., Andrews, R., Gavalas, V., Bachas, L. G. 2004. Aligned multiwalled carbon nanotube membranes. *Science* 303: 62–65.

[48] Holt, J. K., Park, H. G., Wang, Y., Stadermann, M., Artyukhin, A. B., Grigoropoulos, C. P., Noy, A., Bakajin O. 2006. Fast mass transport through sub-2-nanometer carbon nanotubes. *Science* 312: 1034–1037.

[49] Fornasiero, F., Park, H. G., Holt, J. K., Stadermann, M., Grigoropoulos, C. P., Noy, A., Bakajin, O. 2008. Ion exclusion by sub-2-nm carbon nanotube pores. *Proc. Natl. Acad. Sci.* 105: 17250–17255.

[50] Hu, M., Mi, B. 2013. Enabling graphene oxide nanosheets as water separation membranes. *Environ. Sci. Technol.* 47: 3715−3723.

[51] Yang, Z., Guo, H., Tang, C. Y. 2019. The upper bound of thin-film composite (TFC) polyamide membranes for desalination. *J. Membr. Sci.* 590:117297.

7 Desalination Membranes
Characterization Techniques

Rund Abu-Zurayk
Nanotechnology Center
The University of Jordan
Amman, Jordan

Hamdi Mango Center for Scientific Research
The University of Jordan
Amman, Jordan

Mohammed R. Qtaishat
Department of Chemical Engineering
The University of Jordan
Amman, Jordan

Abeer Al Bawab
Hamdi Mango Center for Scientific Research
The University of Jordan
Amman, Jordan

Department of Chemistry
The University of Jordan
Amman, Jordan

7.1 INTRODUCTION

Membrane characterization is vital to understanding the key fundamental physical, chemical, morphological, and mechanical properties that ensure the best membrane performance. An understanding of such membrane structural and morphological properties as porosity, pore size and distribution, surface topography and morphology, surface chemistry, as well as the basic mechanical strength, resistance to deformity, and hydraulic resistance, is essential to control the separation properties and usefulness of the resultant membrane.

The difficulty is in gaining such detailed information about the membrane characteristics, especially when the required dimensions and characteristics could be at atomic scale, which requires the latest analytical and spectroscopic technologies. Thus, this chapter provides a comprehensive overview of available membrane characterization technologies . The first section focuses on physical, chemical, thermal and mechanical characterization techniques, while the second section examines the membrane performance in terms of flux and salt rejection rates.

7.2 MEMBRANE CHARACTERIZATION

7.2.1 SURFACE MORPHOLOGY AND PHYSICAL CHARACTERISTICS

The surface morphology and the physical characteristics are usually the first characteristics to be examined in the prepared membranes; these include studying the porous structure of the membrane, which covers the membrane porosity, the pore shape, the average pore size, and the pore size distribution. These characteristics are considered key information, since they are affected by the preparation approach and affect the membrane performance, Such principal information as the microstructure of the membrane and the surface topography can also be obtained. The microstructure inspection provides information regarding the binding between the different layers, along with the binding between the main matrix and any loaded fillers. Besides, it is important to examine the surface roughness as it affects the membrane performance, either by affecting the water flux through the membrane, or by affecting fouling. In addition, foulant contents and types are required to be characterized.

7.2.1.1 Porous Structure Characterization

Characterizations of the porous structure of the membrane include determination of the overall porosity, the average pore size, and the pore size distribution. These characteristics are considered important information since they are influenced by the preparation techniques and any modification that is made on the membrane; moreover, they affect membrane performance (Capannelli et al. 1983). Many techniques can be used to characterize the porous structure. Usually the gravimetric (Archimedes) method is used to determine the porosity (Hou et al. 2009), using the following equation:

$$\varepsilon = \frac{(w_1 - w_2)/\rho_w}{\frac{w_1 - w_2}{\rho_w} + w_2/\rho_p} \qquad (7.1)$$

where w_1 is the wet membrane weight; w_2 is the dry membrane weight; ρ_w is the density of wetting agent; and ρ_p is the density of polymer. Water is usually used as the wetting agent for hydrophobic membranes, while isopropyl alcohol is used as the wetting agent for hydrophilic membranes.

Another method of porosity evaluation is the mercury intrusion method (Lorente-Ayza et al. 2017). In this method, the intrusion volume of mercury

Desalination Membranes 231

in the membrane pores is recorded. However, this technique uses high pressure, which can cause sample deformation. In addition, in this technique, the pores that determine the flux in the selective layer cannot be distinguished from those in the support (the large pores).

Porosity can also be determined by measuring the polymer material density using isopropyl alcohol that penetrates inside the membrane pores and the density of the membrane using water that does not penetrates through the pores. For density determination a balance and a pycnometer are used. Then, the following equation is used:

$$\varepsilon = 1 - \frac{\rho_m}{\rho_p} \tag{7.2}$$

where ρ_m is the membrane density, while ρ_p is the polymer density (Khayet and Matsuura 2001).

Regarding pore size and pore size distribution, several methods are used. SEM is one approach, in which a software is used on the scanned images of the surface, in order to measure the pores sizes (Abdullah et al. 2014, Wyart et al. 2008). Gas permeation method is another technique (Hou et al. 2009). In this technique, the gas is tested using nitrogen, and a flow meter is used to measure the permeation flux. The permeate flux is plotted versus the pressure, from which the intercept (K_o) and the slope (P_o) are determined. The average pore size is calculated using the following equation:

$$r = \frac{16}{3} \left(\frac{P_0}{K_0} \right) \left(\frac{8RT}{\pi M} \right)^{0.5} x\,\mu \tag{7.3}$$

where R, T, M and μ are the gas constant, the absolute temperature, the molecular weight and the gas viscosity, respectively.

The bubble point method is used for determination of the largest pore size in the membrane (Dickenson 1997). The membrane is put in a container with gas from the bottom side and liquid on the topside of the membrane. The gas increases steadily with time. The bubble point is reached when a constant flow of bubbles is seen on the membrane's topside. The pore size is then determined as the pore diameter (d) using the following equation:

$$d = \frac{4\gamma \cos \theta}{P} \tag{7.4}$$

where γ is the surface tension; θ is the contact angle; and P is the pressure at which the bubbles are formed.

Another approach for determination the pore size is by applying permeability data in mathematical models, such as the Capillary Tube (CT) model (Eygeris et al. 2018). In the CT model, the pore radius is calculated as follows:

$$R^2_{pore} = \frac{8K\tau}{\varepsilon} \tag{7.5}$$

where K is the permeability that can be determined using Darcy's law (Eygeris et al. 2018); τ is the tortuosity; and ε is the porosity of the membrane.

Regarding pores at nanoscale, nanopermporometry is used for the determination of pore sizes in the range of 0.5–50 nm. This technique depends on the capillary condensation of a vapor such as water, hexane, or isopropyl alcohol, and on its ability to block the permeation of noncondensable gases such as He and N_2 permeation (Albo et al. 2014, Tsuru et al. 2003). For small pore size, the vapor condenses at a vapor pressure (P_v) that is lower than the saturation vapor pressure (P_s). The pore radius is then calculated using the Kelvin equation, as follows:

$$RT \ln\left(\frac{P_v}{P_s}\right) = 2v\frac{\sigma \cos\theta}{r_k} \tag{7.6}$$

where v is the molar volume; σ is the surface tension; θ is the contact angle; and r_k is the radius of the pore.

7.2.1.2 Electron Microscopy

Electron microscopy is used to obtain very high-resolution images using a beam of accelerated electrons, which have very short wavelength under vacuum, as the source of illuminating radiation. Two main types of electron microscopy are scanning electron microscope (SEM) and transmission electron microscope (TEM).

7.2.1.2.1 SEM

Scanning electron microscope (SEM) produces images when a focused electron beam is scanned across a specific surface area of the specimen. When the electron beam interacts with the surface, it loses energy, which is converted into other forms that provide signals carrying information about the properties of the surface, such as its composition and topography.

For sample preparation, membranes are usually dried overnight at room temperature (Abdullah et al. 2014). After that, samples are sectioned into small strips and mounted on an SEM stage for skin layer (top view), while they are frozen in liquid nitrogen, and fractured to obtain flakes for cross-section imaging, which give information about the membrane morphology and structure. This is followed by sputter coating by gold, palladium, platinum, osmium, or other metals to enhance the surface electrical conductivity, thereby avoiding the charging effect for SEM observation (Hou et al. 2009, Huang et al. 2019, U.S. Department of the Interior Bureau of Reclamation 2009).

SEM is usually used for characterization of membrane topography and morphology. It can be used to quantitatively characterize the membrane surface roughness and the porosity, the average pore diameter, the pore size distribution, and the pore aspect ratio, which can be determined with the help of such software

Desalination Membranes

products as MATLAB and Image J (Abdullah et al. 2014, Wyart et al. 2008). In one study, (Wei et al. 2013) a thin-film composite nanofiltration hollow fiber membrane was examined using SEM to study the surface morphology and the cross sections of the composite's membranes. It was found that there is a difference in surface morphology between the outer surface and the inner surface of the hollow fiber membrane. The outer surface was found to be smooth, while the inner one was found to be rough surface that was packed with tiny nodules. In addition, the middle and outer edge of the hollow fiber membrane were porous and full of finger-like configurations, as seen in Figure 7.1.

In addition, SEM can be used to examine the foulants' nature at the microscopic level (US Department of the Interior Bureau of Reclamation 2009) and

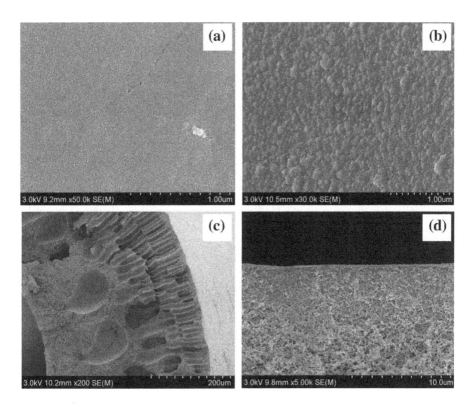

FIGURE 7.1 FE-SEM images of the fabricated composite NF hollow fiber membrane (a), outer surface (b), inner surface (c), cross section x 200 (d), and cross section x 5 K. (Wei et al. 2013).[1]

---

1 Reprinted from *Chemical Engineering Journal* 223. Wei, Xiuzhen, Xin Kong, Chengtian Sun, and Jinyuan Chen, "Characterization and application of a thin-film composite nanofiltration hollow fiber membrane for dye desalination and concentration." 172–182, Copyright (2013), with permission from Elsevier

to make a qualitative analysis of the foulant types that are deposited on the surface of the membranes (Butt et al. 1997). SEM images are capable of showing the different thickness of fouled materials on different membranes, for a variety of membrane materials and water conditions (Sachit and Veenstra 2017).

It can also be used to examine the surface binding in case of composite membranes, or detect any nanoparticles included in the membrane matrix, in addition of determination the nanoparticle size in the polymer matrix (Kadhom and Deng 2019, Khulbe and Matsuura 2017). Khadom et al (Kadhom and Deng 2019) used SEM to measure the size of bentonite in polymeric membranes. SEM images showed that bentonite nanoparticles were around 50nm.

7.2.1.2.2 Transmission Electron Microscopy (TEM)

In TEM, the image is formed as a result of interaction between the electrons and the specimen, which occurs when the focused ion beam is transmitted through the specimen.

Sample preparation depends on the type of the specimen, which is deposited on a support grid in case it is a powdered substance or a nanotube. Other specimens such as biological tissues are embedded in a resin to withstand the high vacuum, before being cut into very thin sections (less than 100 nm thickness), using ultra-microtome. In cases where the specimens can withstand the high vacuum, such as some polymers; it is ultra-microtomed as it is, without being embedded. Sometimes, specimens need to be stained.

TEM is used in membrane characterization to study the morphology and microstructure of the samples (Huang et al. 2019). Xu et al (Xu et al. 2019) used TEM to examine graphene quantum dots incorporated in thin film composite (TFC) membranes. TEM images showed that the membranes had thin skin layers and fingerlike pores, as seen in Figure 7.2. They were able to determine the thickness of the substrate layer, which was found to be 60 nm on average.

In addition, TEM can be used to examine the nanoparticles' shape and size inside the polymer matrix of the membrane. In one study, TEM images showed the bentonite nanoparticles as grey blocks inside the membrane. The effect of the nanoparticles on the structure of the membrane when compared to unfilled thin film membrane was revealed, in which the thickness of the membrane was increased in the presence of nanoparticles from (100–300 nm) to 500 nm (Kadhom and Deng 2019).

7.2.1.3 Atomic Force Microscopy (AFM)

AFM is another common tool for studying surface topography. In AFM, atomic force is used to plot the probe – sample interaction. This interaction between the sample and the probe tip is used to form a three-dimensional image of the sample surface at a very high resolution.

AFM is usually used to examine the roughness of the membrane, and the pore size and distribution, in addition to the phases on the membrane surface (Khulbe and Matsuura 2017).

Desalination Membranes

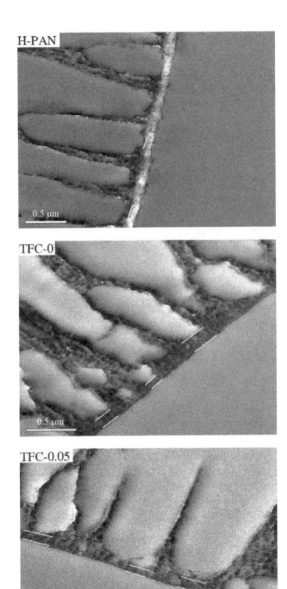

FIGURE 7.2 The cross-sectional TEM images of the H-PAN, the pristine TFC-0membrane, and the TFC-0.05 membrane (Xu et al. 2019).[2]

---

[2] Reprinted from *Desalination* 451. Xu, Shengjie, Feng Li, Baowei Su, Michael Z Hu, Xueli Gao, and Congjie Gao. "Novel graphene quantum dots (GQDs)-incorporated thin film composite (TFC) membranes for forward osmosis (FO) desalination." 219–230, Copyright (2019), with permission from Elsevier.

It is important to check the roughness because it is expected to positively affect the water flux owing to higher specific area as found by Xu et al (Xu et al. 2019) who examined thin film membranes that are incorporated with graphene quantum dots for forward osmosis desalination. They examined the roughness of the membranes using AFM. Those filled with graphene quantum dots showed higher surface roughness and higher flux.

The roughness of the membrane surface is also an important characteristic that affects fouling/antifouling properties. Fouling occurrence increases at surfaces with high roughness (Mahmoud et al. 2015). Zhao et al (Zhao et al. 2013) examined the surface of PVDF and GO/PVDF membranes using AFM. They found that pure PVDF membranes had high roughness, while the GO/PVDF membranes had smoother surfaces, which indicated that addition of GO to PVDF decreases the fouling potential because higher membrane surface roughness leads to contaminants accumulation. Similar results were found by Safarpour et al (Safarpour et al. 2015) who incorporated rGO/TiO$_2$ nanocomposites in a polyamide layer. The roughness of the membranes was determined using AFM. It was found that increasing the content of the nanocomposite in the polymer layer reduced the roughness of the membranes.

In another study, composite polysulfone membranes were examined for the effect of grafting on the roughness of the membranes using AFM. It was found that the grafted polysulfone membranes had rougher surfaces than the untreated polysulfone membranes, which indicated that the graft polymerization was successful, as shown in Figure 7.3. (Akbari et al. 2010).

AFM can be used to calculate the root mean square roughness of reverse osmosis (RO) membranes at different directions over different samples (Ferrero et al. 2011). Moreover, it is used to calculate the pore size and pore size distribution, and surface roughness of the membranes (Shirazi et al. 2013).

Another use of AFM is the understanding the mechanism of membrane formation, as found by Huang et al (Huang et al. 2019) who formed hydrothermal reduction with time for reduced graphene oxide (rGO) membranes. They found that regularly intact rGO membranes that have black appearance were formed under treatment times of 0.5, 1, and 2 h. However, when treated for more than 4h, rGO sheets agglomerated and formed particles instead of films.

7.2.1.4 Fourier-Transform Infrared (FTIR) Microscopy

FTIR microscopy techniques are used to characterize the contaminants that accumulate on the membrane surface. They provide information about fouling distribution on the membrane surface (Ferrero et al. 2011), and are used together with FTIR spectroscopy to allow both visualization of the surface and analysis of the contents accumulating on the surface. Ferrero et al (Ferrero et al. 2011) used FTIR microscopy in the characterization of fouling distribution on the reverse osmosis membrane surface. They were able to conclude that the membrane surface is covered by proteins, polysaccharides, and inorganic species as shown in Figure 7.4.

Desalination Membranes

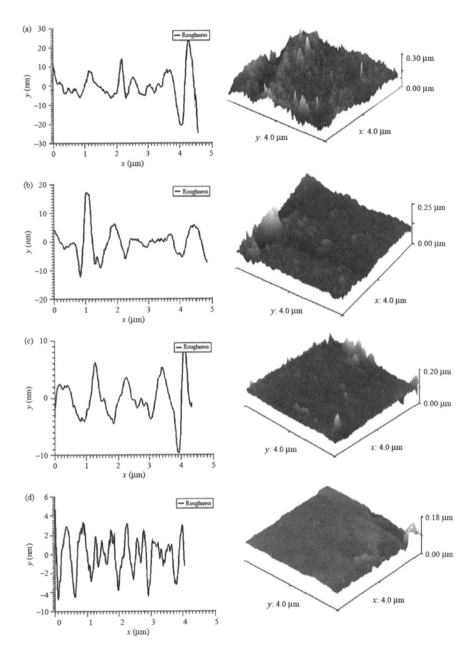

FIGURE 7.3 AFM images of pristine polysulfone (PSF) ultrafiltration (UF) membrane and PSF-grafted-PAA after (a) 1 hour, (b) 2 hours, (c) 3 hours, (d) UV irradiation time, (Akbari et al. 2010).[3]

3 Reproduced from A. Akbari et al. (2010) *Water Science & Technology*, 62.11 (2655–2663) with permission from the copyright holders, IWA Publishing.

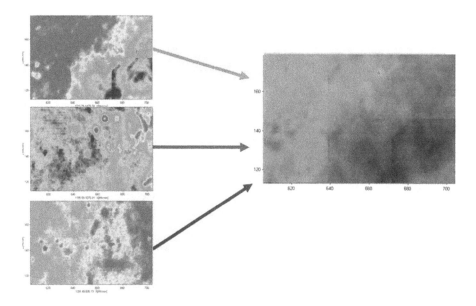

FIGURE 7.4 Infrared images for the main components found in the membrane surface: proteins (1711–1480 cm^{-1}), inorganics (1291–936 cm^{-1}), polysaccharides (1165–1076 cm^{-1}). Combination of the three FTIR images in a RGB graphic: proteins (green), inorganics (red), polysaccharides (blue). (Source: Ferrero et al. 2011).

7.2.2 Surface Chemical Characteristics

It is essential to perform surface chemical characterization, to complement the information that is provided by characterization techniques addressing surface morphology and physical characteristics, so that membrane performance is further understood. Surface chemical characterization techniques include techniques for providing qualitative chemical characterization and elemental compositions for points on the membrane surface such as Energy Dispersive X-ray (EDX) and X-ray Photoelectron Spectroscopy (XPS). Furthermore, the structure and phases of the membrane materials need to be examined. This can be achieved using ATR-FTIR and XRD. In addition, it is necessary to inspect the surface hydrophobicity using the contact angle technique and the surface charge using the zeta potential measurement technique, since they both affect the fouling of the membrane.

7.2.2.1 Energy Dispersive X-ray Spectroscopy (EDX)

The EDX technique is used to deliver a qualitative chemical characterization of specific points on the membrane surface. EDX can be coupled with other characterization techniques such as SEM and TEM (Khulbe and Matsuura 2017). The quantity of energy that is released upon focusing an electron beam onto the sample surface depends on the electron's starting and ending shell.

Desalination Membranes

The generated X-ray is converted into an EDX plot that shows different peaks, each corresponding to different energy levels in the received X-ray. Since each chemical element has its own unique structure, each peak in the EDX plot is identified by comparing it with a peak of a known atomic structure as shown in Figure 7.5 (Ferrero et al. 2011).

Shahabi et al (Shahabi et al. 2019) characterized their membrane of graphitic carbon nitride using EDX. They observed that the membrane consisted of carbon and nitrogen without the presence of any other impurities. In another study, the EDX technique was used for mapping of Ti elements in the membrane surface of GO, to examine the dispersion of TiO_2 nanoparticle in the GO sheets (Safarpour et al. 2015).

7.2.2.2 X-ray Photoelectron Spectroscopy (XPS)

XPS is a quantitative spectroscopic technique that is surface-sensitive. It is able to measure the elemental composition within a film; in addition, it shows the elements they are bonded to. Furthermore, the chemical state, the electronic state and the empirical formula of the elements can also be determined using this technique (Khulbe and Matsuura 2017). In XPS, a beam of X-ray

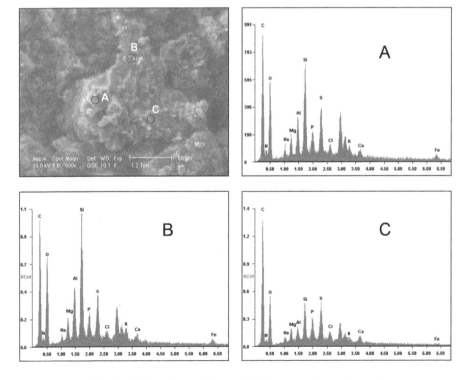

FIGURE 7.5 SEM-EDX analyses. Dry solid from surface of membrane element (Source: Ferrero et al. 2011).

is irradiated onto the material, simultaneously, analysis is made on the kinetic energy and the electrons number that escape from the surface (0–10 nm). High vacuum is required for XPS (Mather 2009). Wei et al (Wei et al. 2013) used XPS for chemical composition analysis for hollow fiber membranes that consisted of a supporting membrane of polysulfone/polyether sulfone and an active layer of polyamide. Figure 7.6 presents the results of the analysis. As shown in Figure 7.6, the two major emissions at 532 eV and 284.8 eV correspond to the binding energies of O_{1s} and C_{1s}, respectively, while the emissions at 167 eV and 230 eV are attributed to sulfur atoms in the sulfone group. As also shown, an emission at 399 eV is only seen in the composite membrane. This indicates that the polyamide layer was successfully incorporated on the support.

7.2.2.3 Attenuated Total Reflectance – Fourier-Transform Infrared (ATR–FTIR)

FTIR is a technique that is used to identify materials. The infrared absorption bands identify the structures and the specific molecular components and give information about the functional groups of the membrane via detecting bonds between atoms. Attenuated total reflectance (ATR) is a sampling tool that is

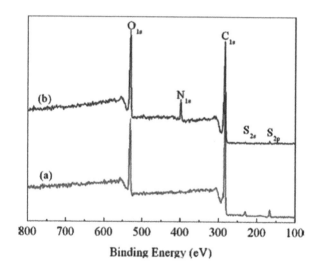

FIGURE 7.6 XPS spectra of (a), the PS/PES-UF supporting membrane and (b), composite NF hollow fiber membrane (Wei et al. 2013).[4]

---

4 Reprinted from *Chemical Engineering Journal* 223. Wei, Xiuzhen, Xin Kong, Chengtian Sun, and Jinyuan Chen. "Characterization and application of a thin-film composite nanofiltration hollow fiber membrane for dye desalination and concentration." 172–182, Copyright (2013), with permission from Elsevier.

Desalination Membranes

used together with FTIR to measure the surface properties of the membrane (Ferrero et al. 2011, Khulbe and Matsuura 2017).

ATR-FTIR makes it possible to assess the efficiency of cleaning methods such as cleaning with NaOH, SDS, and Na-EDTA, by comparing the spectra before and after cleaning. In addition, it can be used to identify the fouling chemical nature (Ferrero et al. 2011, Khulbe and Matsuura 2017, U.S. Department of the Interior Bureau of Reclamation 2009). It can differentiate between different kinds of fouling layers, but cannot determine the thickness of fouling layers (Khulbe and Matsuura 2017). In addition, ATR-FTIR is used to confirm successful preparation or treatment of the filler in the polymer matrix (Xu et al. 2019).

The sample needs to be dried for FTIR. Samples are usually dried in an oven overnight before FTIR measurement (Xu et al. 2019).

In one study, a thin film composite nanofiltration hollow fiber membrane was characterized for chemical structure using ATR-FTIR in order to identify the functional groups, in which the amide group and the carboxylic group were identified (Wei et al. 2013). In another study, polyamide desalination membranes were grafted with MPEG-NH$_2$, and ATR-FTIR were used to confirm grafting to the membrane surface. It was noted that there was an increase in peak intensities in regions referable to PEG. In addition, the grafted membranes had lower nitrogen and higher oxygen contents compared to the ungrafted membrane, which was consistent with a PEG-grafted surface (Van Wagner et al. 2010). Akbari et al (Akbari et al. 2010) prepared a composite of polysulfone membranes for nanofiltration application, and confirmed grafting of polyacrylic acid onto polysulfone using ATR-FTIR, in which the bands corresponding to the CO stretching at 1,713cm^{-1} and OH stretching at 3,396cm^{-1} increased upon grafting.

7.2.2.4 X-ray Powder Diffraction (XRD)

X-ray interacts with electrons of atoms, which causes some photons to be deflected away from their original direction. The diffracted photons are composed of sharp interference peaks, which are related to interlayer spacing according to Brag's Law:

$$n\lambda = 2dsin\theta \qquad (7.7)$$

where n is the order of reflection; λ is the wavelength of x-ray; d is the spacing between planes in the atom; and θ is the angle at which the peak occurs. d-spacing is used to determine an unknown crystal.

Based on the aforementioned, XRD is a technique that is used to identify the phases and the crystal structure of different materials. XRD is usually used for characterization of membranes in order to determine the relative amounts of fouling material on the membrane surface, in addition to the various phases in which fouling material is present (Butt et al. 1997). In addition, XRD can be used to calculate interlayer spacing of the membrane material or filler to determine the degree of exfoliation that can be determined from the peak shifting (Huang et al. 2019).

XRD was used to examine the structure of graphene acid (GA) and graphene oxide (GO) membranes (Khorramdel et al. 2019). Results showed that GA had larger d- spacing compared to GO, which means that the number of layers were lower.

Huang et al (Huang et al. 2019), prepared GO membranes and rGO membranes, and also used XRD to determine d-spacing of different membranes. In addition, they were able conclude from XRD results the coexistence of both GO and rGO phases because of the existence of two peaks.

In another study, in which a hydrophilic modifier of graphitic carbon nitride nanosheets were synthesized, XRD was used to confirm the purity of the nanosheets (Shahabi et al. 2019).

7.2.2.5 Water Contact Angle

Water contact angle is measured between a tangent to the liquid surface, where the vapor – liquid interface meets the surface of the solid, and the plane of the solid surface on which the liquid settles or moves. It is the most common parameter that is used to determine the hydrophobicity of the used membrane. High contact angle (>90°) indicates high hydrophobicity (Khulbe and Matsuura 2017). Other surface properties can be known from the water contact angle, including roughness, homogeneity, and cleanness (Tretinnikov and Ikada 1994).

The goniometric technique is one way of measuring the contact angle with sessile drops. Placing a tangent to the drop plane at the contact point performs a direct measurement. However, this technique has a large error of ±2°, which makes it unsuitable for small angles (Dimitrov et al. 1991). An alternative method is the axisymmetric drop shape analysis (ADSA). This method is based on a drop profile using a computer processing techniques that uses drop height in its calculations. ADSA-CD, however, is based on the measurements of two parameters — the drop volume and the contact diameter. This technique can be applied for the measurements on imperfect solids (Dimitrov et al. 1991).

Another technique that is used to determine the contact angle is the Wilhelmy Method. In this method, the membrane sample is immersed in a liquid. Then, the measured force is used to calculate the contact angle by applying the Wilhelmy equation (Della Volpe et al. 2001)

$$F = \gamma \rho \, cos \, (\theta) \qquad (7.8)$$

where θ is the contact angle; γ is the surface tension of the liquid; and ρ is the wetted perimeter.

Albo et al (Albo et al. 2014) used a goniometer to examine the membrane wettability by sessile water drops. The contact angle was determined by specifying a circle around each drop and then recording the tangent angle that was formed at the surface.

In another study, in which graphene quantum dots were incorporated into thin film membranes, the water contact angle was measured using an automatic contact angle meter. It was noted that incorporation of graphene quantum dots had increased the contact angle because the graphene quantum dots

Desalination Membranes

had larger surface that consisted of hydrophobic aromatic rings, which rendered the membrane surface into a less hydrophilic surface(Xu et al. 2019).

Ren et al (Ren et al. 2015) used the contact angle to determine the effect of grafting of alumina membrane using fluoroalkylsilane. They found that the surface was converted from hydrophilic to hydrophobic with a contact angle of 133°. Furthermore, contact angle measurement was used to examine the effect of carboxylic functionalization of GO membranes on the hydrophobicity/hydrophilicity of the membrane. It was observed that the hydrophilicity was enhanced upon functionalization when the contact angle was decreased (Khorramdel et al. 2019).

7.2.2.6 Zeta Potential Measurement Surface Charge

Zeta potential is the difference in potential between the surface of solid particle and the liquid in which it is immersed. It is an important indicator of stability of colloidal dispersion. High values of zeta potential indicate good stability of the colloid. In membranes, zeta potential is used to determine the surface charge, which is important since it affects rejection of ions through the membranes (Van Wagner et al. 2010), and significantly affects fouling (Xu et al. 2019). Increasing solution's pH changes the zeta potential of the surface from positive to negative, consequently affecting the accumulation of charged particles on the surface of the membrane (fouling) (Khulbe and Matsuura 2017).

Wei et al (Wei et al. 2013) studied the surface charge of their hollow-fiber nanofiltration membrane via surface zeta potential measurement. At pH lower than 6.6, the membrane surface had positive charge, while at pH higher than 6.6, the membrane surface had negative charge.

Moreover, zeta potential can be used to determine the cleaning effect on membrane surfaces. In a study that used both virgin membranes and fouled membranes, different cleaning agents (HCl, NaOH, SDS and mixed agents) were investigated. Zeta potential were measured for all membranes before and after cleaning. Results showed that the surface properties and performance of membranes were affected by cleaning. The permeability of both virgin membranes and fouled membranes, were higher when a cleaning agent is used. It was concluded that both permeability and charge characterization methods were useful in checking the cleaning efficiency to restore the membrane performance without affecting the material of the membrane or its surface charge (Al-Amoudi et al. 2007).

7.2.3 THERMAL AND MECHANICAL PROPERTIES

7.2.3.1 Thermal Gravimetric Analysis (TGA) and Differential Scanning Calorimetry (DSC)

TGA is a thermal analysis technique that shows the thermal behavior of a material under controlled heating conditions. It can be used to study the membrane material specifically to compare different membranes with different chemical compositions, and to relate it to the membrane performance, which helps in examining the membranes morphology/characterization (Khulbe and Matsuura 2017). TGA provides information about the examined membrane, such as initial decomposition temperature, thermal decomposition stages, and percentage of inorganic materials. Such information proves the existence and the percentage of any nanofiller in the membrane polymer matrix, in addition to confirming any grafting that is performed on the membrane material (Kim et al. 2013).

Mohan et al (Mohan and Kullová 2013) studied the relationship between conditions of preparations such as reactant concentrations and the performance of different polyamide thin film composites membranes, that were supported by polyethersulfone (PES). Using TGA, they found that initial decomposition temperatures of the blends of PES and polyamide were not the same for different membranes. They ranged from 20°C to 90°C lower than that of PES. Such differences in thermal behavior suggested differences in chemical composition.

DSC is another thermal analysis technique that examines the crystallization behavior of a material. It depends on measuring the heat that is required to increase the temperature of the sample. A reference sample, with a well-known heat capacity over the range of heating/cooling, is heated at the same rate as the tested sample to compare the difference in heat during melting and crystallization of the tested sample. Similar to TGA, DSC is used to study the membrane materials. It provides information about melting and crystallization onset temperatures, as well as melting and crystallization temperatures. Such information confirms different chemical compositions in addition to the presence of any nanofillers.

Furthermore, glass transition temperature (T_g) can be found using DSC. Higher T_g indicates that the membrane has a looser structure owing to more free volume fraction (Arthanareeswaran et al. 2004)

7.2.3.2 Tensile Strength Test

The tensile strength test is a mechanical test that examines the effect of tension on the tested materials in terms of stress – strain plot. Tensile modulus, tensile strength, yield point and percentage of elongation can be calculated from stress – strain plot. For membrane testing, the tensile test can be used to test the membrane strength, and to study the effects of pores shape, size and density on the mechanical properties of the membrane (Mahmoud et al. 2015). Tensile test is mainly used to test RO membranes that operate at very high pressure (Khulbe and Matsuura 2017).

7.3 MEMBRANE PERFORMANCE

This section focuses on the permeation flux and methods of measuring it, in addition to salt rejection rate measurement. These parameters are considered significant parameters when performing membrane characterization since they directly evaluate the membrane performance.

Desalination Membranes

7.3.1 Permeation Flux

Permeation flux is the volume of fluid that passes through a unit area of membrane per unit time. It is considered a principal indication of membrane performance and fouling, since it declines with time as a result of coverage of membrane surface with foulants.

It can be measured using a crossflow filtration system with constant flow rate. The measurement is performed until constants values of flux are obtained (Wei et al. 2013). It can be calculated using Equation 7.9:

$$f = \frac{V}{A\ \Delta t} \qquad (7.9)$$

where f is the permeate flux, V is the volume of solution permeated through the membrane, A is the cross sectional area of the membrane, and Δt is the operational time of the experiment.

Ferrero et al (Ferrero et al. 2011) recorded data for the permeate and the concentrate flow 15 minutes after the experiment's start-up, then every 15 minutes until one hour of testing was completed. In another study, Wei et al (Wei et al. 2013) used pure water and pure water with added solutes. Results showed that flux for deionized water was higher than those of deionized water with solutes, and that flux of pure water was proportional to the working pressure.

A cross-flow filtration setup can be used to measure the flux, in which the permeate flux can be measured by weighing the water that is accumulated with time via LABVIEW software (Kadhom and Deng 2019).

Flux can also be measured using vacuum filtration system, as in the work that (Huang et al. 2019) performed on their GO/rGO membrane system, in which pure water and 0.1 M NaCl solution were introduced.

For reverse osmosis membrane systems, usually a dead-end cell is used for flux measurement. The following equation is used to calculate the water permeance L_p:

$$L_p = \frac{\Delta v}{A.\Delta t.(\Delta P - \Delta \pi)} \qquad (7.10)$$

where Δv is the permeate volume; A is the membrane area under testing; Δt is the permeation time; Δp is the operating pressure; and $\Delta \Pi$ is the difference in osmotic pressure (Albo et al. 2014).

7.3.2 Membrane Liquid Entry Pressure of Water (LEP_w)

In membrane distillation (MD), the liquid feed, which is in contact with the surface of the membrane, can only penetrate the dry membrane pores when the applied pressure exceeds the *liquid entry pressure of water* (LEP_w). This pressure depends on the membrane's hydrophobicity and the pore size. In order to measure this pressure, the membrane is placed in a cell between the feed side, which is filled with pure water, and the permeate side, which is connected to a flowmeter. First, a slight pressure is applied for 10 min; then, the pressure is increased stepwise. The LEP_w is determined as the pressure at which a steady flow is noted (Khayet and Matsuura 2001).

7.3.3 SALT REJECTION RATE

One important parameter that is used to examine the membrane performance is the rejection rate of the salt. Rejection rate can be calculated using Equation 7.11.

$$R = \left(1 - \frac{C_p}{C_f}\right) \times 100 \ \% \tag{7.11}$$

where R is the percentage of solute rejection; and C_p is the concentration of solute in the permeate, while C_f is the concentration of the solute in the feed. C_p and C_f can be determined by measuring the electrical conductivity of the solution using a conductivity meter (Safarpour et al. 2015, Wei et al. 2013). The conductivity is proportional to the concentration of salt; hence, the salt rejection can also be determined in terms of conductivity rejection, using Equation 7.12 (Álvarez-Sánchez et al. 2018):

$$Conductivity \ rejection = \left(1 - \frac{\Omega_{60}}{\Omega_0}\right) \times 100\% \tag{7.12}$$

where Ω_0 is the feed conductivity, while Ω_{60} is the permeate conductivity after 60 minutes of permeation.

Conductivity can be converted to salt concentration using relations that are derived by linear interpolation between table values for salts, which are usually given by the producer of the conductivity standards (US Department of the Interior Bureau of Reclamation 2009).

Specific ions rejection rate can be determined using Equation 7.11. Vaseghi et al (Vaseghi et al. 2016) compared the performance of two types of membranes — reverse osmosis membranes and nanofiltration membranes — by studying the effect of specific ion rejection rates. They studied the rejection rates of sodium, potassium, calcium, magnesium, chloride, fluoride, sulfate, bicarbonate, nitrate, and silica.

7.4 CONCLUSIONS

A comprehensive characterization of the membrane system helps in enhancement of its performance in terms of preparation optimization and use. Several characterization techniques are used to provide a clear picture about the membranes. These include physical, chemical, thermal, and mechanical properties, in addition to the characterization techniques that examine the membrane performance. Physical properties characterization techniques focus mainly on the

Desalination Membranes

porous structure of the membrane, roughness, and the foulants on the membrane surface. They include electron microscopy, atomic force microscopy, and FTIR microscopy. Rather, chemical properties characterization techniques address the chemical composition of the membrane material, their phases, and their surface charge. Such techniques include EDX, XPS, ATR-FTIR, XRD, contact angle, and zeta potential measurements. As for thermal properties, TGA and DSC are used mainly to determine the effect of composition on the thermal behavior of membranes. Concerning mechanical properties, the tensile test is usually used to examine the strength of membrane materials. To complete the picture, membrane performance assessment is significant, and is mainly achieved by determination of flux and salt rejection rates.

REFERENCES

Abdullah, Syed Z, Pierre R Bérubé, and Derrick J Horne. 2014. "SEM imaging of membranes: Importance of sample preparation and imaging parameters." *Journal of Membrane Science* 463: 113–125.

Akbari, Ahmad, Maryam Homayoonfal, and Vahid Jabbari. 2010. "Synthesis and characterization of composite polysulfone membranes for desalination in nanofiltration technique." *Water Science and Technology* 62(11): 2655–2663.

Al-Amoudi, Ahmed, Paul Williams, Steve Mandale, and Robert W Lovitt. 2007. "Cleaning results of new and fouled nanofiltration membrane characterized by zeta potential and permeability." *Separation and Purification Technology* 54 (2): 234–240.

Albo, Jonathan, Hideaki Hagiwara, Hiroshi Yanagishita, Kenji Ito, and Toshinori Tsuru. 2014. "Structural characterization of thin-film polyamide reverse osmosis membranes." *Industrial & Engineering Chemistry Research* 53 (4): 1442–1451.

Álvarez-Sánchez, Jesús, Griselda Evelia Romero-López, Sergio Pérez-Sicairos, German Eduardo Devora-Isiordia, Reyna Guadalupe Sánchez-Duarte, and Gustavo Adolfo Fimbres-Weihs. 2018. "Development, characterization, and applications of capsaicin composite nanofiltration membranes." In *Desalination and Water Treatment*, 256–268. IntechOpen. DOI: 10.5772/intechopen.76846

Arthanareeswaran, Gangasalam, Palanisamy Thanikaivelan, K. Srinivasn, Doraisamy Mohan, and Munnuswamy Rajendran. 2004. "Synthesis, characterization and thermal studies on cellulose acetate membranes with additive." *European Polymer Journal* 40 (9): 2153–2159.

Butt, FH, Faizur Rahman, and Uwais Baduruthamal. 1997. "Characterization of foulants by autopsy of RO desalination membranes." *Desalination* 114 (1): 51–64.

Capannelli, Gustavo, Fernando Vigo, and Stelio Munari. 1983. "Ultrafiltration membranes—characterization methods." *Journal of Membrane Science* 15 (3): 289–313.

Della Volpe, Claudia, Devid Maniglio, Stefano Siboni, and Marco Morra. 2001. "An experimental procedure to obtain the equilibrium contact angle from the Wilhelmy method." *Oil & Gas Science and Technology – Revue d'IFP Energies nouvelles, Institut Français du Pétrole*, 56 (1): 9–22.

Dickenson, T Christopher. 1997. *Filters and Filtration Handbook*. Elsevier Advanced Technology The Boulevard, Kidlington, Oxford, UK.

Dimitrov, Antony S., Peter Kralchevsky, Alex Nikolov, Hideaki Noshi, and Mutsuo Matsumoto. 1991. "Contact angle measurements with sessile drops and bubbles." *Journal of Colloid and Interface Science* 145 (1): 279–282.

Eygeris, Yulia, Emily V White, Qiaoyi Wang, John E Carpenter, Michael Grünwald, and Ilya Zharov. 2018. "Responsive nanoporous membranes with size selectivity and charge rejection from self-assembly of polyelectrolyte "hairy" nanoparticles." *ACS Applied Materials & Interfaces* 11 (3): 3407–3416.

Ferrero, Enrique, Susana Navea, Carme Repolles, Jordi Bacardit, and Jorge Malfeito. 2011. Analytical Methods for the characterization of reverse osmosis membranes fouling. *IDA World Congress – Perth Convention and Exhibition Centre (PCEC), Perth, Western Australia September*, 4–9.

Hou, Deyin, Jun Wang, Dan Qu, Zhaokun Luan, and Xiaojing Ren. 2009. "Fabrication and characterization of hydrophobic PVDF hollow fiber membranes for desalination through direct contact membrane distillation." *Separation and Purification Technology* 69 (1): 78–86.

Huang, Hsin-Hui, Rakesh K Joshi, K Kanishka H De Silva, Rajashekar Badam, and Masamichi Yoshimura. 2019. "Fabrication of reduced graphene oxide membranes for water desalination." *Journal of Membrane Science* 572: 12–19.

Kadhom, Mohammed, and Baolin Deng. 2019. "Thin film nanocomposite membranes filled with bentonite nanoparticles for brackish water desalination: A novel water uptake concept." *Microporous and Mesoporous Materials* 279: 82–91.

Khayet, Mohamed, and Takeshi Matsuura. 2001. "Preparation and characterization of polyvinylidene fluoride membranes for membrane distillation." *Industrial & Engineering Chemistry Research* 40 (24): 5710–5718.

Khorramdel, Hasan, Erfan Dabiri, Farshad Farshchi Tabrizi, and Mohammad Galehdari. 2019. "Synthesis and characterization of graphene acid membrane with ultrafast and selective water transport channels." *Separation and Purification Technology* 212: 497–504.

Khulbe, KC, and T Matsuura. 2017. "Recent progresses in preparation and characterization of RO membranes." *Journal of Membrane Science and Research* 3 (3): 174–186.

Kim, Sang Gon, Jeong Hwan Chun, Byung-Hee Chun, and Sung Hyun Kim. 2013. "Preparation, characterization and performance of poly (aylene ether sulfone)/modified silica nanocomposite reverse osmosis membrane for seawater desalination." *Desalination* 325: 76–83.

Lorente-Ayza, Maria-Magdalena, Olga Pérez-Fernández, Raquel Alcalá, Enrique Sánchez, Sergio Mestre, Joaquin Coronas, and Miguel Menéndez. 2017. "Comparison of porosity assessment techniques for low-cost ceramic membranes." *Boletín De La Sociedad Española De Cerámica Y Vidrio* 56 (1): 29–38.

Mahmoud, Khaled A, Bilal Mansoor, Ali Mansour, and Marwan Khraisheh. 2015. "Functional graphene nanosheets: The next generation membranes for water desalination." *Desalination* 356: 208–225.

Mather, Robert Rhodes. 2009. *Surface modification of textiles by plasma treatments.* Woodhead Publishing Series in Textiles. Editor Q. Wei, 296–317. UK.

Mohan, Dodda Jagan, and Lucie Kullová. 2013. "A study on the relationship between preparation condition and properties/performance of polyamide TFC membrane by IR, DSC, TGA, and SEM techniques." *Desalination and Water Treatment* 51 (1–3): 586–596.

Ren, Chunlei, Hong Fang, Jianqiang Gu, Louis Winnubst, and Chusheng Chen. 2015. "Preparation and characterization of hydrophobic alumina planar membranes for water desalination." *Journal of the European Ceramic Society* 35 (2): 723–730.

Sachit, Dawood, and John Veenstra 2017. "Foulant analysis of three ro membranes used in treating simulated brackish water of the Iraqi Marshes." *Membranes* 7 (2): 23.

Safarpour, Mahdie, Alireza Khataee, and Vahid Vatanpour. 2015. "Thin film nanocomposite reverse osmosis membrane modified by reduced graphene oxide/TiO2 with improved desalination performance." *Journal of Membrane Science* 489: 43–54.

Shahabi, Soulmaz Seyyed, Najmedin Azizi, and Vahid Vatanpour. 2019. "Synthesis and characterization of novel g-C3N4 modified thin film nanocomposite reverse osmosis membranes to enhance desalination performance and fouling resistance." *Separation and Purification Technology* 215: 430–440.

Shirazi, Mohammad Mahdi A, Dariush Bastani, Ali Kargari, and Meisam Tabatabaei. 2013. "Characterization of polymeric membranes for membrane distillation using atomic force microscopy." *Desalination and Water Treatment* 51 (31–33): 6003–6008.

Tretinnikov, Oleg N, and Yoshito Ikada. 1994. "Dynamic wetting and contact angle hysteresis of polymer surfaces studied with the modified Wilhelmy balance method." *Langmuir* 10 (5): 1606–1614.

Tsuru, Tsuru, Yuko Takata, Hiroyasu Kondo, Fumi Hirano, Tomohisa Yoshioka, and Masashi Asaeda. 2003. "Characterization of sol–gel derived membranes and zeolite membranes by nanopermporometry." *Separation and Purification Technology* 32 (1–3): 23–27.

US Department of the Interior Bureau of Reclamation. 2009. "Characterization of membrane foulants in seawater reverse osmosis desalination." Desalination and Water Purification Research and Development Program Report No.14.

Vaseghi, Ghazaleh, Abbas Ghassemi, and Jim Loya. 2016. "Characterization of reverse osmosis and nanofiltration membranes: Effects of operating conditions and specific ion rejection." *Desalination and Water Treatment* 57 (50): 23461–23472.

Van Wagner, Elizabeth Marie, Benny D Freeman, Mukul M Sharma, Michael A Hickner, and Susan J Altman. 2010. "Polyamide desalination membrane characterization and surface modification to enhance fouling resistance." University of Texas, Sandia National Laboratories, SAND2010-5540.

Wei, Xiuzhen, Xin Kong, Chengtian Sun, and Jinyuan Chen. 2013. "Characterization and application of a thin-film composite nanofiltration hollow fiber membrane for dye desalination and concentration." *Chemical Engineering Journal* 223: 172–182.

Wyart, Yvan, Gaelle Georges, Carole Deumie, Claude Amra, and Philippe Moulin. 2008. "Membrane characterization by microscopic methods: Multiscale structure." *Journal of Membrane Science* 315 (1–2): 82–92.

Xu, Shengjie, Feng Li, Baowei Su, Michael Z Hu, Xueli Gao, and Congjie Gao. 2019. "Novel graphene quantum dots (GQDs)-incorporated thin film composite (TFC) membranes for forward osmosis (FO) desalination." *Desalination* 451: 219–230.

Zhao, Chuanqi, Xiaochen Xu, Jie Chen, and Fenglin Yang. 2013. "Effect of graphene oxide concentration on the morphologies and antifouling properties of PVDF ultrafiltration membranes." *Journal of Environmental Chemical Engineering* 1 (3): 349–354.

8 Carbon Molecular Models for Desalination

Georgia Karataraki and Anastasios Gotzias
NCSR – Demokritos Institute of Nanoscience and
Nanotechnology (INN)
Agia Paraskevi
Greece

Elena Tocci
CNR Institute on Membrane Technology (IT)
Rende, Italy

8.1 INTRODUCTION

Surface and underground freshwater resource scarcity have rendered water desalination one of the most important means of obtaining fresh water. (Azoulay and Houngbo 2018). Traditional desalination approaches involve either the process of distillation, which needs a large amount of energy, or the filtration approach, which uses polymeric membranes to achieve both high salt rejection rate and high freshwater flux. Membrane flux is the main challenge in every filtration approach. The high-flux membrane is highly desirable not only for reducing the membrane area, but also for increasing productivity (Chen et al. 2018).

To address such challenges, the interest has been turned towards nanostructured carbon (NC)-based membranes, that promise higher water treatment effectiveness and productivity while maintaining their low cost and easy manufacturing. Carbon-based nanomaterials such as carbon nanotubes (CNTs), graphene, and graphene oxide have been further examined owing to their appealing features such as large surface area, extended range of porosity, superior thermal and electrical conductivity, and exceptional mechanical strength and stiffness (Teow and Mohammad 2019).

Membrane based technologies that are commonly used for water treatment can be categorized on the basis of their driving force: pressure difference [microfiltration (MF), ultrafiltration (UF), nanofiltration (NF), reverse osmosis (RO)]; concentration difference [forward osmosis (FO)]; electric potential difference [electrodialysis (ED), electrodeionization (EDI)]; and temperature differences (membrane distillation). The computational technique called molecular dynamics (MD) is broadly employed to improve the understanding of membranes and separation mechanisms (Ebro et al. 2013).

8.2 EXISTING MODELS

8.2.1 Graphene and Graphene Oxide (GO)

Molecular dynamics simulations are broadly employed to investigate water desalination through functionalized nanoporous graphene membranes. Yunhui Wang's et al. (Wang et al. 2017) examined six different functionalized nanopores terminated by hydrogen or hydroxyl functional groups, each having a different pore size. For the simulation of the systems, membranes were located at the center of the simulation box, which was filled with NaCl aqueous solution, corresponding to a salt concentration close to that of seawater (see Figure 8.1).

Firstly, the energy was minimized in each system, and afterward, velocities were assigned according to the Maxwell-Boltzmann distribution at 300 K. Finally, nonequilibrium MD simulation was conducted at 298 K. For all runs, the membrane was held flexible, except the carbon atoms at the edge of the graphene sheet, while Na$^+$ and Cl$^-$ ions were allowed to move freely during the MD simulations.

Results showed that both water flux and salt rejection were sensitive to the pore diameter and chemical functionalization. The water flux increases with increasing external pressure as a result of the decrease in the energy barrier of water molecules. Besides, the largest the pore, the higher the water flux; however, salt rejection decreases. Furthermore, compared to existing technology, water permeance that is two or three orders of magnitude higher is obtained.

Water desalination through fluorine-functionalized nanoporous graphene oxide membranes was also investigated by Mostafa Hosseini et al. (Hosseini et al. 2019). The simulated system was a Fluorinated Nanoporous Graphene Oxide (F-NPGO) and Fluorinated Nanoporous Graphene (F-NPG) membrane, placed in the center of the simulation box and a pristine graphene, placed at a certain distance, because the barrier sheet for the molecules permeating the membrane to do not re-enter the feed (see Figure 8.2).

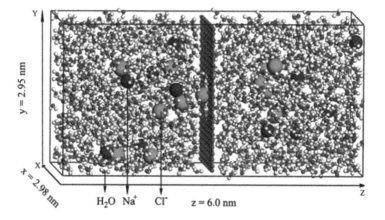

FIGURE 8.1 Schematic diagram of simulation cell for the hydrogenated membrane with largest pore size system (Wang et al. 2017).

Carbon Molecular Models for Desalination

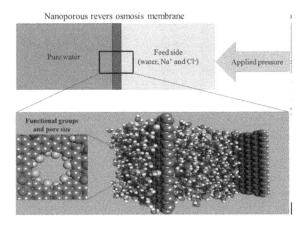

FIGURE 8.2 Simulation system of fluorinated nanoporous graphene oxide membrane (Hosseini et al. 2019).

The range of the applied pressure was 5–150 MPa, and the salt rejection was studied through the calculation of the potential of the mean force (PMF), which indicates that fluorine atoms in the pore provide a high energy barrier for the ions flow through the membrane pore, especially for Cl⁻ ions. MD simulations results showed that hydrophilic groups on the surface render the NPGO an ultra-permeable membrane, which is a good indicator for achieving high water flux.

Another way of achieving high water flux and the high salt rejection rate is applying an external electric field to the system, instead of applying pressure. Jafar Azamat (Azamat 2016) studied a system in which two graphene nanosheets were put in the simulation box, and ions were placed between them. Both sheets had a functionalized pore in their centers, one with fluoride and one with hydrogen atoms, so that there was preferential selectivity to Na^+ and Cl^- respectively (Figure 8.3). Such behavior was justified by the potential mean force calculations though MD simulations.

Functionalization of graphene nanopores have brought into light a new nanodevice based on graphene that gives promising results in ion separation, and could be further examined so that it also applied in water desalination processes (see Figure 8.4). Yaojia Chen et al. (Chen et al. 2019) used MD simulations to examine the selectivity of K^+/Na^+ using the so-called 18-crown-6-like graphene nanopore (18C6) with different layers.

Nanopores were obtained by removing carbon atoms from the central region of graphene, and the carbon atoms around the pore edge were then replaced by six oxygen atoms according to the arrangement of oxygen atoms in 18C6 molecules. The carbon atoms in the outer edge of graphene were fixed, whereas the other carbon and oxygen atoms were flexible. Results showed that both the interlayer spacing and the layer number are critical structural parameters that could affect the K^+/Na^+ selectivity.

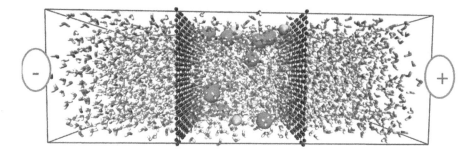

FIGURE 8.3 Schematic representation of the simulation system. The pore of left graphene was functionalized with the fluoride (F-pore), and the pore of right graphene was functionalized with the hydrogen atoms (H-pore). The direction of the applied electric field was shown also (yellow: Na^+; green: Cl^-; red: oxygen; white: hydrogen; black: carbon) (Azamat 2016).

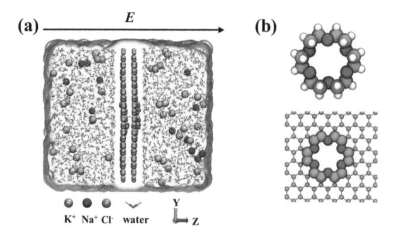

FIGURE 8.4 (a) Lateral view of the simulation box that consists of bilayer 18C6 graphene nanopores in the center and the reservoirs on both sides of the nanopore; (b) Top view of 18-crown-6 (up) and the monolayer 18C6 graphene nanopore (down). The black arrow represents the direction of the employed electric field (Chen et al. 2019).

Instead of only functionalizing the graphene sheets, there is also another efficient way of increasing salt rejection. Adding flocculant chemicals into the salt water solution can improve the efficiency of nanoporous membranes for water desalination. Mateus H. Köhler et al. (Köhler et al. 2019) added ferric chloride so that Na^+ ions aggregate, forming large clusters and finally reinforcing the ion blockage at the nanopore interface. They created a simulation box, built with a graphene sheet at one end acting as a rigid piston that applies an external

Carbon Molecular Models for Desalination

pressure on the ionic solution. The pressure gradient forces the solution against the 2D nanopore at the membrane located at the center of the box as depicted in Figure 8.5.

MD simulations showed that the increase of $FeCl_3$ in the solution leads to an increase in the desalination rate. In order to clarify whether the clusterization results from the confinement or from the nonequilibrium nature of the flow, three different cases were examined: bulk, confined between graphene sheets, and

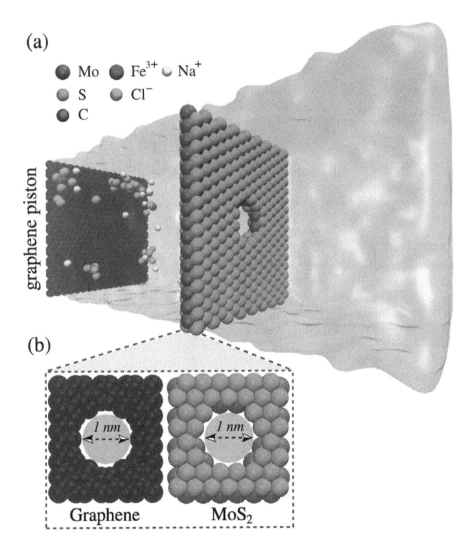

FIGURE 8.5 (a) Simulation framework. The piston (graphene) on the left side pressing the ionic solution of water against the nanopore membrane; (b) Pore diameter of each membrane (Köhler et al. 2019).

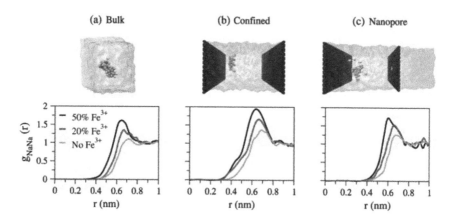

FIGURE 8.6 (a) Simulation framework. The piston (graphene) on the left side pressing the ionic solution of water against the nanopore membrane; (b) Pore diameter of each membrane (Köhler et al. 2019).

nonequilibrium nanopore case for water desalination simulations. The results presented in Figure 8.6 revealed that the ionic clusterization phenomenon is a result of the ionic inversion effect associated with the addition of $FeCl_3$ to the solution.

8.2.2 Carbon Nanotube (CNT)

Carbon nanotube-based membranes can differ widely in terms of their size, orientation, and density, as well as in terms of the presence of chemical functional groups on the nanotubes. Michael Thomas et al. (Thomas and Corry 2016) used MD simulations to determine the permeability and salt rejection capabilities for a wide range of CNT-based filtration membranes, examining all of the aforementioned factors.

Each system was composed of a 4×3 array of hexagonally packed, armchair type nanotubes which were bounded by an upper saltwater reservoir and a lower freshwater reservoir, as depicted in Figure 8.7. The system was periodic in all directions, and the diameter of the nanotubes was reported as the distance between centers of opposing carbon atoms. The CNT/water/ion systems were first minimized and then equilibrated under an NPT ensemble. The pressure was introduced as a force on the oxygen atoms of the water molecules in the upper half of the upper reservoir and the lower half of the lower reservoir. While the pressure was applied, an NVT ensemble was used, and the dimensions of the cell are set at the final cell dimensions of the equilibration simulations.

Some of the main results showed that permeability increases with nanotube diameter (see Figure 8.8). Regarding salt rejection, it increases as the pressure decreases; however, all nanotubes simulated in this work showed a modest ion rejection rate. Finally, the functionalization of the endings of CNTs decreases

Carbon Molecular Models for Desalination 257

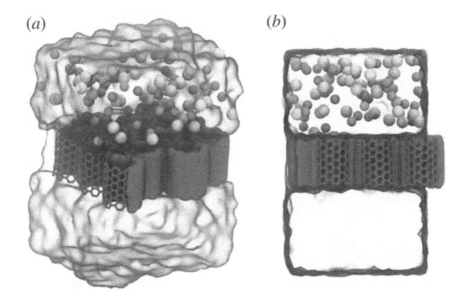

FIGURE 8.7 (a) Angled; and (b) side view of the system of a 4Ã U3 array of CNTs used in this study. The nanotubes are represented in red, Na^+ in yellow, Cl^- in cyan and the water as the transparent surface (Thomas and Corry 2016).

the permeability of the membranes. The electrostatic interaction between the polar water and charged functional forces tend to slow water transport.

Quingsong Tu et al. (Tu et al. 2016) designed a device that consists of a rotating carbon nanotube membrane filter (RCNT-MF) for reverse osmosis water desalination, which improves desalination efficiency. The key features are the centrifugal force that propels the process and the porous carbon nanotube where the separation of ions and water molecules takes place.

The device's design concept is illustrated in Figure 8.9. A CNT motor at the top of the proposed desalination device provides power to keep the CNT filter rotating. The functional part of RCNT-MF is a partial double-wall CNT system, and when an electric voltage is applied to the outer CNT, it may exert a torque to the inner CNT. For the purpose of desalination, many small holes with less than 1 nm diameter exist on the wall of the inner CNT. For the chiral type CNTs, the rotating motion generates a negative pressure that draws the salt water into the top entrance of the RCNT-MF. Once salt water enters into the CNT membrane channel, the centrifugal force generated by CNT rotation throw the fresh water molecules out from the pores on the CNT wall. Simultaneously, NaCl ions are kept inside the rotating CNT pipeline if the pore size on the CNT wall is carefully selected.

All-atom MD simulations were performed to simulate the aforementioned device. Initially, there were water molecules outside the CNT and salt water

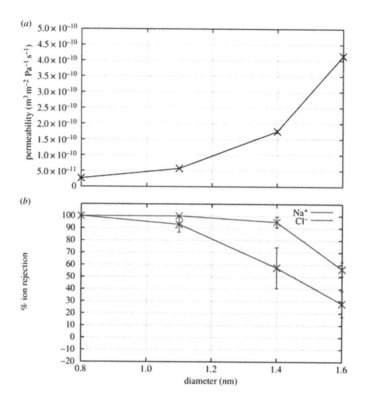

FIGURE 8.8 (a) Permeability; (b) Ion rejection of pristine CNT membranes of various pore diameter in 600 mM NaCl feed solution. Ion rejection is determined at the lowest simulated pressure for each diameter (Thomas and Corry 2016).

inside the CNT wall. Pores with different diameters were created, and the dangling carbon atoms at the edge of them were decorated with hydrogen atoms. The simulation was done in three steps: first, the system was relaxed to minimize the free energy; an NVT ensemble simulation was then performed to achieve equilibration; and finally, the desalination simulation was conducted.

Solutions showed that the rotating system of the device provides a Coriolis force that is acting at the tangential direction of the CNT wall and pushes off ions that are stuck near the site of nanoscale pores, which is an excellent way of antifouling (Figure 8.10). Moreover, the self-propelled mechanism allows the salt water to flow through the membrane without the need for extra pressure, enforcing the desalination process. Also, the large diameter CNTs that are used take advantage of both CNT and graphene membranes; thus, desalination efficiency is increased.

Michelle P. Aranha et al. (Aranha and Edwards 2018) used MD simulations to investigate ion transport through a single-walled carbon nanotube under the influence of electric fields and a fixed surface charge. Results of the simulation

Carbon Molecular Models for Desalination 259

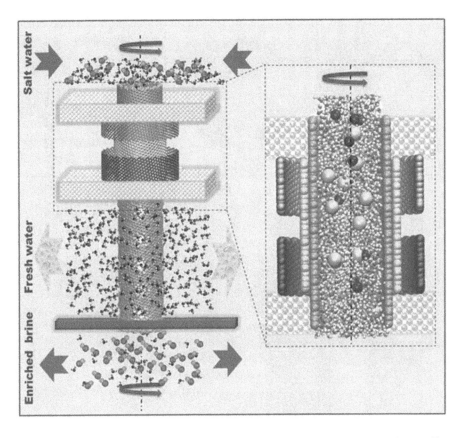

FIGURE 8.9 Illustration of the design concept of the rotating CNT membrane filter and its desalination mechanism based on the working principle of the double-wall-rotating CNT (Tu et al. 2016).

showed that the higher the electric field strength, and the higher the magnitude of the negative fixed charges applied to the tube, the higher was the cation selectivity. However, the latter seemed to be increased in a much lesser extent compared to the former.

The simulation box consisted of a carbon nanotube, connected with two bulk reservoirs, one containing a sodium-chloride solution and one containing pure water, as depicted in Figure 8.11. A separate NPT simulation was performed, in which the CNT was solvated in a water bath. The two reservoirs were also first equilibrated when containing only water molecules. Afterward, one of the reservoirs was specified at a 1M electrolyte concentration by replacing the water molecules with sodium-chloride ions.

The reservoirs were bounded by rigid graphene walls perpendicular to the axial direction of the CNT. Periodic boundary conditions were applied in the other directions. No salt solute was present in the nanotube at the beginning of

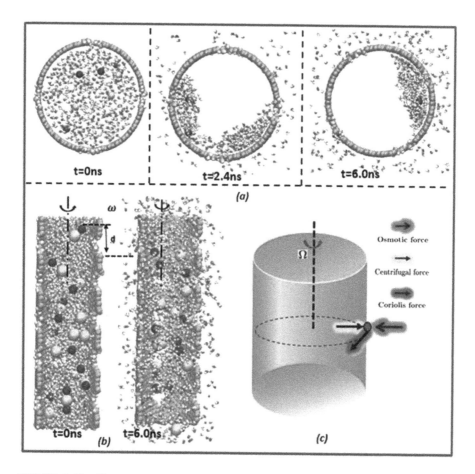

FIGURE 8.10 Time sequences from MD simulation and snapshots, taken from a top view (left) and half-section side views (right). (a) Perspective in cross-section view; (b) Perspective in longitudinal view, and (c) Schematic illustration of force diagram with the directions of the centrifugal force, the Coriolis force, and the osmotic pressure (Water molecules that are initially outside the CNT membrane are hidden.) (Tu et al. 2016).

the simulation, while ions could pass from one reservoir to the other only via the CNT.

Carmen Rizzuto et al. (Rizzuto et al. 2018) used MD simulations to study the effect of different numbers of walls on permeability in water desalination. The simulated MWCNT (Multiwalled Carbon Nanotube) system contained three sections (Figure 8.12). The first section was a reservoir filled with water molecules and sodium and chloride ions, simulating the water salinity. The second section was a multiwall carbon nanotube membrane, wherein three perpendiculars to two regular graphene sheets carbon nanotubes were placed; CNTs were modeled in single-walled, double-walled, triple-walled, or six-walled

Carbon Molecular Models for Desalination

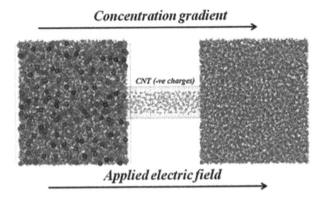

FIGURE 8.11 Schematic representation of the simulated CNT/reservoir system (water molecules (cyan), sodium ions (red); chloride ions (blue); and carbon atoms (black)) (Aranha and Edwards 2018).

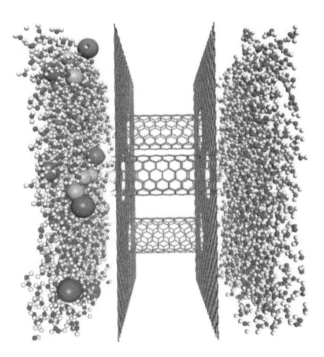

FIGURE 8.12 MWCNT membrane with SW (6,6). Armchair CNTs and two graphene sheets compose the membrane (Carbon atoms: grey; oxygen atoms: red; hydrogen atoms: white; Cl^- ions: green; Na^+: violet (Rizzuto et al. 2018).

configurations. The third and last section was a reservoir containing only water molecules.

All systems were energy minimized and then equilibrated in an NVT ensemble. Nonequilibrium MD simulations were carried out to simulate the water and ions permeation under hydrostatic pressure. Pressure difference was induced by applying an external force to the oxygen atoms of the water molecules so that they can pass through the CNT membrane. Results showed that by increasing the number of walls in the system, water conductance also increased, whereas the system performed relatively well in salt rejection and selectivity.

8.2.3 Biomimetic

Qing Li et al. (Li et al. 2016) examined the performance of biomimetic carbon nanotubes with different charged and polar groups added in their interior. The (10,10) CNTs were built to imitate the biological aquaporin system. The system contained four hexagonally packed CNTs, periodic boundaries were applied in order to create the membrane and two layers of water on each side with Na^+ and Cl^- ions randomly placed (Figure 8.13). Four types of functional groups (carboxylate anion COO^-, amine cation NH_3^+, hydroxyl OH, and amide $CONH_2$ were added to the interior of the CNTs.

We performed an energy minimization for all the systems and then equilibration under constant ambient pressure and temperature. Afterward, the simulation was performed under hydrostatic pressure of 200 MPa by applying a constant force to water layers on each side across the membrane.

The results showed that even the lowest water flux of the modified (10,10) CNTs is higher than that of a traditional RO membrane and unmodified (8,8) CNTs. When four -COO^- or -$CONH_2$ groups were added in the interior 100%, water desalination was achieved, while water conductance was not affected.

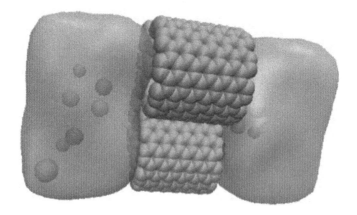

FIGURE 8.13 A model of simulating the system, which is formed by hexagonally packing 4 CNTs in a periodic cell and immersing in a solution of NaCl in water (Li et al. 2016).

Charged groups ($-NH_3^+$ or $-COO^-$) added in interior blocked the same charged ions outside while attracting opposite charged ions into the pore. When four $CONH_2$ groups were added in the interior of the CNTs, both Na^+ and Cl^- ions were unable to enter the CNTs. CNTs modified in the interior, rather than those modified at the entrance, appeared to have lower water flux by time; however salt, selectivity seemed to increase.

Chang Seon Lee et al. (Seon Lee et al. 2018) investigated the performance of functionalized graphene oxide-based membranes with a synthetic peptide motif designed to mimic the water-selective filter of natural aromatic/arginine (Ar/R) aquaporins. The octapeptide sequence RFRFRFRF (Arg-Phe-Arg-Phe-Arg-Phe-Arg-Phe), named RF8, was immobilized on the GO surface to produce GO-RF8 (where Phe = phenylamine). Results showed that the GO-RF8 membrane achieved ultrafast water permeation while maintaining the same ion rejection rate (Figure 8.14).

For comparison, GO was functionalized with two other octapeptides, Copep and random, to generate GO-Copep and GO-random. For each peptide model, energy minimization was conducted, and an equilibration simulation was then performed with a periodic boundary condition and isothermal-isobaric (NPT) ensemble. Each GO model system consisted of one large upper GO layer and two small lower GO layers located in the simulation box. A constant force was applied to each water molecule along the z-direction to produce hydrostatic pressure across the membrane. The simulations were performed under the canonical (NVT) ensemble (Figure 8.15).

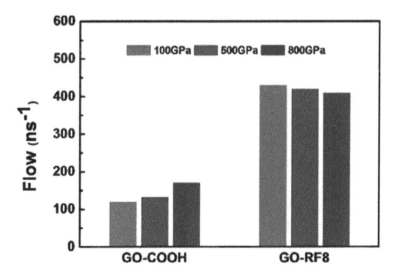

FIGURE 8.14 MD simulation results of GO-COOH and GO-RF8, water flow comparison (Seon Lee et al. 2018).

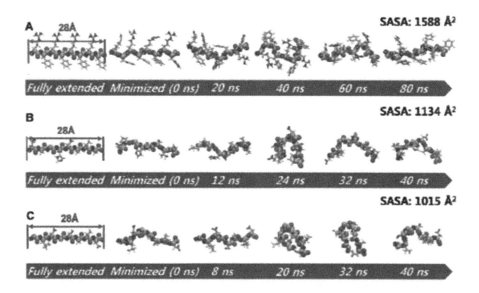

FIGURE 8.15 MD simulation results of octapeptides in water. Conformational changes in both backbone and side chains of: (a) RF8; (b) Copep; and (c) random peptides. Average solvent-accessible surface area for 40 ns is calculated (Seon Lee et al., 2018).

High cost and low stability of natural water channels have led the way to artificial water channels, made mainly out of carbon incorporated in lipid or polymer membranes. Daohui Zhao et al. (Zhao et al. 2019) examined porous organic cages (POCs) embedded in a lipid membrane as a synthetic membrane for water desalination by reverse osmosis (RO). The POC examined was the prototypical CC3 ($C_{72}H_{84}N_{12}$), which is synthesized by a single step condensation of 1,3,5-triformylbenzene with 1,2-diaminocyclohexane (DPPC) lipid. Four different CC3 channels were considered: straight, tilted, tilted with 3-openings, and crystalline.

The simulation system contained two chambers, and the membrane was placed between them. The first one was a feed chamber with NaCl aqueous solution of concentration close to that of seawater and a permeate chamber with pure water (Figure 8.16). Two graphene layers were also added to the system, and hydrostatic pressures were applied on them to simulate the reverse osmosis process.

Each system was initially subjected to energy minimization using the steepest descent method. Then, the desalination simulation took place, in which velocities were assigned according to the Boltzmann distribution. Lipid atoms were harmonically restrained by a constant force to prevent them from inserting into the CC3-embedded cages.

Results showed that the alignment, opening, and packing of the channels are essential in water permeation. The straight channel displayed a higher water

Carbon Molecular Models for Desalination

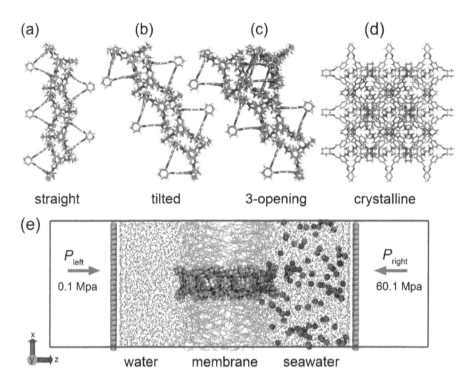

FIGURE 8.16 Types of CC3 channels: (a) straight; (b) tilted; (c) tilted with 3-openings; (d) crystalline; and (e) a simulation system for water desalination. An aqueous NaCl solution and pure water are on the right and left chambers of CC3 embedded lipid membrane. Two grapheme layers outside the two chambers are exerted under pressures *Pleft* and *Pright*, respectively (Zhao et al. 2019).

flow rate than the tilted channel owing to its vertical alignment to the membrane surface. The crystalline channel possesses the highest water flow because water molecules travel easily through its 3D interconnected morphology. The wetting – dewetting transition occurs in most of the CC3 channel types: straight, tilted, and 3-opening.

8.3 CONCLUSIONS

We have presented an indicative set of molecular simulation studies which model membrane desalination using porous carbons or carbon nanostructures. Every study follows a slightly different approach considering the definition of the models and the design of the relevant simulation workflow. In general, most of the studies use the same universal forcefield for the molecular interactions and same dynamic properties for the faces the edges and the polyhedra of the carbon nanostructures. The simulation boxes varied from elementary ones containing single pore models such as carbon nanotubes or graphene oxide sheets, to highly complex designs with discrete permeate and retendate sections on both sides of the membrane and complex membrane models with substrate structures and lipid bilayers with solvated carbon units. The simulation workflow in the presented studies usually begins with a simple minimization, in order to remove any plausibly overlapping molecules or unit fragments of the membrane models from the initial configurations. The minimization is then followed by quick system relaxation. This is accomplished through a submission of fast time NVT and NPT simulations. Notably, the production runs differ substantially from one experiment to another. In some cases the studies report lengthy simulations of the NVT type on the nanosecond and sometimes microsecond time scale. Others consider more elaborate simulation schemes like umbrella sampling where they perform multi submissions along a reaction coordinate in which they pull the salt substances through the pore channels of the membrane, from the permeate to the retendete sections of the simulation box.

REFERENCES

Aranha, Michelle P, and Brian J Edwards. 2018. "Ion Transport through Single-Walled Carbon Nanotubes: Effects of Electric Field and Fixed Surface Charge." *Chemical Physics Letters* 712: 95–101. https://doi.org/10.1016/j.cplett.2018.09.072.

Azamat, Jafar. 2016. "Functionalized Graphene Nanosheet as a Membrane for Water Desalination Using Applied Electric Fields: Insights from Molecular Dynamics Simulations." *The Journal of Physical Chemistry C* 120 (41): 23883–91. https://doi.org/10.1021/acs.jpcc.6b08481.

Azoulay A and G F Houngbo. 2018. *The United Nations World Water Development Report 2018: Nature-Based Solutions for Water. UN Water Repor*t. https://doi.org/ https://unesdoc.unesco.org/ark:/48223/pf0000261424.

Chen, Wei, Shuyu Chen, Tengfei Liang, Qiang Zhang, Zhongli Fan, Hang Yin, Kuo-Wei Huang, Xixiang Zhang, Zhiping Lai, and Ping Sheng. 2018. "High-Flux Water Desalination with Interfacial Salt Sieving Effect in Nanoporous Carbon Composite Membranes." *Nature Nanotechnology* 13 (4): 345–50. https://doi.org/10.1038/s41565-018-0067-5.

Chen, Yaojia, Yudan Zhu, Yang Ruan, Nana Zhao, Wei Liu, Wei Zhuang, and Xiaohua Lu. 2019. "Molecular Insights into Multilayer 18-Crown-6-like Graphene Nanopores for K+/Na+ Separation: A Molecular Dynamics Study." *Carbon* 144: 32–42. https://doi.org/10.1016/j.carbon.2018.11.048.

Ebro, Hannah, Young Mi Kim, and Joon Ha Kim. 2013. "Molecular Dynamics Simulations in Membrane-Based Water Treatment Processes: A Systematic Overview." *Journal of Membrane Science* 438: 112–25. https://doi.org/10.1016/j.memsci.2013.03.027.

Hosseini, Mostafa, Jafar Azamat, and Hamid Erfan-Niya. 2019. "Water Desalination through Fluorine-Functionalized Nanoporous Graphene Oxide Membranes." *Materials Chemistry and Physics* 223: 277–86. https://doi.org/10.1016/j.matchemphys. 2018.10.063.

Köhler, Mateus H, José R Bordin, and Marcia C Barbosa. 2019. "Ion Flocculation in Water: From Bulk to Nanoporous Membrane Desalination." *Journal of Molecular Liquids* 277: 516–21. https://doi.org/10.1016/j.molliq.2018.12.077.

Li, Qing, Dengfeng Yang, Jinsheng Shi, Xiang Xu, Shihai Yan, and Qingzhi Liu. 2016. "Biomimetic Modification of Large Diameter Carbon Nanotubes and the

Desalination Behavior of Its Reverse Osmosis Membrane." *Desalination* 379: 164–71. https://doi.org/10.1016/j.desal.2015.11.008.

Rizzuto, Carmen, Giovanni Pugliese, Mohammed A Bahattab, Saad A Aljlil, Enrico Drioli, and Elena Tocci. 2018. "Multiwalled Carbon Nanotube Membranes for Water Purification." *Separation and Purification Technology* 193: 378–85. https://doi.org/10.1016/j.seppur.2017.10.025.

Seon Lee, Chang, Moon-ki Choi, Ye Young Hwang, Hyunki Kim, Moon Ki Kim, and Yun Jung Lee. 2018. "Facilitated Water Transport through Graphene Oxide Membranes Functionalized with Aquaporin-Mimicking Peptides." *Advanced Materials* 30: 1705944. https://doi.org/10.1002/adma.201705944.

Teow, Yeit Haan, and Abdul Wahab Mohammad. 2019. "New Generation Nanomaterials for Water Desalination: A Review." *Desalination* 451: 2–17. https://doi.org/10.1016/j.desal.2017.11.041.

Thomas, Michael, and Ben Corry. 2016. A Computational Assessment of the Permeability and Salt Rejection of Carbon Nanotube Membranes and Their Application to Water Desalination. *Philisophical Transactions of the Royal Society A: Mathematical, Physical and Engineering Sciences* 374: 20150020.

Tu, Qingsong, Qiang Yang, Hualin Wang, and Shaofan Li. 2016. "Rotating Carbon Nanotube Membrane Filter for Water Desalination." *Scientific Reports* 6: 26183 EP. https://doi.org/10.1038/srep26183.

Wang, Yunhui, Zhongjin He, Krishna M Gupta, Qi Shi, and Ruifeng Lu. 2017. "Molecular Dynamics Study on Water Desalination through Functionalized Nanoporous Graphene." *Carbon* 116: 120–27. https://doi.org/10.1016/j.carbon.2017.01.099.

Zhao, Daohui, Jie Liu, and Jianwen Jiang. 2019. "Porous Organic Cages Embedded in A Lipid Membrane for Water Desalination: A Molecular Simulation Study." *Journal of Membrane Science* 573: 177–83. https://doi.org/10.1016/j.memsci.2018.11.053.

9 Virtual Material Design

A Powerful Tool for the Development of High-Efficiency Porous Media

Stéphan Barbe and Aron Kneer

9.1 CHARACTERIZATION OF MEMBRANE STRUCTURE

Major developments in polymeric porous membranes are nowadays strongly driven by the growing economic impact of separation techniques based on membrane technology (e.g., reverse osmosis, nanofiltration, ultrafiltration, and microfiltration). In this context, modules with integrated high performance polymeric membranes are implemented in large scale processes in order to separate impurities, salts, micro-organisms, and viruses from different bulks (Barbe et al. 2006).

The performance of a membrane involved in a separation process strongly relies on its structure. Currently, numerous methods are available for the characterization of membrane structure such as Scanning Electron Microscope (SEM), Confocal Laser Scanning Microscope (CLSM) and X-Ray Tomography. SEM (Figure 9.1) and CLSM (Figure 9.2) are commonly used to investigate the topography of a membrane (pore size, homogeneity, asymmetry). Environmental Scanning Electron Microscopy (ESEM) is a great modification of conventional SEM allowing the visualization of membrane structures under wet conditions, as well as the investigation of swelling/shrinking behaviors (Figures 9.3 and 9.4).

While SEM and CLSM investigations require the achievement of cut using e.g., microtomes, X-ray tomography enables the full, three-dimensional visualization of porous samples without affecting the sample integrity (Barbe et al. 2008). Figure 9.4 shows the structure of a pore from Q-Sartobind membrane in a wet state. This image was obtained in Figure 9.5: Radiographs of a Q-Sartobind® (with water saturated) membrane at different depths using X-ray microtomography (carried on the ID 19 beamline at the ESRF, Grenoble, France), X-ray energy: 15 keV, pixel size: 0.56µm. Zero depth corresponds to the gas side of the membrane. manufacturers (Barbe et al. 2007).

ESEM mode by using a large field detector. Figure 9.5 shows radiographs of a Q-Sartobind® membrane saturated with deionized water at different

270   Membrane Desalination

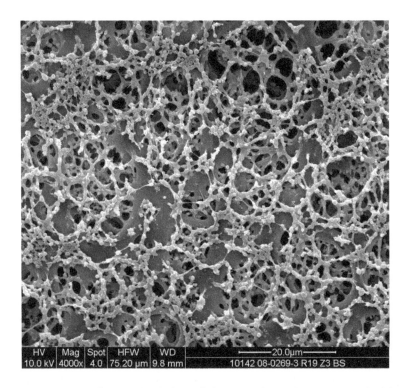

FIGURE 9.1 SEM micrograph of nitrocellulose membrane (sputter coated with gold, WD detector).

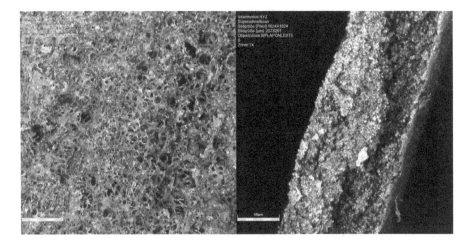

FIGURE 9.2 CLSM micrographs of a membrane from reconstructed wood.

Virtual Material Design

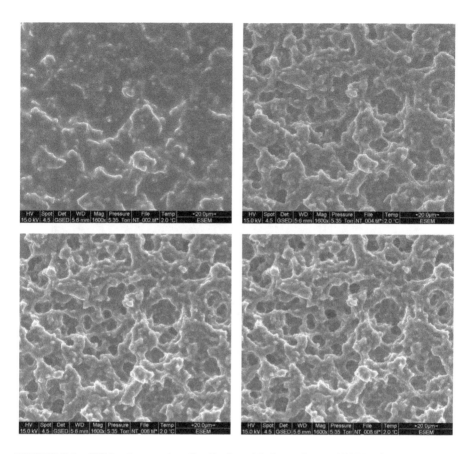

FIGURE 9.3 SEM micrographs of a Q- Sartobind membrane (GSED detector, ESEM mode, magnification – 1600x) during drying, water pressure has been progressively reduced from A to D.

depths (zero depth corresponds to the gas side of the membrane). Barbe et al. (2007) obtained these images by X-ray microtomography, which was carried on the ID 19 beamline at the ESRF (Grenoble, France). A progressive deterioration of the porous homogeneity could be observed along the membrane depth. At zero depth, coarse and fine porosities could be recognized and both categories seemed to be homogeneously distributed, resulting in an apparent homogeneous pore scale. In the middle of the active membrane layer (Depth = 32 μm), the radiograph revealed the first signs of heterogeneity. At this depth, polyester fibers and pores with a higher diameter scale became visible. Near the interface between the active layer and polyester fleece (Depth = 52 μm), a very heterogeneous and wide pore size distribution appeared.

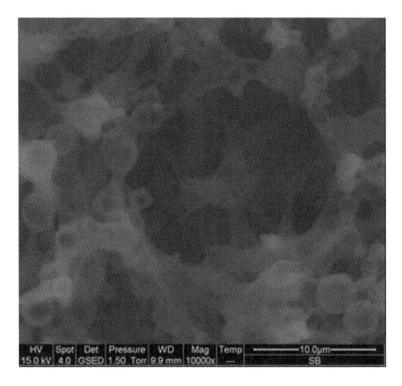

FIGURE 9.4 SEM micrograph of a Sartobind Q membrane using a Large Field Detector (LFD) detector (Barbe et al. 2011).

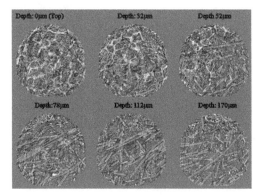

FIGURE. 9.5 Radiographs of a Q-Sartobind® (with water saturated) membrane at different depths using X-ray microtomography (carried on the ID 19 beamline at the ESRF, Grenoble, France), X-ray energy: 15 keV, pixel size: 0.56µm. Zero depth corresponds to the gas side of the membrane. manufacturers (Barbe et al. 2007).

Virtual Material Design

9.2 COMPUTATIONAL FLUID DYNAMICS IN POROUS MEDIA

Computational Fluid Dynamics (CFD) is a technique that aims to solve complex two-dimensional or three-dimensional fluid mechanics problems by approximation using numerical methods. Conventional equations used in CFD are Navier-Stokes equations and Euler equations. For this purpose, the fluid flow volume has to be meshed by means of a numerical grid (Figure 9.6) that defines the location's *cells* at which the variables are to be calculated (Ferziger & Peric 2002).

Subsequently, the governing partial differential equations are converted to a large system of nonlinear algebraic equations by using an appropriate discretization scheme (finite differences, finite volumes, or finite elements) and boundary conditions are defined (inlet, outlet, wall, symmetry plane, Dirichlet, Neumann). Finally, a dedicated solver allows for calculating the parameter fields (pressure, flow velocity) as well as streamlines. The set of algebraic equations can be expanded with equations related to mass and heat transfer to calculate scalar fields such as temperature or concentration distribution (Figure 9.7). In this regard, there is a very large variety of methods for the calculation of heat and mass transfer coefficients.

Since the structure of porous membranes is usually complex and not easy to digitalize, it is common practice in CFD to model the three-dimensional flow resistance within a porous medium by introducing a permeability tensor

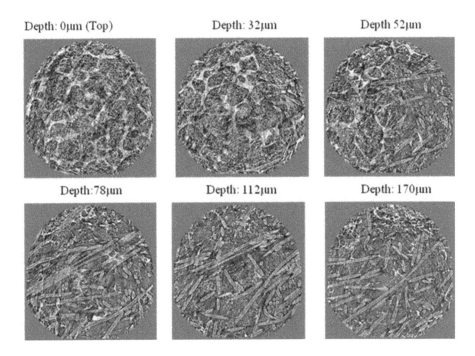

FIGURE 9.6 Fluid flow volume (blue) meshed by means of polyhedrons using Star CCM+.

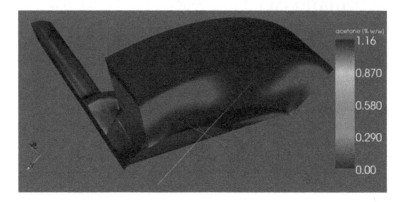

FIGURE 9.7 Field of acetone concentration in an inlet distributor after 2.4 s as a response to a Dirac pulse of acetone (Barbe et al. 2009). Liquid flow rate was set to 0.68 L/min, calculations performed with Code_Saturne (EDF, France, Paris) and visualization was achieved with Paraview.

(Pastor & Barbe 2011). As a result, a good approximation of the pressure and flow distribution can be obtained, but accurate prediction of heat and mass transfer cannot be properly achieved.

The existence of dead-end pores in most porous media has a great impact on mass transfer. A numerical approach based on dual porosity was introduced by Barbe et al. (2009) to achieve a better prediction of the distribution of bovine serum albumin membrane adsorbers (Figure 9.8). According to this model, the

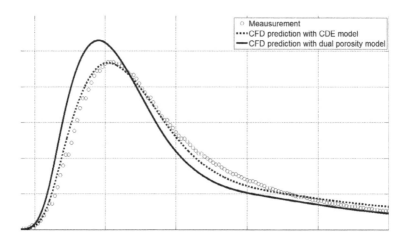

FIGURE 9.8 Comparison between measured and predicted residence time distributions of bovine serum albumin in a Q-Sartobind® Membrane Adsorber System (Barbe et al. 2009).

liquid phase is partitioned into a mobile (flowing) region and an immobile (stagnant) region, and solute exchange between the two regions is modeled as first order process. This requires some modifications of the calculation scheme for diffusion within the CFD code. This work was achieved with the Open Source CFD Code_Saturne (EDF, France).

9.3 VIRTUAL MATERIAL DESIGN (VMD)

In order to design a novel porous structures, it is necessary to use an algorithm that decomposes a three-dimensional space into n-volumes that may be named *pores* via so-called Voronoi decomposition.

A Voronoi decomposition is a decomposition of the n-dimensional space R_n into areas that are defined by a predetermined set of points, primarily n = 2 (a two-dimensional plane) or n = 3 (the usual three-dimensional space). An area is determined by exactly one point from the set of points (the center of the area) and consists of all points of the space that are closer to the center of the Euclidean distance range than at all other points in the set of points. An area is regarded as the catchment area of the center and is also called a Voronoi cell. An example of a level with a score of 14 points is shown in Figure 9.9. The Voronoi cell is bounded by straight line segments with points having exactly the same distance to the center of the cell as to one of the other centers. These straight line segments are thus parts of the midperpendicular on the connecting line between two centers. Closely related to the Voronoi decomposition is the Delaunay triangulation, which is defined as a triangular network with certain advantageous properties. A Delaunay triangulation combined with a Voronoi decomposition is shown in Figure 9.10. Originally created in of mathematics

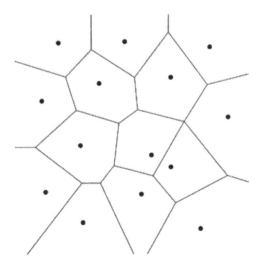

FIGURE 9.9 Voronoi-decomposition.

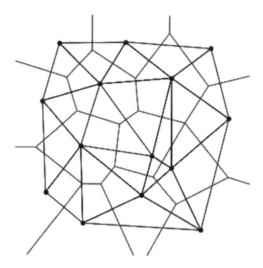

FIGURE 9.10 Delauny-triangulation.

and algorithmic geometry, there are now many applications of Voronoi decomposition in other fields such as architecture, biotechnology (membranes), chemistry, crystallography, materials science, and ecology.

PACE3D (Parallel Algorithms for Crystal Evolution in 3D) is a software package developed at the Institute for Digital Materials Science (Karlsruhe University of Applied Sciences, Karlsruhe, Germany) that embodies modules for the solution of numerous applications, including:

- phase-field models for microstructure formations in multicomponent and multiphase materials;
- CFD solvers for modelling fluid flow processes based on the Navier-Stokes equations and on the Lattice-Boltzmann method;
- solid mechanics;
- micromagnetism;
- electrical fields; and
- grand chemical potential and grand elastic potential.

PACE3D is based on the phase field model of Nestler et al. (2005). The ability of its preprocessing methods for the generation of CFD suitable porous microstructure has been confirmed (Barbe et al. 2011; Kneer et al. 2018, 2019). Figure 9.11 depicts the workflow of the preprocessing module of PACE 3D for the generation of porous microstructures.

This workflow consists of the three following steps:

1. The volume is filled with a dense bead package as basis of the porous structure. The Bead radius determines the pore volumes.

Virtual Material Design

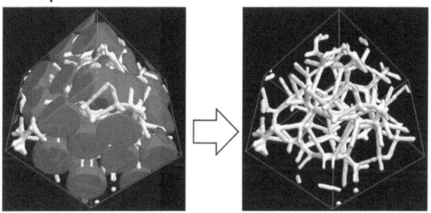

FIGURE 9.11 Workflow of the preprocessing package of PACE3D for the generation of porous microstructure.

2. A 3D-Voronoi-Diagramm is generated from the location of the bead centers.
3. Connection balks are generated where at least three Voronoi cells have a common point.

Such microstructures can be imported in the preprocessing of conventional CFD codes such as Star-CCM+, where the solid porous network, as well as the fluid flow volume, can be extracted via Boolean operations. In this case,

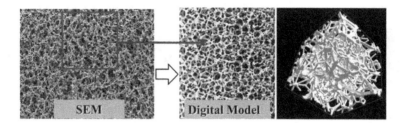

FIGURE 9.12 Reconstruction of the porous microstructure of a cellulose acetate membrane with the PACE3D. The structure was generated by overlapping a coarse structure and fine structure with different balk shapes (Barbe et al. 2011).

VMD allows for investigating heat and mass transfer in porous media by using state of the art CFD codes. Barbe et al. (2011) used the preprocessing module of PACE3D in order to reconstruct a cellulose acetate membrane and run CFD simulations with Star-CCM+ and Code_Saturne. The reconstruction was achieved by overlapping a coarse structure with a fine structure exhibiting different balk shapes (Figure 9.12).

9.4 SYNTHETIC MEMBRANE STRUCTURES

The previous example shows that porous structures can be reconstructed. However, this technique can be used to design novel structures that are not production-ready and exhibit particular properties such as specific permeability. Figure 9.13 shows three different small parts of a virtually generated membrane. Each part has porosity of 83%, but the surface areas differ, and therefore, the friction force of any fluid flowing through.

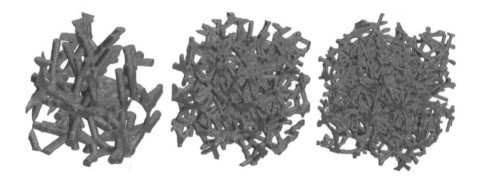

FIGURE 9.13 Generated synthetic membrane structures.

Virtual Material Design

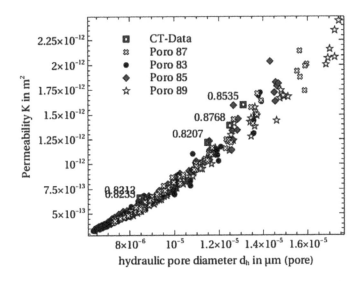

FIGURE 9.14 Permeability for different synthetic membrane types and real membranes (Kneer et al. 2018).

Attempts to change the inner surface area may have a great impact on the wall shear stress, and the flow inside the membrane might change entirely. If a capillary effect should be enabled, the pore size should be low enough and the wetting surface high enough. If pressure drop plays an important role (dialysis), but at the same time the membrane must be able to separate even very small molecules, the membrane needs an optimized structure for such an application. Comparing the permeability of virtual membranes with a porosity of 83% to real membranes (CT-Data), Figure 9.14 shows a similar dependence of permeability on hydraulic pore diameter. Overall, it should be highlighted that any kind of membrane property can be virtually generated, but the production of those membranes is still a challenge.

9.5 VMD AND CFD SIMULATIONS

Virtual porous structures can be used for CFD simulations e.g., in order to predict streamlines (Figure 9.15). The preprocessing module of the software package Star-CCM+ was used to mesh the reconstructed structure (solid porous network) depicted in Figure 9.12. It is important to note that the triangulation of the surface mesh generated for the structure was used as the base for the mesh of the fluid flow volume. This, in turn, leads to a perfect match at the interface of both meshes (Figure 9.16). The resulting CFD-ready meshes were placed in a virtual square channel (Figure 9.17). The generated mesh for the porous fluid flow volume was used as a base for the meshing of both the inlet section and the outlet section of the channel (so-called interface for mesh

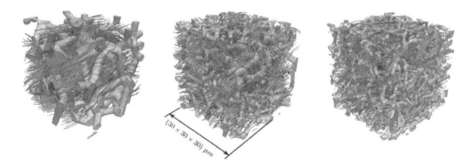

FIGURE 9.15 Computed streamlines through the synthetic membranes.

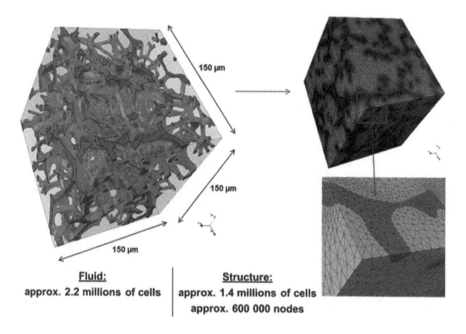

FIGURE 9.16 Meshing of the structure and the fluid flow volume of a digitalized cellulose acetate membrane. The mesh was generated by the preprocessing module of the Star-CCM + software package.

junction). The flow of a gaseous mixture (90.5% w/w nitrogen, 8.0% w/w isopropanol and 1.5% w/w water) entering the channel with a flow velocity of 8.8 m/s (temperature = 20°C; pressure = 1 atm) was simulated by using two different CFD codes: Star CCM+ and Code_Saturne.

Virtual Material Design

281

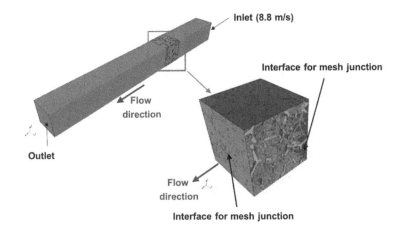

FIGURE 9.17 Integration of the meshed virtual porous medium in a square channel.

Convergence was achieved with both codes and Figure 9.18 shows the calculated and complex velocity fields in a middle cross section of the reconstructed porous medium. Compared to Code_Saturne, a slightly higher maximal flow velocity spot was predicted by Star CCM+; both codes predicted very similar flow directions and dead volumes could be clearly identified. The flow field obtained with Code_Saturne after convergence was frozen and further used to predict the residence time distribution after the injection of a Dirac pulse of acetone, as shown in Figure 9.19.

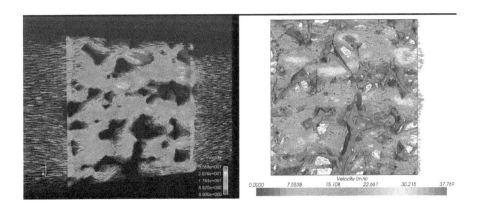

FIGURE 9.18 Simulation of the flow of a gaseous mixture (90.5 % w/w nitrogen, 8.0 % w/w ispropanol and 1.5 % w/w water) entering a square channel with a flow velocity of 8.8 m/s (temperature = 20°C, pressure = 1 atm) containing a virtual reconstructed porous medium was simulated by using two different CFD codes: Star CCM+ and Code_saturne.

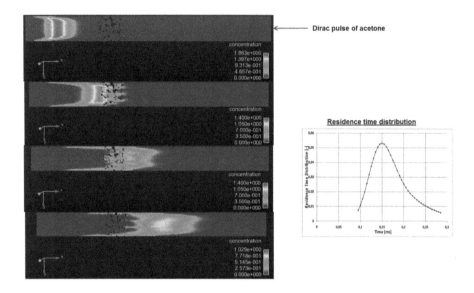

FIGURE 9.19 Prediction of the residence time distribution in reconstructed membrane after the injection of a Dirac pulse of acetone.

9.6 CONCLUSIONS

The ability of Voronoi diagrams to virtually reconstruct real porous media from imaging methods such as SEM, X-Ray, or VMD tomography is very promising for the development of desalination membranes. Veroni diagrams offer an attractive way to analyze CFD in membranes and estimate the impact of CFD on the pressure and velocity fields. Furthermore, they enable investigation of the impact of porous media on mass transfer and the prediction of local concentration gradients that may greatly influence the performance of desalination membranes. They are also the basis of the systematic characterization of membranes under extreme conditions (e.g., high pressure and high temperature). Finally, macroscopic parameters related to porous media (e.g., permeability tensor, pressure loss) can be derived from the simulations (Kneer 2014).

REFERENCES

S. Barbe, E. Boller, R. Faber, V. Thom, T. Scheper, The use of x-ray microtomography for the characterization of macroporous membranes. *Journée annuelle de la Socété Française de Métallurgie et de Matériaux*, Ecole Nationale Supérieure des Mines de Saint Etienne, France (2007).

S. Barbe, A. Kneer, M. Roelle, A. August, B. Nestler, The use of PACE 3D models for the generation of CFD suitable macroporous structures. *Code_Saturne User Meeting edition 2011*, Paris, France (2011).

Virtual Material Design

S. Barbe, A. Kneer, M. Wirtz, T. Scheper, Prediction of mass transport phenomena in Membrane Adsorber Systems (MAS) using a dual porosity model. *Code_Saturne User Conference 2008*, R&D EDF Paris, France (2008).

S. Barbe, A. Kneer, M. Wirtz, T. Scheper, *Fluid Dynamics in Membrane Adsorber Systems*, DECHEMA Working Party Membrane Technology, DECHEMA HAUS, Frankfurt am Main, Frankfurt am Main, Germany (2009).

S. Barbe, D. Nussbaumer, W. Demmer, A. Weiss, R. Faber, T. Scheper, Presentation of a Tandem–Sartobind pilot plant as an approach for large scale membrane chromatography, *Desalination* **200** (2006): 480–482.

J. H. Ferziger, M. Peric, *Computational Methods for Fluid Dynamics*, 3rd Edition, Springer, Heidelberg (2002).

A. Kneer, *Numerische Untersuchung des Wärmeübertragungsverhaltens in unterschiedlichen porösen Medien*, Vol. **42** Schriftreihe des Instituts für Angewandte Materialien, KIT Scientific Publishing, Karlsruhe (2014).

A. Kneer, M. Wirtz, P. Altschuh, S. Barbe, *Virtual material design: New generation of porous media*, STEPsCON 2018 International Scientific Conference on Sustainability and Innovation, Leverkusen, Germany (2018).

A. Kneer, M. Wirtz, S. Yurtsever-Kneer, S. Barbe, A. August, Modern times need enlightened innovation and sophisticated materials, *Galvanotechnik* **110** (2019): 712–719.

B. Nestler, H. Garcke, B. Stinner, Multicomponent alloy solidification: phase-field modelling and simulations, *Physical Review E* **71** (2005): 1–8.

A. Pastor, S. Barbe, Disposable chromatography for large-scale biomanufacturing in Single-Use Technology. In *Biopharmaceutical Manufacturing*, Edited by Regine Eibl and Dieter Eibl.ch25, Wiley (2011): 301–310. doi: 10.1002/9780470909997.

10 Estimation of Process Parameters in Industrial Membrane Manufacture Using Computational Fluid Dynamics

Stéphan Barbe, Fei Wang, Aron Kneer, Michael Metze, Christian Wenning, Patrick Altschuh, and Britta Nestler

10.1 INTRODUCTION

In recent decades, membrane desalination have become an important key operation for the treatment of seawater and brackish water. The heavy-duty requirements for desalination membranes have become sharper and considerable efforts are dedicated to a more accurate understanding of membrane formation. Indeed, the functional behavior of a polymeric membrane is intimately tied to its microstructure, and current performance limitations are strongly related to the difficult control of microstructure evolution during large scale membrane manufacturing. Achieving a targeted microstructure in a precise and reproducible way is still challenging for engineers and membrane scientists. Furthermore, membranes involved in desalination modules are manufactured with production rates up to several thousand square meters per day, and process fluctuations usually affect the lot-to-lot consistency and herewith the associated reachable specifications.

The high performance of polymeric membranes mostly relies on their asymmetric structure, which consists of a thin, fine pored skin atop a much thicker and coarse porous substructure. The skin provides selectivity in filtration, while the substructure acts as a depth filter and a mechanical support. One of the most common techniques for the manufacture of asymmetric polymeric membranes is the so called dry cast (Metze et al. 2017), also referred to as evaporative casting. In this process, a homogeneous polymer solution is cast as a thin film on a moving endless steel belt (substrate). In most cases, the considered polymer solution is a ternary or a quaternary system, which consists of one polymer, one highly volatile solvent, and one or two less volatile nonsolvents. Then, the cast film moves through several chambers, where different drying conditions are applied. The preferential evaporation of more volatile solvent over nonsolvent(s) causes the film to phase separate (liquid – liquid demixing) in a polymer lean phase and a polymer rich phase (Musa et al. 2018). The latter builds the solid matrix of the membrane while the polymer lean phase builds the pores. The final membrane microstructure results from the evolution and maturation of this heterogeneous system via coalescence, ripening, or hydrodynamic flow. The resulting microstructure is finally fixed by drying at high temperature and low partial pressures (Metze et al. 2017).

The microstructure of polymeric filtration membranes may be controlled by:

- Maturation time: coarse microstructures require long maturation times to be developed.
- Position in the phase diagram where demixing occurs: the higher the interfacial tension of the resulting heterogeneous system is, the coarser the microstructure would be.
- Demixing mechanism: nucleation is the most observed demixing mechanism during the formation of polymeric membranes.

In this regard, Figure 10.1 shows the dramatic impact of the film composition at demixing on the resulting structure of nitrocellulose membranes (Metze 2014). The current progress achieved in the field of computational fluid dynamics (CFD), cryogenic scanning electron microscopy (cryo-SEM), small angle neutron scattering (SANS), and modeling of phase separation are reviewed in this chapter. The intelligent combination of these techniques to build up a novel approach for the development of high performance membranes involved in desalination is discussed.

10.2 INVESTIGATION OF LIQUID – LIQUID DEMIXING IN POLYMER SOLUTIONS AND MICROSTRUCTURE EVOLUTION

The experimental determination of binodal curves by means of cloud point determination is a common method for the thermodynamic investigation of phase equilibrium in polymer solutions (Metze et al. 2012, 2013). However, binodal curves can also be computed by combining the Flory-Huggins theory with group contribution theories (Figure 10.2) as demonstrated by Wenning et al. (2018a) for the ternary system: poly(ethylene oxide) (PEO), poly(propylene oxide) (PPG), and 1,6-hexamethylene diisocyanate (HDI). Furthermore, the evidence of Liquid – Liquid demixing during the synthesis of bisoft segment isocyanate-terminated polyurethanes (ITPUs) was reported by Wenning et al. (2018b).

For the purpose of this investigation, the authors developed an experimental setup based on coupled in-line reaction monitoring via UV/Vis and FTIR techniques (Figure 10.3). It is so possible to simultaneously monitor phase separation and reaction progress (expressed as isocyante conversion % NCO conversion)

Computational Fluid Dynamics 287

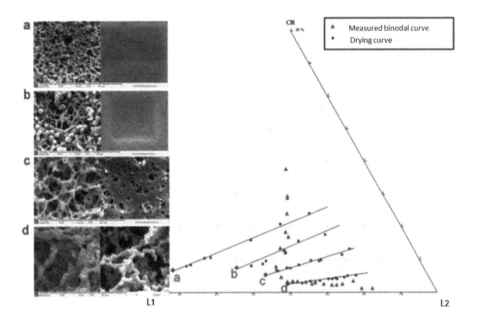

FIGURE 10.1 Impact of demixing composition on the structure of nitrocellulose membranes. L1 = solvent, L2 = non-solvent. Micrographs (left) = belt side, micrographs (right) = air side. Metze (2014).

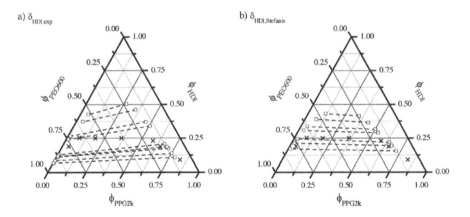

FIGURE 10.2 Predicted binodal curves and connodes (white squares) and experimentally determined cloud points of the ternary mixture PPG2k/PEO600/HDI (black crosses). The solubility parameter for 1,6-hexamethylene diisocyanate (HDI) was experimentally determined and calculated by using group contribution theories of Stefanis, Wenning et al. (2018a). Reused with permission, copyright 2018 Wiley.

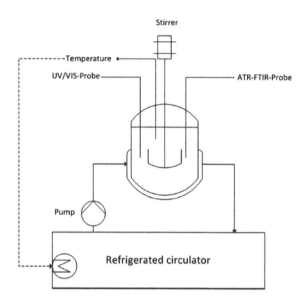

FIGURE 10.3 Experimental set-up developed and used by Wenning et al. (2018b) for the investigation of liquid – liquid demixing phenomena during bisoft segment polyurethane prepolymerization. (Wenning et al. 2018b). Reused with permission, copyright 2018 Springer.

during a prepolymerization reaction using 1,6-hexamethylene diisocyanate and a polyether polyol mixture (Figure 10.4).

The development of an experimental setup dedicated to the kinetic analysis of the very first minutes of membrane formation process via SANS measurements was reported by Metze et al. (2017). Core part of the described setup was a neutron transparent flow-through cell (NTFT-Cell, Figure 10.5) which was constructed and used for the investigation of liquid – liquid – demixing in a quaternary system (cellulose nitrate – methyl acetate – isopropanol – deuterium oxide).

The increase of nonsolvent proportion during dry cast operations and the corresponding crossing of the binodal curve were simulated by adding nonsolvent into a homogeneous casting solution which consisted of solvent, nonsolvent and polymer.

The dope continuously circulated through the NTFT-Cell, which was placed in the neutron beam of the D11 instrument at Institut Laue-Langevin (ILL, Grenoble, France). D11 is a pinhole geometry SANS instrument (Figure 10.6), as well as the world's longest SANS instrument, comprising a 40 m collimation section and a 40 m maximum detector distance (Metze et al. 2017). Additionally, a probe for light transmission was placed in the stirred reactor, and allowed for monitoring the turbidity of the casting solution during SANS measurements (Figures 10.7 and 10.8).

Figure 10.9 shows some SANS measurements obtained with the described polymer solution; well-defined peak was observed at about $2 \cdot 10^{-3}$ Å^{-1}

Computational Fluid Dynamics 289

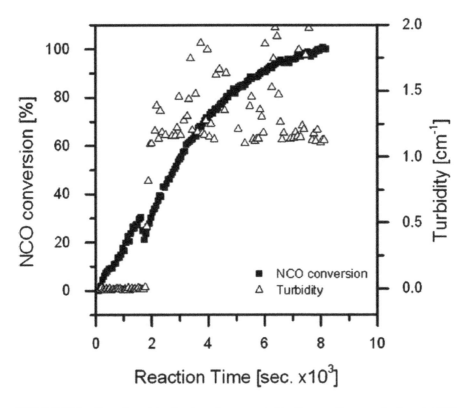

FIGURE 10.4 Experimental setup based on coupled in-line reaction monitoring via UV/Vis and FTIR technique developed by Wenning et al. (2017). Reused with permission, copyright 2017 Springer.

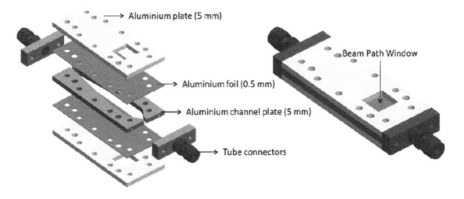

FIGURE 10.5 Design and construction of the Neutron-Transparent Flow-Through-Cell (NTFT-Cell) used and developed by Metze et al. (2017) for SANS investigations.

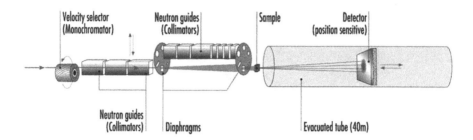

FIGURE 10.6 Description of D11 Instrument at Institut Laue Langevin (ILL, Grenoble, France).

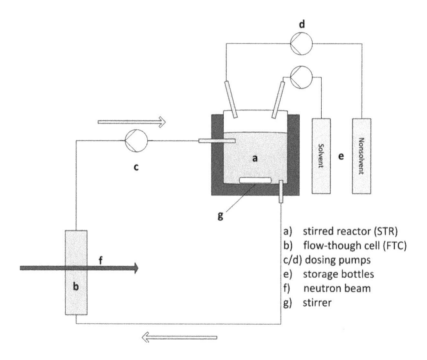

FIGURE 10.7 Flow sheet of the experimental set-up developed and used by Metze et al. (2017) for the investigation of liquid – liquid-demixing in a quaternary system (cellulose nitrate – methyl acetate – isopropanol – deuterium oxide) via SANS.

resulting from liquid – liquid demixing. After its arrival in the neutron beam, the scattered intensity passed two maxima. The intensity significantly dropped between 25 and 35 min owing to the formation of larger structures, sizes of which were outside the detection range of SANS. This experimental setup is well-suited for kinetic investigation of the casting process, and opens the possibility to explore and improve the production parameters.

Computational Fluid Dynamics 291

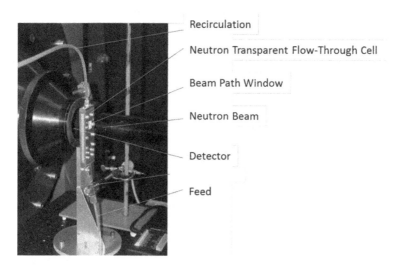

FIGURE 10.8 Photograph of the Neutron-Transparent Flow-Through-Cell (NTFT-Cell) used and developed by Metze et al. (2017).

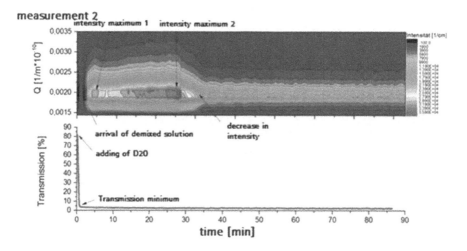

FIGURE 10.9 SANS and transmission measurement achieved with a Neutron Transparent Flow Through Cell (NTFT-Cell) and a polymer solution (cellulose nitrate – methyl acetate – isopropanol – deuterium oxide), Metze et al. (2017).

Cryogenic Scanning Electron Microscopy (Cryo-SEM) is an attractive technique for the investigation of membrane formation kinetics. After casting, the film is frozen in liquid nitrogen and subsequently lyophilized at different time intervals. The film lyophilisates are then investigated by SEM.

It allows for retracing and resolving the evolution of microstructure evolution along the film depth. Metze (2010) investigated the formation of a macroporous nitrocellulose membrane via Cryo-SEM. The corresponding micrographs are depicted in Figure 10.10 and show the progressive formation of porous network, especially on the belt side.

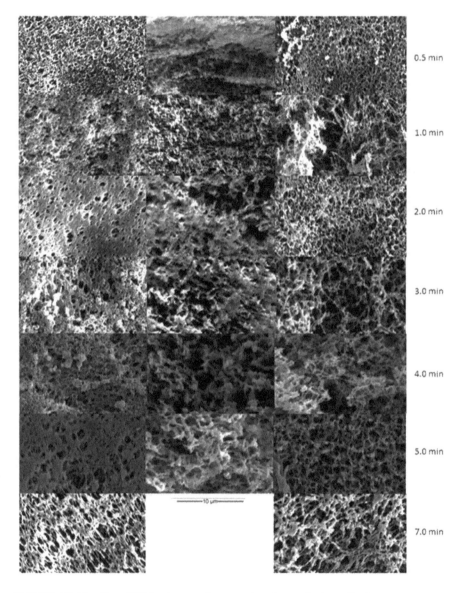

FIGURE 10.10 Cryo-SEM micrographs retracing the evolution of microstructure evolution in a nitrocellulose membrane (Metze 2010).

10.3 NUMERICAL PREDICTION OF HEAT AND MASS TRANSFER IN CASTING MACHINES

CFD became a powerful tool for the investigation of flow dynamics in membrane modules (Barbe et al. 2008, 2009, Pastor & Barbe 2011, Kneer et al. 2019). Its main objective is the simulation of flow dynamics and the prediction of heat and mass transfer in complex 3D geometries. Nonetheless, it can also be utilized for the design of dry cast equipment. As shown in Figure 10.11, the highly versatile and flexible CFD package TinFlow (Tinnit Technologies GmbH, Karlsruhe, Germany) has been successfully used for the prediction of the velocity field and solvent concentration field (methylene chloride) in the flowing gaseous phase, as well as the evolution of film thickness along the machine during dry cast operations. The polymeric film was coated with a wet thickness of 3.15 mm on a moving stainless steel belt. In this case, a slow countercurrent stream of nitrogen was used to evaporate the solvent and induce liquid – liquid demixing within the film. The model properly predicts the accumulation and acceleration of heavy methylene chloride on the belt of the machine. The k-ε turbulence model was selected in this study and the turbulent kinetic energy was used for the calculation of the local mass transfer coefficient.

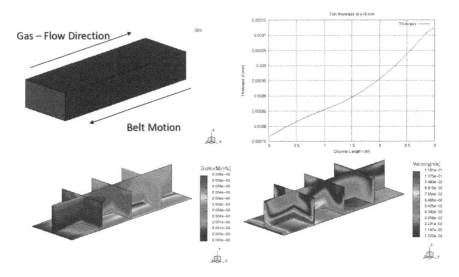

FIGURE 10.11 Prediction of the velocity field and solvent concentration field (methylene chloride) in the flowing gaseous phase, as well as the evolution of film thickness along the machine during dry cast operations using the CFD package TinFlow (Tinnit Technologies GmbH, Karlsruhe, Germany).

10.4 MODELING OF PHASE SEPARATION

Membranes are produced primarily from polymer solutions, in which the liquid solution decomposes into a polymer-rich phase and a polymer-lean phase, as mentioned in the previous sections. Numerical simulations of this demixing process can shed light on the kinetic process of morphological evolution, which provides additional insight into the experimentally observed microstructures of membranes. The premise of the modeling and simulation on this demixing process is based on thermodynamic databases of the polymer solutions. There is a paucity of experimental thermodynamic databases for polymer solutions forming membranes. The modeling of phase separation in polymer solutions is mostly based on the theoretical model, the Flory-Huggins theory.

A simple case is the binary mixture with a polymer species and solvent, where the free energy density is expressed as (Flory 1941)

$$f(\phi) = k_b T \left[\frac{\phi}{N} \ln \phi + (1 - \phi) \ln(1 - \phi) + \chi \phi (1 - \phi) \right] \tag{10.1}$$

Here, k_b is the Boltzmann constant; T represents the temperature; ϕ stands for the volume concentration of the polymer species; N denotes the degree of polymerization (DP); and χ is the Flory parameter. For upper critical temperature systems, the Flory parameter is inversely proportional to the temperature as $\chi = Z \Delta \varepsilon / k L_b T$, where Z is the coordination number and $\Delta \varepsilon$ is the difference between the polymer – solvent interaction energy ε_{PS} and the average interaction energy of the polymer – polymer ε_{PP} and the solvent – solvent ε_{SS}, namely, $\Delta \varepsilon = \varepsilon_{PS} - (\varepsilon_{PP} + \varepsilon_{SS})/2$. For lower critical temperature systems, the dependence of the Flory parameter on the temperature may be modeled as $\chi = \chi_1 T$, with χ_1 being a constant.

By using the free energy density in Equation (10.1), the binary phase diagram can be produced according to the common tangent construction method:

$$\partial_\phi f \vert_{\phi = \phi_p} = \partial_\phi f \vert_{\phi = \phi_s}, \tag{10.2}$$

$$\left(f - \partial_\phi f \phi \right) \vert_{\phi = \phi_p} = \left(f - \partial_\phi f \phi \right) \vert_{\phi = \phi_S} \tag{10.3}$$

Here, ϕ_p and ϕ_s represent the equilibrium volume concentrations in the polymer-rich and polymer-lean phases, respectively. Applying this common tangent construction method at different temperatures, we obtain the miscibility gap of the system. As an example, Figures 10.12(a) and (b) depict the miscibility gap for an upper and lower critical temperature system, respectively, with different DPs. The dashed lines are the spinodal lines which are defined by the locus of $\partial^2 f / \partial \phi^2 = 0$.

As shown in Figure 10.12(a), the miscibility gap, as well as the spinodal lines, both shift upward and leftward with an increase in DP. For such a system, there are two ways to model the phase separation in the polymer solutions. The first of these is thermally induced phase separation (TIPS) (Kim

Computational Fluid Dynamics

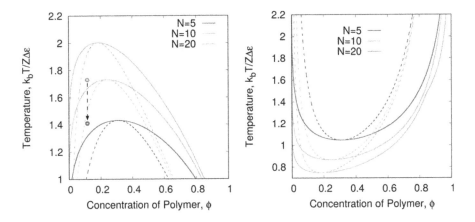

FIGURE 10.12 Phase diagram of binary polymer solutions constructed from the Flory-Huggins theory for an upper (a); and lower (b) critical temperature system. The solid curves depict the binodal lines and the dashed curves are the spinodal lines. The Flory parameters in (a) and (b) are $\chi_1 = 1.5$ and $\chi_1 = 1.0$, respectively. Different colors correspond to distinct DPs.

et al. 2016), which is achieved by quenching the system from a high-temperature state (yellow point) to a low temperature state (red point). The system enters the spinodal region via such a thermal quench, engendering phase separation. A second mechanism is polymerization-induced phase separation (PIPS) (Wang et al. 2019). As denoted by the yellow point in Figure 10.12(a), the initial state is above the critical temperature of the miscibility gap with $N = 10$ and the polymer solution is homogeneous. Because of the cross-link reaction, DP increases with time $N(t)$. When the DP is sufficiently large, e.g., $N = 20$, the state of the system moves into the spinodal region and phase separation starts. This indicates that PIPS can spontaneously occur at a constant temperature without quenching.

Formation of membranes from polymer solutions by phase separation has been widely modeled by the Cahn-Hilliard model. The starting point is the free energy functional which is expressed as (Cahn 1961)

$$F = \int_V \left[f(\phi) + \kappa(\nabla\phi)^2 \right] dx \qquad (10.4)$$

where κ is the gradient energy coefficient, which is determined by the interfacial tension σ of the polymer-rich and polymer-lean phases as $\kappa = \sigma/2 \int_{\infty}^{\infty} \left[\frac{d\phi}{dx}\right]^2 dx$. The time evolution of the volume concentration ϕ is such that it reduces the free energy functional of the system, following the variation approach as (Wang et al. 2019)

$$\partial_t \phi = \nabla[M(\phi)\nabla(\delta F/\delta \phi) + \xi] \qquad (10.5)$$

where the operator $\delta F/\delta \phi$ denotes the functional derivative and $M(\phi)$ is a volume concentration dependent mobility. The term ξ represents the random fluctuations.

Figure 10.13(a)–(c) show the time evolution of the phase separation morphologies with an initial monomer concentration $\phi_0 = 0.3$ for PIPS. Initially, noises obeying Gaussian distribution are introduced to perturb the homogenous polymer mixture and the isosurface of the polymer-rich phase after perturbation is illustrated by the cyan color. In the simulations, DP increases with time, following the expression $N = 1 + kt$, where k is the reaction rate. Arising from the variation in DP, the phase diagram shifts upward and leftward, as in Figure 10.12(a). After a specific period of time, the system moves into the spinodal region and the solution decomposes into two separate phases, forming a bicontinuous porous structure, portrayed in Figure 10.13(b). Here, the interface of the polymer-rich and the polymer-lean phase is represented by the cyan color; the yellow surface is an isosurface standing inside the polymer-rich phase. With a further increase in time, the ligaments of the bicontinuous structure get coarser because of Ostwald ripening. With a decrease in the initial monomer concentration to 0.2, 0.1, and 0.05 (Figure 10.13(d), (e), and (f)), the following

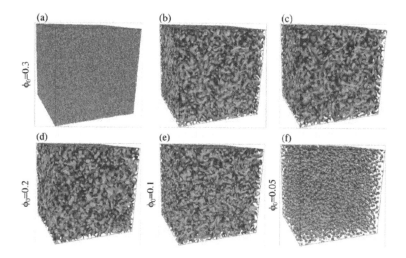

FIGURE 10.13 Porous microstructures resulting from polymerization-induced phase separation for the phase diagram shown in Figure 10.12 (a): (a)–(c) time evolution of the microstructure with an initial polymer concentration $\phi_0 = 0.3$(d), (e), and (f). The microstructures with distinct initial polymer concentrations, $\phi_0 = 0.2$, 0.1 and 0.05, respectively, at the same time. The polymer-rich phase is depicted by the cyan color and the yellow surface is an isosurface standing inside the polymer-rich phase. The simulations are performed on a cubic domain with a size of 400×400×400.

morphological transition is observed: bi-continuous structure → a structure consisting of polymer particles and polymer ligaments → polymer particles. The transition is the results the change in the volume fraction of the polymer-rich phase as a result of the variation in the initial monomer concentration.

Figure 10.14 shows the morphological evolution for TIPS with a linear concentration gradient in the z-dimension (Mino et al. 2015), where $\phi_0 = 0.25$ at the bottom ($z = 0$), and $\phi_0 = 0.35$ at the top ($z = 128$). The simulations consider the phase diagram of the PMMA/cyclohexanol system with a temperature of 50°C. Ascribing to the concentration gradient, we observe an anisotropic membrane structure with bicontinuous morphologies at the bottom and polymer droplets at the top. The cross section of the 3D porous structure [Figure 10.14(e)] demonstrates that a higher ϕ_0 leads to a lower porosity. These results show qualitative agreement with previous experiments (Matsuyama et al. 2000). During the production of membranes, a concentration gradient can be established because of solvent evaporation. In reality, the concentration gradient is likely to be non-linear, in lieu of the linear distribution considered in Figure 10.14. Thus, the anisotropic effect of the membrane structures is actually more pronounced.

Modeling of the phase separation in polymer solutions with more than two components is much more complex than in binary solutions. One challenge is the unknown thermodynamic database. As for binary polymer solutions, the phase diagram for multicomponent polymer solutions can be constructed by using the Flory-Huggins free energy density

$$f\left(\vec{\phi}\right) = k_b T \left[\sum_{i=1}^{K} \frac{\phi_i \ln \phi_i}{N_i} + \sum_{i=1, j=1, i \neq j}^{K,K} \chi_{ij} \phi_i \phi_j\right] \quad (10.6)$$

where $\vec{\phi} = (\phi_1, \phi_2, \cdots, \phi_K)$ is a vector with ϕ_i describing the volume concentration of the i-th species. The parameters K and χ_{ij} depict the number of components and the Flory interaction parameter between the i-th and j-th components, respectively.

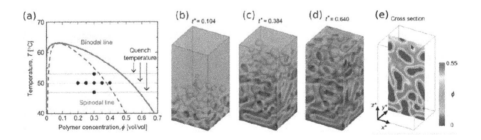

FIGURE 10.14 Anisotropic porous microstructures of thermally induced phase separation for the PMMA/cyclohexanol system (Mino et al. 2015): (a) Phase diagram of PMMA/cyclohexanol.; (b)–(d) Time evolution of the phase separation structure with a linear concentration gradient in the z-dimension. (e) Cross section of the microstructure in (d). The polymer concentration is indicated by the color bar. Reuse with permission, copyright 2015 Elsevier.

Figure 10.15(a) shows the phase diagram of a ternary system: polymer-solute-nonsolvent with $\chi_{ps} = \chi_{ns} = 0$ and $\chi_{pn} = 1.213$, where p, n and s denote polymer, nonsolvent and solvent, respectively. The miscibility gap and the spinodal line are represented by the solid and dashed lines, respectively. The initial setup is that a nonsolvent bath is in contact with a polymer solution (bottom panel, Figure 10.15(a)). The initial polymer concentrations in the polymer solutions are shown by the blue, violet, and green points in the ternary phase diagram. These compositions are outside the miscibility gap, and phase separation cannot take place. Engendering by the diffusion between the nonsolvent phase and the polymer solution, the concentration of the nonsolvent species in the polymer solution increases. As a consequence, the polymer solution moves into the spinodal region and phase separation is induced. This process is called as nonsolvent-induced phase separation. Figure 10.15(b), (c), and (d) illustrate typical phase separation microstructures for different initial concentrations. These microstructures are classified into three categories: nonsolvent droplets, alternating lamellar structures, and polymer droplets, corresponding to the initial concentrations represented by the blue, violet, and green dots, respectively, in the phase diagram. The bottom panel presents the 3D simulation results for these three typical microstructures.

Phase separation in membranes is a strongly time-dependent process in which the morphologies of the forming microstructure continuously change (Wang et al. 2019). Since the microstructure's properties are massively affected

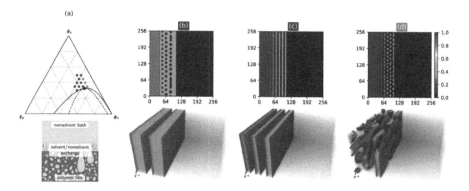

FIGURE 10.15 Phase separation in a ternary system (Tree et al. 2019). (a) Top: Phase diagram for a polymer-solvent-nonsolvent system with the Flory parameter: $\chi_{ps} = \chi_{ns} = 0$ and $\chi_{pn} = 1.213$. Bottom: Sketch for the initial setup in which a polymer solution is in contact with a nonsolvent phase. The miscibility and spinodal lines are depicted by the solid and dashed lines, respectively. The blue, violet, and green circles denote different initial polymer compositions, which lead to three typical microstructures: (b) nonsolvent droplets; (c) alternating lamella; and (d) polymer droplets, respectively. Reused with permission, copyright 2019 Royal Society of Chemistry.

Computational Fluid Dynamics

by its morphology, quantitative characterization methods are highly required when phase separation and complex microstructures in general are investigated. This characterization of forming microstructures reveals fundamental insights and leads to a deeper understanding of both the involved formation mechanism and the resulting material properties. By connecting this knowledge to processing technologies, the linkage between fabrication processes, microstructure morphologies, and effective properties can be realized and an essential contribution to the efforts of designing materials can be made.

Commonly applied characterization methods for membranes act on a broad length scale (Bostanabad et al. 2018). For instance, experimental methods such as capillary flow porometry operate on the macroscopic length scale and enable extraction of the materials properties by simplified theoretical models, such as the Young-Laplace equation. Since no information about pore location and corresponding pore sizes are accessible, it is not surprising, that the information content regarding the morphology of the materials is very limited. In contrast, computational materials science provides powerful opportunities for extracting the morphology of microstructures on the pore-scale by obtaining digital representations of the microstructure and applying sophisticated computer-aided characterization methods (Bostanabad et al. 2018). A full description of the morphology of structures requires geometrical and topological information (Xiong et al. 2016). Geometrical characterization methods include such concepts as the medial axis, which reduces the pore space by projecting the topology of the microstructure on a central skeleton (Blunt et al. 2013), as shown in Figure 10.16. Other methods focus on Euclidean distance maps (EDM) while also combining the advantages of both skeleton and EDM methods is presented in literature (Ley et al. 2018). By classifying the elements of the skeletons and applying graph theoretical models, pores and pore throats can be distinguished, located, and put in neighborly relations at the same time while their pore and pore throat sizes are assigned correspondingly (Yin et al. 2018). The resulting graphs serve as a basis for comprehensive studies on the morphology of microstructures when time-dependent data can be also used as input.

Another very potential approach includes statistical characterization techniques, which are motivated by the fact that heterogeneous materials such as membranes embody a certain degree of randomness (e.g., bicontinuous cluster, orientation of structure ligaments) (Bostanabad et al. 2018). The methods are based on statistical functions, in which two-point correlations are very commonly used. They provide a very comprehensive picture of the salient features, but consequently create an extremely large feature space. Hence, further processing of the correlation function results is conducted by applying principal component analysis (PCA) in order to reduce dimensionality and extract main characteristics of the microstructure in both static and time-dependent data in an unsupervised fashion (Kalidindi & De Graef 2015, Altschuh et al. 2017, Bostanabad et al. 2018).

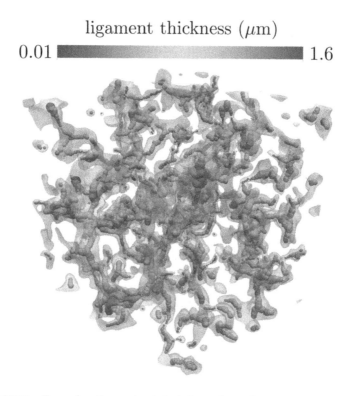

FIGURE 10.16 Example of an extracted skeleton from the structure space of reconstructed X-ray computed tomography data. The presented microstructure is from a cellulose nitrate membrane. The color bar shows the range of detected ligament thicknesses.

10.5 CONCLUSIONS

The recent progress achieved in the thermodynamic and kinetic investigation of liquid – liquid-demixing in polymer solutions as well as the possibility to mimic dry cast operations with a NTFT-Cell and analyze demixing mechanisms via SANS offers a considerable support to membrane scientists and engineers. The combination of these techniques with state-of-the-art microtomography and SEM allows the establishment of a relationship between structural information and the drying process in the form of microstructural maps that are representations of parameters related to microstructure evolution on phase diagrams using contour lines. These parameters (growth rates, demixing mechanisms, droplet size at demixing point) can be acquired by exploring process relevant regions of the phase diagram via SANS experiments. Furthermore, the calculation of drying process paths in dry cast equipment by using CFD calculations and plotting these paths on microstructural maps may allow membrane scientists and engineers to predict the progression of liquid – liquid demixing and model microstructure evolution in polymeric films during industrial evaporative

Computational Fluid Dynamics 301

casting. The whole methodology should be regarded as an opportunity to virtualize the membrane manufacturing process, which is clearly in the scope of Industry 4.0. The ultimate goal is to learn how to design and operate industrial evaporative casting processes in order to control liquid – liquid demixing within polymeric films by accurately achieving the targeted porous structure over the whole membrane area.

REFERENCES

P. Altschuh, Y. C. Yabansu, J. Hoetzer, M. Selzer, B. Nestler, S. R. Kalidindi, Data science approaches for microstructure quantification and feature identification in porous membranes, *J. Membrane Sci.* **540** (2017): 88–97.

S. Barbe, A. Kneer, M. Wirtz, T. Scheper, Prediction of mass transport phenomena in Membrane Adsorber Systems (MAS) using a dual porosity model, Code_Saturne User Conference 2008, R&D EDF Paris, France (2008).

S. Barbe, A. Kneer, M. Wirtz, T. Scheper, *Fluid Dynamics in Membrane Adsorber Systems, DECHEMA Working Party Membrane Technology*, DECHEMA HAUS, Frankfurt am Main, Germany (2009).

M. J. Blunt, B. Bijeljic, H. Dong, O. Gharbi, S. Lglauer, P. Mostaghimi, A. Paluszny, C. Pentland, Pore-scale imaging and modelling, *Adv. Water Resour.* **51** (2013): 197–216.

R. Bostanabad, Y. Zhang, X. Li, T. Kearney, L. C. Brinson, D. W. Apley, W. K. Liu, W. Chen, Computational microstructure characterization and reconstruction: Review of the state-of-the-art techniques, *Prog. Mater. Sci.* **95** (2018): 1–41.

J. W. Cahn, On spinodal decomposition, *Acta Metall.* **9** (1961): 795–801.

P. J. Flory, Thermodynamics of high polymer solutions, *J. Chem. Phys.* **9** (1941): 660–661.

S. R. Kalidindi, M. De Graef, Materials data science: Current status and future outlook, *Annu. Rev. Mater. Res.* **45** (2015): 171–193.

J. F. Kim, J. H. Kim, Y. M. Lee, Thermally induced phase separation and electrospinning methods for emerging membrane applications: A review, *AIChE J.* **62** (2016): 461–490.

A. Kneer, M. Wirtz, S. Yurtsever-Kneer, S. Barbe, A. August, Modern times need enlightened innovation and sophisticated materials, *Galvanotechnik* **110–4** (2019): 712–719.

A. Ley, P. Altschuh, V. Thom, M. Selzer, B. Nestler, P. Vana, Characterization of a macro porous polymer membrane at micron-scale by confocal-laser-scanning microscopy and 3D image analysis, *J. Membrane Sci.* **564** (2018): 543–551.

H. Matsuyama, M. Yuasa, Y. Kitamura, M. Teramoto, D. R. Lloyd, Structure control of anisotropic and asymmetric polypropylene membrane prepared by thermally induced phase separation, *J. Membrane Sci.* **179** (2000): 91–100.

M. Metze, Untersuchung zum Entmischungsverhalten und zur Membranbildung von Cellulosenitratmembranen, Diplomarbeit, Göttingen University (2010).

M. Metze, Untersuchungen zur Bildung von porösen Membranen aus Cellulosederivaten nach dem Verdunstungsverfahren, Dissertation, Hannover University (2014).

M. Metze, S. Barbe, A. Reiche, A. Kesting, R. Schweins, A Neutron-Transparent Flow-Through Cell (NTFT-CELL) for the SANS investigation of microstructure evolution during industrial evaporative casting, *J. Neutron Res.* **19** (2017): 177–185.

M. Metze, S. Barbe, A. Reiche, T. Scheper, The application of flory–huggins–thermodynamics for industrial membrane manufacture via evaporative casting, *Procedia Eng.* **44** (2012): 1460.

M. Metze, A. Reiche, D. Melzner, S. Barbe, The influence of surfactants on the formation of macroporous membranes via evaporative casting, Deutsche Physikalische Gesellschaft, Regensburg, Germany (2013).

Y. Mino, T. Ishigami, Y. Kagawa, H. Matsuyama, Three-dimensional phase-field simulations of membrane porous structure formation by thermally induced phase separationin polymer solutions, *J. Membrane Sci.* **483** (2015): 104–111.

S. Musa, O. Richter, M. Balsam, A. Kneer, S. Barbe, Macroporous films from acetylated lignin and cellulose as precursors for smart coatings based on regenerated wood, *Eur. J. Wood Wood Prod.* **76** (2018): 1363–1366.

A. Pastor, S. Barbe, Disposable chromatography for large-scale biomanufacturing in single-use technology in biophamaceutical manufacturing, DOI: 10.1002/9780470909997.ch25, Wiley, Hoboken, NJ (2011).

D. R. Tree, L. F. Dos Santos, C. B. Wilson, T. R. Scott, J. U. Garcia, G. H. Fredrickson, Mass-transfer driven spinodal decomposition in a ternary polymer solution, *Soft Matter* **15** (2019): 4614–4628.

F. Wang, P. Altschuh, L. Ratke, H. Zhang, M. Selzer, B. Nestler, Progress report on phase separation in polymer solutions, *Adv. Mater.* **31** (2019): 1–14.

C. Wenning, S. Barbe, D. Achten, A. M. Schmidt, M. C. Leimenstoll, Prediction of initial miscibility for ternary polyurethane reaction mixtures on basis of solubility parameters and flory-huggins theory, *Macromol. Chem. Phys.* **219** (2018a): 1–8.

C. Wenning, J. Noe, S. Barbe, M. C. Leimenstoll, Evidence of liquid-liquid demixing during bi-soft segment polyurethane prepolymerization, *Macromol. Res.* **26** (2018b): 1–4.

Q. Xiong, G. B. Todor, P. J. Andrey, Review of pore network modelling of porous media: Experimental characterisations, network constructions and applications to reactive transport, *J. Contam. Hydrol.* **192** (2016): 101–117.

X. Yin, H. Aslannejad, E. T. de Vries, A. Raoof, S. M. Hassanizadeh, Droplet imbibition into paper coating layer: Pore-network modeling simulation, *Transport Porous Med.* **125** (2018): 239–258.

11 Towards Energy-Efficient Reverse Osmosis

Mohamed T. Mito
School of Engineering and Applied Science
Aston University, Birmingham, UK

Philip Davies
School of Engineering, University of Birmingham
Edgbaston, Birmingham, UK

11.1 INTRODUCTION

Reverse Osmosis (RO) represents the state of the art in desalination technology today. Owing to its affordability and relatively low energy consumption, it has clearly overtaken thermal desalination technologies such as multistage flash (MSF) and multiple effect distillation (MED) (Weaver, Howells, & Brown, 2018). The energy usage of a desalination process is measured in terms of its specific energy consumption (SEC) $[kWh/m^3]$, which is defined as the energy consumed per volume of water produced. Though the SEC of RO desalination is much lower than that of thermal desalination, it is still several times higher than for other types of water treatment such as standard treatment of surface water by coagulation and filtration (Voutchkov, 2018a). The energy usage of RO desalination remains significant because it accounts for about half the cost of the product water. Moreover, since the energy to drive desalination is typically obtained from burning of fossil fuels, RO has a high carbon footprint compared to conventional water treatment. For these reasons, there is much interest in reducing the energy consumption of RO desalination to make it even more affordable and to reduce its environmental impact.

It is possible to define an ideal minimum SEC in the thermodynamic sense for any desalination process, including RO. The thermodynamic minimum SEC is independent of the technology used to carry out the process, applying in principle to thermal (e.g. MED, MSF), pressure-driven (e.g., RO), and electrical technologies (e.g. electrodialysis) alike. No matter how much effort is put into developing these technologies, they will never beat the thermodynamic ideal minimum SEC_{ideal} – they can only approach it. Although any one of these technologies could in theory approach the minimum, in practice some are able to approach it much more closely than others. In particular, RO technology has been quite successful in gradually approaching the ideal minimum SEC.

The textbooks on chemical thermodynamics describe the concept of chemical potential as the work needed to remove a unit of a certain species from a solution (Smith, 2004). When this general concept is applied to the case of removing water from a solution of salt in water, chemical potential becomes equivalent to SEC_{ideal}. Moreover, there is a fundamental relation between chemical potential and osmotic pressure, such that the SEC_{ideal} is also equivalent to the osmotic pressure of the salt solution to be desalinated. The only difference lies in the units conventionally used to measure these two quantities. Thus, for standard seawater containing 3.5% salt, the osmotic pressure is 26 bar=2.6 MPa resulting in a SEC_{ideal} of 2.6/3.6=0.72 kWh/m^3. This means that no future desalination technology will ever be able to extract a cubic meter of freshwater from the sea while consuming less than 0.72 kWh of energy. An exception arises if the seawater is taken from an area of the sea, such as an estuary, that is less salty than standard seawater. This is because the osmotic pressure (and thus SEC_{ideal}) varies more or less in proportion to the salt concentration. Therefore, the SEC_{ideal} for estuarine water may be less than 0.72 kWh/m^3.

There are many reasons why SEC exceeds the ideal minimum in real desalination plants. A fundamental reason relates to the recovery ratio (r), i.e., the fraction of freshwater that is recovered from the incoming seawater. The theoretical concept of chemical potential applies to a hypothetical situation in which a small volume of freshwater is recovered from an infinite reservoir of seawater, such that r is virtually zero. This is not the case in practice for several reasons. For example, the seawater has to be pumped onshore to reach the desalination plant. This incurs costs of intake pipework and power supplied to the pump. To avoid pumping an excessive amount of seawater, the recovery ratio of most desalination plants is kept below 50%. As water is extracted from a finite amount of seawater, the osmotic pressure must go up and the corresponding SEC_{ideal} for a recovery ratio $r>0$ increases according to (Qiu & Davies, 2012a):

$$SEC_{ideal} = SEC_{ideal_{r0}} \left(\frac{1}{r} \right) \ln \left(\frac{1}{1-r} \right) \tag{11.1}$$

where $SEC_{ideal_{r0}}$ is the value of SEC_{ideal} at $r = 0$. (This equation is obtained by a process of mathematical integration and is based on the assumption that the osmotic pressure increases in proportion to the salt concentration, which is quite accurate, except at high concentrations). For a recovery ratio of 50%, Equation 11.1 gives $SEC_{ideal} = 1$ kWh/m^3, whereas (as shown subsequently) real RO desalination plants have an SEC of more than twice this value.

A second reason for this disparity between real and ideal SEC relates to the membrane area and capital cost. In the RO process, permeate water is driven through a selective RO membrane by means of a transmembrane pressure Δp that must exceed the osmotic pressure $\Delta \pi$. The difference ($\Delta p - \Delta \pi$), called the net driving pressure, determines the flux J_w of water through the RO membrane (i.e., the flow per unit membrane area) according to the equation:

$$J_w = A(\Delta p - \Delta \pi) \tag{11.2}$$

Here, A is the permeability – a property of the membrane that depends on the technology used in its fabrication. The value of A is always finite, such that the net driving pressure has to be above zero to maintain a certain flux. The flux should not be too small, otherwise this results in a large area of membrane being needed for a desired flow of product water, with associated increase in the cost of the desalination plant. On the other hand, whenever ($\Delta p - \Delta \pi$) exceeds zero substantially, there is an energy penalty in SEC because the thermodynamic ideal case is to apply a pressure across the membrane *only just* exceeding $\Delta \pi$. This shows that there is a trade-off between minimizing SEC and minimizing the cost of the RO plant.

To obtain a better trade-off in this respect, one approach would be to select a membrane with higher value of permeability A. Indeed, RO membrane manufacturers have gradually increased A, such that fluxes from RO membranes today are in practice nearly twice the values possible 20 years ago. Here, however, there is another trade-off to consider that relates to the ability of the RO plant to reject salt. The flux J_s of salt through an RO membrane is, unlike the flux of water, virtually independent of the transmembrane pressure. It depends instead on the difference Δc in concentration across the membrane.

$$J_s = B\Delta c \tag{11.3}$$

where B is the salt diffusion coefficient which, like A, is a property of the membrane. It is not desirable for J_s to be too high, because this would result in too much salt reaching the product water, making it unsuitable for drinking. We would then say that the RO plant has not rejected enough salt. The difficulty in selecting membranes with higher A, however, is that such membrane also tend to have higher B. As such, they are not sufficiently selective in excluding salt and allowing water to pass through. This is referred to as the selectivity-permeability trade-off that is a general feature of membrane separation processes (Werber, Deshmukh, & Elimelech, 2016).

In summary, we have introduced the idea of an ideal minimum SEC for a RO desalination plant and highlighted some fundamental trade-offs to consider when aiming towards this ideal minimum. Not only do we wish to minimize SEC of the plant, we also wish to minimize its capital cost, increase freshwater recovery, and increase salt rejection. These four objectives tend to conflict. In the rest of this chapter, we explore in more detail the reasons for nonideal SEC and the advances that are being made to lower SEC in the various areas of technology that are used in desalination plants. In the next section, we take a closer look at the SEC of real RO plants.

11.2 ENERGY CONSUMPTION OF RO PLANTS

The SEC of RO plants is constantly decreasing. In the 1990s, it ranged from 5 to 10 kWh/m^3 (Amy et al., 2017). Today, the industry average SEC has been reduced to 3.1 kWh/m³, with modern plants achieving between 2.5 and 2.8 kWh/m³ (Voutchkov, 2018a). Although these values are much better than those of the 1990s, they are still more than twice the ideal value of 1 kWh/m³. It is therefore important to investigate where and how energy is consumed in a typical RO plant, as this will help in understanding what further progress can be made to reduce the SEC.

11.2.1 ENERGY BREAKDOWN

The energy consumption of a RO plant depends on its overall design and the design of its components. Other factors such as age, mode of operation, and feed water conditions also have an influence. An RO plant consumes energy in the form of electricity. The following breakdown of electrical energy consumption has been provided by Qasim et al. for a typical seawater RO desalination plant (Qasim, Badrelzaman, Darwish, Darwish, & Hilal, 2019). Figure 11.1 shows that the RO system uses 71% of the total energy, while the remaining 29% is used by the intake, pretreatment, delivery, and other auxiliary systems.

11.2.1.1 Intake Systems

The intake system normally consists of a low-pressure pump drawing seawater through a submarine pipe and inlet grating. It has to collect seawater without ingesting wildlife and litter – and without endangering marine traffic or bathers. The inlet should be sufficiently distant from outlet pipes of other plants or the desalination plant itself, otherwise it would suck in contaminated seawater. Typically, the intake pump consumes about 5% of the total plant power. This depends on the pump selected and the pipe design, which in turn is influenced by the siting of the plant and local bathymetry.

11.2.1.2 Pretreatment

Pretreatment is the second most energy intensive process in the RO plant, after the RO system itself. It can include conventional methods such as

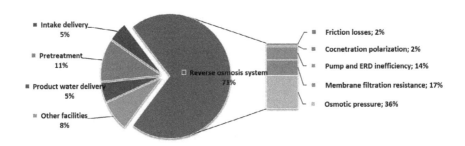

FIGURE 11.1 Energy consumption distribution of a typical RO plant (Qasim et al., 2019).

coagulation, flocculation, sedimentation, and media filtration or advanced methods such as ultrafiltration. Pretreatment uses about 11% of the total plant energy consumption to get the water to the required standard prior to admission into the RO system. The pretreatment system is crucial to prevent excessive fouling, membrane damage, and to ensure membrane longevity.

11.2.1.3 Product Water Delivery

The permeate water must be post-treated to meet the drinking water standards and then pumped into the water distribution network. Post-treatments may include boron removal, remineralization, and chlorine removal. This process requires approximately 5% of the entire plant energy consumption.

11.2.1.4 Other Processes and Services

Commercial RO plants have a compound design that relies on multiple subsystems and processes that contribute to the water production cost and overall energy consumption. These processes, which include chemical dosing, membrane cleaning, concentrate discharge, waste disposal, as well as plant lighting, heating, instrumentation, and routine maintenance activities, could make up nearly 8% of the overall RO plant energy consumption.

11.2.1.5 RO System

The major energy load in the desalination system is the high-pressure pump (HPP) that drives the RO process. The HPP normally provides a pressure in the range 50–70 bar. Around 50% of this energy is required to overcome essential osmotic pressure difference; 20% is lost to inefficiencies in the HPP and the energy recovery device (ERD); 2.5% is lost through flow friction in the concentrate and permeate channels; and 2.5% is lost owing to concentration polarization. The remaining 25% is consumed by the HPP to overcome:

- Pressure losses corresponding to hydrodynamic losses in the membrane. This energy depends on the permeability A (see section 11.1). A higher value of A would result in lower energy being required.
- Pressure losses corresponding to the increase of salt concentration, and therefore osmotic pressure, from the inlet to the outlet of the RO modules. Because the applied pressure must at all positions exceed the maximum osmotic pressure occurring at the outlet, there is an excess pressure at the inlet. The resulting wastage of energy is discussed further in section 11.4.1.

Because RO systems consume the most energy, the rest of this chapter examines more closely the different factors affecting their energy consumption, beginning with RO membranes and the prospects to improve plant performance through use of new and better membrane technology (Karabelas, Koutsou, Kostoglou, & Sioutopoulos, 2018).

11.3 RO MEMBRANES AND PLANT ENERGY CONSUMPTION

Any improvement in membrane permeability and water flux should decrease the pressure required by the HPP, thus reducing the energy requirement of the RO plant. Membranes used in RO plants are nonporous, semipermeable membranes that separate fresh water from seawater. Seawater is pressurized above the osmotic pressure, which leads to freshwater permeating through the membrane from the high-salinity (brine) side to the low-salinity (permeate) side. As such, salt is retained on the high-salinity side and seawater concentration increases along the membrane. Accordingly, two streams are formed – a high-salinity feed water (concentrate or brine) stream and a low-salinity fresh water (permeate) stream (Voutchkov, 2013).

RO membranes are efficient in blocking all suspended solids; however, dissolved solids could permeate along with fresh water through the membrane. Ideally, water transport rate through the membrane is higher than salt passage rate by several orders of magnitude, which is how seawater is desalted. The solution–diffusion model for nonporous semipermeable membranes governs the mechanism of water and salt transport. The rate of salt and water passage through the membrane defines the performance of a RO membrane.

The performance of an RO membrane depends on its material, which determines performance characteristics such as permeability, salt rejection, fouling, and reactivity (Goh, Matsuura, Ismail, & Hilal, 2016). Research and development for RO membranes target increasing permeability and salt rejection by using innovative materials for the active layer, which would directly reduce the applied hydraulic pressure, and thus reduce the energy requirement and water production cost. Other than decreasing the energy consumption for the same permeate production, increasing membrane permeability would lead to increased flux and recovery for the same energy consumption.

Increasing membrane permeability tends to increase salt passage owing to the trade-off between polymer chain stiffness and interchain spacing, which presents the selectivity–permeability trade-off as an inherent problem for RO membrane development (Goh et al., 2016). For single-stage RO plants, the maximum energy saving is constrained by the brine osmotic pressure, which dictates the minimum hydraulic pressure and the minimum SEC. A recent study quantified the drop in SEC resulting from increased membrane permeability. The SEC was reduced by 3.7% for an increase in water permeability from the current commercial average of 2 L/m^2 h bar to 10 L/m^2 h bar. If the permeability increased further from 10 L/m^2 h bar to 100 L/m^2 h bar, an additional 1% reduction in energy consumption would occur (Park, Kamcev, Robeson, Elimelech, & Freeman, 2017). This suggests that there is a saturation in SEC reduction with increasing permeability, and that further increase in permeability could be considered as a negative aspect, owing to the eventual drop in membrane selectivity and salt rejection. Moreover, higher membrane flux would increase the effect of concentration polarization and induce higher membrane fouling. Accordingly, targeting membrane development, as a means of reducing energy consumption requires a compromise that considers the selectivity-permeability trade-off, membrane longevity and durability, membrane fouling, and capital cost. Despite technical development of RO membranes, the water permeability is limited to approximately 2 L/m^2 h bar for a commercial seawater RO membrane (Tang, Wang, & Hélix-Nielsen, 2016).

Common membrane types include the thin-film composite and thin-film nanocomposite membranes. They use either polyamide or cellulose acetate for the thin-film active layer.

11.3.1 Thin-Film Composite Membranes

Currently, thin-film composite membranes are widely used in the desalination industry. They consist of three parallel layers: 1) a nonporous semipermeable thin film made of either cellulose acetate or aromatic polyamide, 2) a microporous polyethersulfone support layer, and 3) a reinforcing fabric (Khorshidi, Thundat, Fleck, & Sadrzadeh, 2016). The structure of a thin-film composite membrane is illustrated in Figure 11.2.

The thin-film semipermeable layer is the active layer of the RO membrane, which gives the membrane its salt rejection characteristics. It has a nonporous random molecular structure, such that water and solute molecules permeate through the membrane active layer by diffusing and traveling on a multidimensional curvilinear path through the polymer matrix. The two support layers beneath the thin-film polyamide layer support the membrane structure, integrity, and durability. For the first generation of RO membranes, the thin-film active layer was composed of cellulose acetate. However, polyamide based thin-film offers much higher permeability, better salt rejection, and stability over a wide pH range (2–11). Consequently, they have mostly superseded the cellulose acetate membranes.

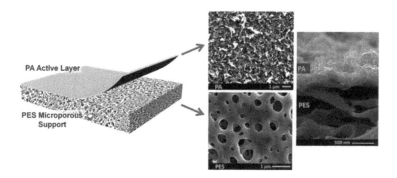

FIGURE 11.2 Schematic illustration of a thin-film composite structure and micrographs of the top and cross section – Reused with permission (Khorshidi et al., 2016).

11.3.2 Nanocomposite Membranes

Nanocomposite membranes are categorized into two types. The first type is thin-film nanocomposite membranes that include inorganic nanoparticles embedded into a conventional thin-film polymer layer, shown in Figure 11.3 (a). The second type is the carbon nanotube membrane, which consists of a structured porous film that includes a densely packed array of nanotubes, shown in Figure 11.3 (b).

Nanocomposite membranes offer a high specific permeability compared to conventional thin-film composite membranes, which translates to higher permeate flow rate for the same surface area and pressure. In addition, nanocomposite membranes offer similar or lower fouling rates than thin-film composite for the same operating conditions. Research in RO membrane is heading towards transforming membrane structures to be made completely of tubes having uniform size; this has the potential to increase water production per unit area up to 20 times compared to current commercial RO membranes (Voutchkov, 2013).

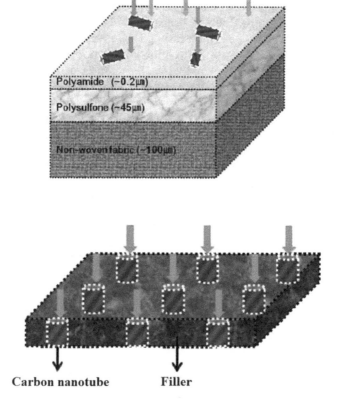

FIGURE 11.3 a) Schematic illustration of a thin-film nanocomposite membrane; and b) a vertically aligned carbon nanotube membrane – Reused with permission (Ahn et al., 2012).

11.3.3 Emerging Membrane Types

Innovative technologies such as biomimetic and graphene-based membranes show potential to deliver high permeability and salt rejection that could represent a breakthrough in decreasing the RO process SEC (Tang et al., 2016). Biomimetic membranes take inspiration from the biological water channel aquaporin. Aquaporin is a membrane protein structured from tetramers of four identical monomers, each monomer containing a water pore. Aquaporin is proposed for RO membranes owing to its high permeability, which originated from its structure. It has an hourglass structure with a wide entrance and a narrow hydrophobic centre (see Figure 11.4 (a)). The mechanism of rejection

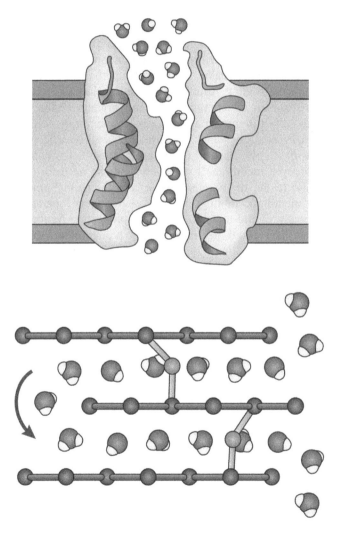

FIGURE 11.4 a) Aquaporin protein embedded in an impermeable film; and b) Graphene based framework – Reused with permission (Werber et al., 2016).

for aquaporin membranes is based on size and charge selection. The channel excludes molecules larger than 0.28 nm in diameter, which is ideal for passing water molecules (0.275 nm in diameter). In addition, the positively charged arginine side chain prevents the permeation of charged solutes (Werber, Osuji, & Elimelech, 2016).

Graphene-based membranes are an emerging technology that has yet to be developed commercially, owing to manufacturing challenges. Graphene membranes are structured as nanoporous graphene or graphene-based frameworks; both offer selectivity based on size selection. Currently, graphene-based frameworks shown in Figure 11.4 (b) are suggested for nanofiltration and not recommended for RO application.

11.3.4 Membrane Fouling

Membrane fouling, the build-up of substances on the membrane surface, has a direct effect on membrane performance and energy requirements. Membrane fouling can be identified as scaling resulting from the accumulation of precipitated salts; biological fouling resulting from admission and growth of microorganisms that form an adherent biofilm; and organic fouling resulting from adsorption of dissolved organic matter. Excessive membrane fouling affects permeability and salt rejection, which, in turn, has a direct effect on the energy requirement owing to the necessity of higher pressure to maintain the same water flux. Moreover, membrane fouling leads to the need for excessive chemical cleaning, shortening the membrane life, and increasing running cost. Factors that affect membrane fouling are feed water quality, operation parameters (i.e., flux and pressure) and membrane surface properties. As feed water quality is site dependent and could be improved by pretreatment processes, current research is focused on developing membrane materials with low fouling propensity to minimize the effect of fouling on energy consumption. Propensity to fouling can be reduced by increasing the membrane surface hydrophilicity, using smoother membrane surfaces, and introducing strong electrostatic repulsion between membrane surface and charged foulants. RO membranes with smooth, hydrophilic surfaces have shown resistance to biofouling and biofilm formation, leading to lower energy requirements and requiring less frequent chemical cleaning (Goh et al., 2016).

11.4 CONFIGURATIONS OF RO SYSTEM TO MINIMIZE SEC

Although improvements to RO membranes have helped to increase plant output and reduce SEC over the years, future prospects to reduce SEC through better membrane technology may be bottlenecked by the configuration of the RO plant (Elimelech & Phillip, 2011). Innovations in plant configurations are important to reduce SEC and to take advantage of emerging membrane technologies, as discussed in this section. Improvements in RO plant configuration can help optimize the plant energy consumption, maximize operation flexibility, and reduce capital cost. This section describes alternative configurations that target the reduction of RO plant SEC while simultaneously benefiting the other areas.

11.4.1 MULTISTAGE RO SYSTEMS

Theoretically, the minimum energy required for desalination is the energy needed to raise the feed water pressure to the osmotic pressure across the membrane. In each pressure vessel, fresh water is gradually removed from the seawater side along the RO module. This leads to a salt concentration gradient along the length of the membrane, whereby concentration can nearly double from the inlet value (Qiu & Davies, 2012a). Thus, the minimum feed inlet pressure has to match the outlet osmotic pressure of the concentrate to ensure seawater permeates at all points along the membrane. Figure 11.5 (a) shows that a significant fraction of the inlet pressure is lost owing to the longitudinal concentration gradient.

Multistage configurations are used to reduce the energy loss caused by the longitudinal concentration gradient. The concentrate of each stage is boosted by an inter-stage pump and fed to the subsequent stage. As shown in Figure 11.5 (b), the first stage operates at a low pressure, equivalent to its concentrate salinity. Then, the concentrate of the first stage is fed to the subsequent stage

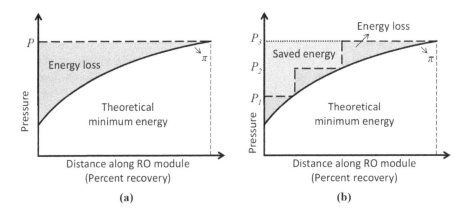

FIGURE 11.5 a) The relationship between pressure energy and osmotic pressure along the length of the membrane module. The theoretical minimum energy required for permeation along the module is represented by the area under the osmotic pressure line π. Since the inlet pressure must match the osmotic pressure at the vessel outlet, the total energy consumption is represented by the area of the rectangle. The difference between the two shapes is the energy lost owing to the longitudinal concentration gradient. b) A multistage RO system with interstage boosting is presented. The first stage is operated at a low pressure that matches the feed salinity of the first stage outlet. This provides a significant energy saving compared to a single stage system. [Adapted with permission from (Qiu & Davies, 2012a)].

at a higher pressure according to the increased salinity by using a booster pump. This eliminates the need for the first stage pump to raise the pressure to the final osmotic pressure and reduces energy lost because of the longitudinal concentration gradient (Qiu & Davies, 2012a).

By using multistage configuration, the system recovery can be maximized without exceeding the maximum recovery limit per RO element. This reduces the total volume of concentrate produced by the RO process and increases production capacity for the same volume of seawater supplied.

To maintain a uniform feed flow for each stage, the number of parallel pressure vessels for each stage decreases downstream. Traditionally, a 2:1 staging ratio is used for a two-stage system (Voutchkov, 2013). Although multiple RO stages improve the overall system recovery and SEC, additional stages increase the plant capital cost. The optimum number of stages is different for each plant, depending on the source water quality, required permeate quality, production capacity, and cost of energy and equipment.

11.4.2 Multipass RO Systems

A multiple-pass RO configuration (Figure 11.6) is used when the source water salinity is relatively high, such that the target permeate quality cannot be achieved in a single pass. It is designed to treat the permeate multiple times, with each RO pass improving the permeate until the target quality is reached. However, multipass systems have high capital cost and high SEC; and they produce less permeate than a single-pass system processing the same volume of feed water. In a conventional two-pass system, the feed water is initially treated through a SWRO train; then, the permeate water from the first pass is treated though a second pass of brackish water RO trains. The main drawback of multipass systems is the reduced system recovery and increased SEC compared to single-pass systems, specifically because a portion of the permeate of the first pass is transformed to concentrate in the subsequent pass.

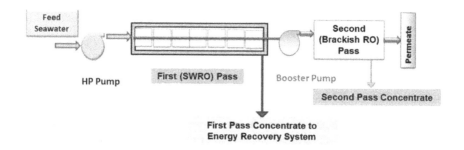

FIGURE 11.6 Schematic of a two-pass RO system. The permeate of the first pass is processed by the brackish water second pass to achieve the target permeate quality [Adapted with permission from (Voutchkov, 2018b)].

Incorporating high-productivity RO modules would enhance the SEC of multipass systems. By installing high productivity modules in the first pass, the required feed pressure and energy consumption are reduced significantly. Owing to the low salt rejection of these modules, a second permeate pass would be required to achieve the target quality. Although the second permeate pass would require more modules to handle the increased flow, the reduction in energy consumption outweighs the increase in capital cost (Peñate & García-Rodríguez, 2012).

11.4.3 Split-Partial Two-Pass Systems

The split-partial two-pass configuration represents an upgrade to conventional multipass systems in terms of cost management and energy reduction. In the split-partial two-pass system, the second pass handles around 50 to 75% of the permeate produced by the first pass, such that the remainder of the permeate produced by the first pass is blended with the permeate of the second pass without additional treatment (Figure 11.7). Because the flowrate to the second pass pump is reduced, the energy consumption of the booster pump is reduced by 25 to 50%, compared to a conventional two-pass system (Voutchkov, 2018a). Accordingly, this configuration improves the SEC, lowers capital costs by reducing the size of the second pass and help achieves the target water quality.

Moreover, the concentrate of the second pass has a significantly lower salinity than the initial seawater feed. This represents an opportunity to recirculate the

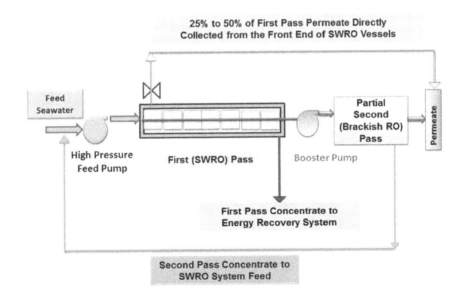

FIGURE 11.7 Split partial two-pass system – Reused with permission (Voutchkov, 2018a).

concentrate and mix it with the feed of the first pass, which would reduce feed salinity, increase system recovery, and decrease required pressure. Generally, the split partial two pass system requires 12% to 20% less energy than a conventional two-pass system.

11.4.4 Hybrid Membrane Configuration

Hybrid membrane configuration is a design strategy for combining RO modules of different productivity and salt rejection within the same pressure vessel to distribute evenly the permeate flux and optimize the energy available for each module along the pressure vessel.

In conventional SWRO systems, feed water is admitted at the front of the pressure vessel, where the concentrate and permeate are collected from the opposite end. As a result, the first membrane element is exposed to the entire flowrate and pressure delivered by the HPP, thus it operates at a higher recovery ratio and higher flux compared to the following membrane elements. This causes an uneven flux distribution. Ideally, if an even flux distribution were achieved for a standard seven-element vessel, each element would produce 14.3% of the vessel production capacity. In reality, the first element produces around 25% of the vessel production capacity, with the last element producing only 6% to 8% (Figure 11.8). The uneven distribution of permeate production is a result of the increasing feed salinity and associated osmotic pressure, as explained previously in Section 11.4.1.

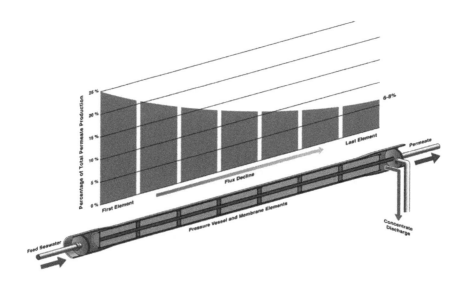

FIGURE 11.8 Permeate flux distribution for a conventional pressure vessel packed with seven RO modules of the same type – Reused with permission from (Voutchkov, 2018a).

Since the energy dissipated in each element is proportional to the flow and pressure, and the first RO element produces around 25% of the permeate flow, this means it uses 25% of the pressure energy available for the entire vessel. This affects the energy efficiency of the subsequent modules as they are underworked and not operating at their optimum efficiency.

Regarding membrane fouling, the first element handles the largest portion of feed flow, making it more prone to fouling than subsequent elements owing to higher exposure to particulate and organic foulants. Thus, its permeability decreases, such that the concentrate from the first element is directed to the second element. As a result, the second element receives higher salinity feed at lower pressure compared to the first element, which reduces its permeate flux and increases the concentration polarization across its surface. The subsequent modules are subjected to higher feed salinity, lower pressure and higher concentration polarization, decreasing overall water production. Mineral scaling caused by supersaturation and crystallisation of sparingly soluble salts (such as calcium carbonate) tends to be more severe in the tail elements.

Higher energy efficiency and more favorable operation conditions, i.e., lower fouling rate and better membrane longevity, could be achieved if the feed flow were distributed evenly across the seven RO modules. Hybrid-membrane configuration targets obtaining an even flux distribution across the membrane by using three different types of RO modules in the same pressure vessel, each with different permeability. As shown in Figure 11.9, the first element in the pressure vessel is a low permeability element with high salt rejection. The low permeability element produces 14% to 18% of the pressure vessel capacity compared to 25% for a conventional system. Thus, higher output is obtained from subsequent modules along the pressure vessel. The second RO element delivers average permeability and salt rejection, producing around 14% to 16% of the

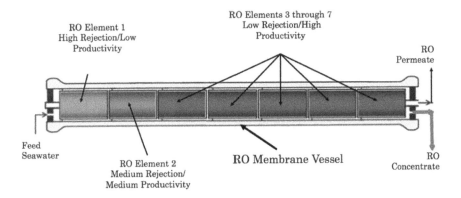

FIGURE 11.9 Hybrid membrane configuration. The different types of RO modules used are shown in terms of their position in the pressure vessel – Reused with permission from (Voutchkov, 2018a).

total permeate flow. The subsequent five modules are the standard high permeability and lower salt rejection modules. This configuration, referred to as 1-1-5 combination, delivers a more even pressure and flux distribution along the pressure vessel, reducing the fouling rate across the membrane and reducing energy consumption by 5% to 15%.

11.4.5 TIME-VARYING PRESSURE SYSTEMS – BATCH RO SYSTEMS

As discussed in section 11.4.1, multistage configurations reduce energy loss caused by the longitudinal concentration gradient. This is achieved by spatially dividing the RO process into multiple stages that operate at different pressures to reduce energy loss. Batch RO desalination shares the same objective of multistaging; however, it uses temporal separation to eliminate the longitudinal concentration gradient and the energy loss associated with it. In batch RO systems, a single vessel is operated at time-varying pressure, synchronized with the increase in salinity and osmotic pressure. This process allows operation at the theoretical minimum energy for desalination by providing only the required pressure to overcome the osmotic pressure across the membrane. This could also be achieved with multistaging; however, an infinite number of stages would be required (Qiu & Davies, 2012a). Batch desalination has not yet been tested for seawater desalination; however, it holds potential for SWRO if high recovery is required. Two approaches have been described for time-varying brackish water RO: Batch RO, and *Closed Circuit Desalination* (CCD, also referred to as semi-batch RO).

The first of the aforementioned approaches, batch RO, was introduced by Davies (Davies, 2011). The operating principle of batch RO is presented in Figure 11.10. The system consists of a pressure vessel containing saline water, whereby a cylinder piston pressurizes it gradually, providing a work input. The piston pushes the fresh water through the membrane until the permeate is expelled. Owing to the increasing concentration on the membrane surface, a 'stirrer' is included to minimize the effect of concentration polarization. The basic operating principle is shown in Figure 11.11 (Qiu & Davies, 2012b). A circulation pump is included to feed the pressurized concentrate back to the

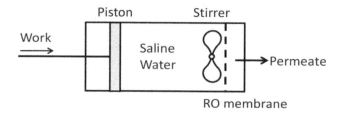

FIGURE 11.10 Operation principle of a batch RO system – Adapted from (Qiu & Davies, 2012a).

Towards Energy-Efficient Reverse Osmosis

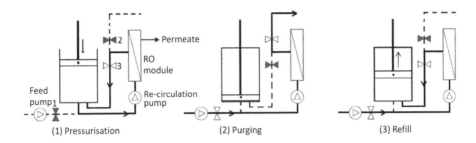

FIGURE 11.11 The three stages of a batch RO system: 1) pressurisation, 2) purging, and 3) refill. These three stages are engaged by controlling the valves. The solid lines indicate water flow and the dashed lines indicates a blocked path – Adapted from (Qiu & Davies, 2012b).

cylinder. As such, the batch mode operation is very useful to provide energy recovery from the concentrate without using an ERD. Although Figure 11.11 shows a piston driven by a rod, in practical arrangements, a free piston that is driven by feed water from a high-pressure pump (HPP) may be used instead (Ali et al., 2019).

A similar brackish water system named *Closed Circuit Desalination* (CCD, also referred to as semi-batch RO) was proposed by Efraty (Efraty, 2012). Its operating principle (Figure 11.12) is generally similar to the aforementioned batch RO system. However, the feed is pressurized and recirculated through the vessel using a circulation pump. Once the target recovery ratio is reached, the process stops, and the vessel is decompressed through a valve allowing the brine to be replaced by a new batch. The efficiency of CCD is predicted to be lower than of batch RO because the feed is mixed with the recirculated brine

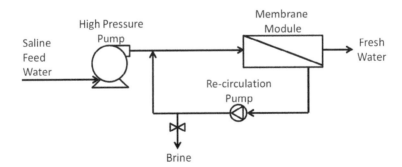

FIGURE 11.12 Schematic of a RO system incorporating Closed Circuit Desalination – Adapted from (Qiu & Davies, 2012a).

320 Membrane Desalination

11.5 PUMPS

As the HPP provides all the power to the RO system, its efficiency has a direct effect on SEC. Every effort should be made to maximize HPP efficiency. This goal is mainly the responsibility of pump and motor manufacturers. Still, the responsibility of RO plant designers and developers is to select the commercial equipment that optimally suits the needs of their system. Two common selection approaches are described next.

11.5.1 HIGH FLOW CAPACITY CENTRIFUGAL PUMPS

Currently, a common approach used by RO plant developers to reduce energy consumption and water production costs is to use a large capacity centrifugal pump to feed multiple RO trains instead of using several smaller pumps to operate individual trains. This approach is referred to as *pressure center design*. Larger pumps save energy because the efficiency of a multistage centrifugal pump increases with flow capacity. The HPP efficiency for a conventional system that includes an individual pump for each RO train usually ranges between 80% and 83%. By comparison, if a large centrifugal pump is used to operate two identical RO trains, the efficiency reaches 85%. This operating principle is implemented at Ashkelon seawater desalination plant in Israel, in which two high-pressure pumps are used to operate 16 SWRO trains. This design is reported to have 88% efficiency (Voutchkov, 2018a).

11.5.2 POSITIVE DISPLACEMENT PUMPS

The required operating pressure for RO plants can vary, based on several parameters such as feed water temperature, salinity, target permeate recovery, membrane fouling, and membrane condition. Selecting a pump to be efficient at a specific operating point could result in inefficiencies when conditions vary. Consequently, there is a move toward using positive displacement instead of centrifugal pumps. Positive displacement pumps deliver high operation efficiency of up to 97% over a wider operation range compared to centrifugal pumps. In addition, positive displacement pumps are superior in terms of operation flexibility, since they deliver consistent efficiency at varying flowrates and offer a linear relationship between feed flow rate and pump speed (Zarzo & Prats, 2018).

11.6 ENERGY RECOVERY DEVICES

Recovering energy from the brine stream is a well-established practice in RO plant design. The incorporation of ERDs in RO plants has reduced their SEC by as much as 60% and paved the way for a SEC below 5 kWh/m^3 (Stover, 2007). Since their introduction in the early 1980s, ERDs have been based on centrifugal machines such as Pelton wheels and turbochargers. Their integration in a RO system is presented in Figure 11.13. A Pelton wheel is used by

Towards Energy-Efficient Reverse Osmosis

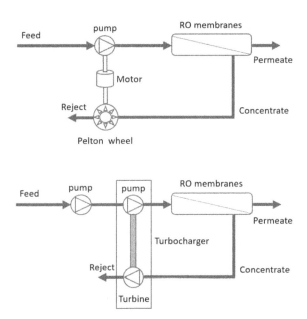

FIGURE 11.13 Schematic diagram of: a) a Pelton wheel; and b) a turbocharger incorporated in an RO system – Adapted from (Mito, Ma, Albuflasa, & Davies, 2019).

directly coupling it to the HPP shaft, where it assists the motor in delivering the driving power. In this case, the energy recovery efficiency is a function of the Pelton wheel, pump, and motor efficiency. If a turbocharger is used, it acts as a booster pump to raise the feed stream pressure after the HPP. Although centrifugal ERDs are favored for high flow rate applications, they have several drawbacks such as a limited range of capacity and a maximum energy efficiency of around 82%. In addition, as a centrifugal machine, they deliver their best performance only at a specific operating point, making them unsuited for flexible operation (Stover, 2007).

Currently, isobaric ERDs are gaining popularity and are becoming the standard choice for commercial RO plants. Isobaric ERDs are positive displacement machines that directly exchange the pressure between the brine and feed stream. Their incorporation into the RO system requires splitting the feed stream between the HPP and the ERD according to the plant recovery ratio (Figure 11.14). A booster pump is needed after the ERD to make up for the pressure lost in the RO membranes and the ERD. Isobaric ERDs deliver a high energy efficiency of up to 97%, and result in the use of a smaller capacity HPP when compared to a system using a centrifugal ERD, since the HPP only handles the permeate portion of the flow (Stover, 2007). In addition, isobaric ERDs deliver consistent efficiency over their operational range, making them very suited for varying the operating point to overcome changes in feed temperature or the required permeate recovery.

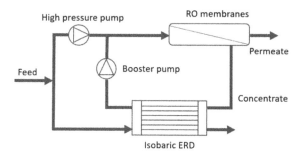

FIGURE 11.14 RO plant configuration incorporating isobaric energy recovery devices – Adapted from (Mito et al., 2019).

Unlike centrifugal devices, isobaric ERDs offer operational flexibility, lower HPP energy consumption, and higher efficiency. Their widespread use has reduced the cost of water production by 10% to 15% over the last 10 years (Voutchkov, 2018a). Many commercial RO plants that originally used centrifugal ERDs are being retrofitted with isobaric devices, providing a significant reduction in SEC.

11.7 SUMMARY AND CONCLUSIONS

Improving the energy efficiency of RO is an important aspect of RO desalination development, as energy contributes nearly half the cost of water production and is directly associated with the carbon footprint and sustainability of RO desalination. However, there are several design trade-offs that must be considered to maximize energy efficiency without compromising other aspects. In addition to minimizing SEC, increasing water production and improving salt rejection are also important, as is reducing capital cost. This chapter has described the status and breakdown of RO plant energy consumption, and outlined the technical advances needed to minimize it.

- Currently, the industry average SEC for commercial SWRO plants is 3.1 kWh/m^3, with some plants operating in the 2.5–2.8 kWh/m^3 range. Although the current SEC is impressive, it is still more than double the ideal SEC. The main reasons for the difference between the ideal and actual SEC are the energy lost because of equipment inefficiency, concentration polarization, membrane filtration resistance, and friction losses. These losses account for 35% of the total plant consumption and represent nearly half of the energy used for the RO process.
- The current standard membrane remains the thin-film composite with polyamide active layer. Innovative designs such as nanocomposite, biomimetic, and graphene-based membranes, are being pursued for their potential to deliver higher permeability with increased salt rejection. However, the contribution of new membrane materials is constrained by

the nonlinear relationship between permeability and SEC. Other improvements in RO plant configuration and component efficiency should also be targeted.

- In addition to innovation in membrane design, improvements in RO plant configuration can directly affect the energy consumption and capital cost. Design approaches such as concentrate and permeate staging are becoming standard in commercial RO plants to reduce SEC and achieve the targeted water quality. In addition, hybrid membrane configurations offer higher energy efficiency and less fouling by achieving an even pressure and flux distribution along the pressure vessel. Other innovative technologies, such as time-varying pressure systems (batch and semi-batch configurations), offer potential to reduce energy consumption by varying the hydraulic pressure as needed to overcome osmotic pressure.

- Since the centrifugal pump efficiency tends to increase with flow capacity, the current trend in pump selection favors high-flow capacity pumps to operate multiple RO trains, instead of using small-capacity pumps for individual RO trains. Another trend is to use positive displacement pumps rather than centrifugal pumps, because they tend to have higher efficiency and more consistent performance over a wider flow range. For the same reasons, there is a shift towards using isobaric energy recovery devices in place of the centrifugal Pelton wheel and turbocharger. This shift has reduced water production costs by 10% to 15% over the last 10 years.

REFERENCES

Ahn, C. H., Baek, Y., Lee, C., Kim, S. O., Kim, S., Lee, S., ... Yoon, J. (2012). Carbon nanotube-based membranes: Fabrication and application to desalination. *Journal of Industrial and Engineering Chemistry*, 18(5), 1551–1559. doi:10.1016/j.jiec.2012.04.005

Ali, H. A., Baronian, M., Burlace, L., Davies, P. A., Halasah, S., Hind, M., ... Naughton, T. (2019). Off-grid desalination for irrigation in the Jordan Valley. *Desalination and Water Treatment*, 168, 143–154. doi:10.5004/dwt.2019.24567

Amy, G., Ghaffour, N., Li, Z., Francis, L., Linares, R. V., Missimer, T., & Lattemann, S. (2017). Membrane-based seawater desalination: Present and future prospects. *Desalination*, 401, 16–21. doi:10.1016/j.desal.2016.10.002

Davies, P. A. (2011). A solar-powered reverse osmosis system for high recovery of freshwater from saline groundwater. *Desalination*, 271(1–3), 72–79. doi:10.1016/j.desal.2010.12.010

Efraty, A. (2012). Closed circuit desalination series no-3: High recovery low energy desalination of brackish water by a new two-mode consecutive sequential method. *Desalination and Water Treatment*, 42(1–3), 256–261. doi:10.1080/19443994.2012.682970

Elimelech, M., & Phillip, W. A. (2011). The future of seawater desalination: Energy, technology, and the environment. *Science*, 333(6043), 712–717. doi:10.1126/science.1200488

Goh, P. S., Matsuura, T., Ismail, A. F., & Hilal, N. (2016). Recent trends in membranes and membrane processes for desalination. *Desalination*, 391, 43–60. doi:10.1016/j.desal.2015.12.016

Karabelas, A. J., Koutsou, C. P., Kostoglou, M., & Sioutopoulos, D. C. (2018). Analysis of specific energy consumption in reverse osmosis desalination processes. *Desalination*, 431, 15–21. doi:10.1016/j.desal.2017.04.006

Khorshidi, B., Thundat, T., Fleck, B. A., & Sadrzadeh, M. (2016). A novel approach toward fabrication of high performance thin film composite polyamide membranes. *Scientific Reports*, 6, 22069. doi:10.1038/srep22069

Mito, M. T., Ma, X., Albuflasa, H., & Davies, P. A. (2019). Reverse osmosis (RO) membrane desalination driven by wind and solar photovoltaic (PV) energy: State of the art and challenges for large-scale implementation. *Renewable and Sustainable Energy Reviews*, 112, 669–685. doi:10.1016/j.rser.2019.06.008

Park, H. B., Kamcev, J., Robeson, L. M., Elimelech, M., & Freeman, B. D. (2017). Maximizing the right stuff: The trade-off between membrane permeability and selectivity. *Science*, 356, 6343. doi:10.1126/science.aab0530

Peñate, B., & García-Rodríguez, L. (2012). Current trends and future prospects in the design of seawater reverse osmosis desalination technology. *Desalination*, 284, 1–8. doi:10.1016/j.desal.2011.09.010

Qasim, M., Badrelzaman, M., Darwish, N. N., Darwish, N. A., & Hilal, N. (2019). Reverse osmosis desalination: A state-of-the-art review. *Desalination*, 459, 59–104. doi:10.1016/j.desal.2019.02.008

Qiu, T., & Davies, P. A. (2012a). Comparison of configurations for high-recovery inland desalination systems. *Water*, 4(3), 690–706. doi:10.3390/w4030690

Qiu, T. Y., & Davies, P. A. (2012b). Longitudinal dispersion in spiral wound RO modules and its effect on the performance of batch mode RO operations. *Desalination*, 288, 1–7. doi:10.1016/j.desal.2011.11.054

Smith, E. B. (2004). *Basic Chemical Thermodynamics* (Vol. 35). London, UK: Imperial College Press.

Stover, R. L. (2007). Seawater reverse osmosis with isobaric energy recovery devices. *Desalination*, 203(1), 168–175. doi:10.1016/j.desal.2006.03.528

Tang, C. Y., Wang, Z., & Hélix-Nielsen, C. (2016). Chapter 14 – Biomimetic membranes for water purification and wastewater treatment. In N. P. Hankins & R. Singh (Eds.), *Emerging Membrane Technology for Sustainable Water Treatment* (pp. 359–369). Boston, MA: Elsevier.

Voutchkov, N. (2013). *Desalination Engineering Planning and Design*. USA: McGraw Hill.

Voutchkov, N. (2018a). Energy use for membrane seawater desalination – Current status and trends. *Desalination*, 431, 2–14. doi:10.1016/j.desal.2017.10.033

Voutchkov, N. (2018b). Membrane desalination – Process selection, design, and implementation. In V. G. Gude (Ed.), *Sustainable Desalination Handbook* (pp. 3–24). Oxford, UK: Butterworth-Heinemann.

Weaver, R., Howells, M., & Brown, H. (2018). *IDA Water Security Handbook 2018–2019*. International Desalination Association: Global Water Intelligence.

Werber, J. R., Deshmukh, A., & Elimelech, M. (2016). The critical need for increased selectivity, not increased water permeability, for desalination membranes. *Environmental Science & Technology Letters*, 3(4), 112–120. doi:10.1021/acs.estlett.6b00050

Werber, J. R., Osuji, C. O., & Elimelech, M. (2016). Materials for next-generation desalination and water purification membranes. *Nature Reviews Materials*, 1, 5. doi:10.1038/natrevmats.2016.18

Zarzo, D., & Prats, D. (2018). Desalination and energy consumption. What can we expect in the near future? *Desalination*, 427, 1–9. doi:10.1016/j.desal.2017.10.046

12 Membrane Fouling and Scaling in Reverse Osmosis

Nirajan Dhakal, Almotasembellah Abushaban, Nasir Mangal, Mohanad Abunada, Jan C. Schippers, and Maria D Kennedy

12.1 INTRODUCTION

The main "Achilles heel" for the efficient and smooth operation of the membrane-based desalination systems is membrane fouling (Flemming et al. 1997). Fouling occurs in all membrane processes such as reverse osmosis, as well as nano-, ultra-, and microfiltration. Membrane fouling results from complex physical and chemical interactions between the fouling constituents in water and between these constituents and membrane surfaces, leading to the attachment, accumulation, and/or adsorption of these constituents onto the membrane surface. As a result, water transport is hindered because of the formation of a fouling layer on the membrane surface, which eventually causes a decline in the quantity and quality of the produced water (Guo et al. 2012). The problems associated with membrane fouling are decreased membrane permeability, increased operating pressure, and increased frequency of chemical cleaning and membrane damage (Matin et al. 2011).

12.2 EFFECT OF FOULING AND SCALING

Fouling and scaling may manifest in three ways, which are detailed next.

12.2.1 Increasing Differential Pressure in Cross - Wound Fibers, or of Spacer Spiral Wound Elements

Particles tend to deposit and bacteria tend to grow on fibers and spacers, which results in clogging of fiber bundles and spacers. Two main mechanisms involved in fouling the membrane surface itself are identified namely; pore blocking and cake/gel formation resulting in a lower permeability of the membranes. Clogging of the spacers and bundles results in higher differential pressure (head loss). The consequences of this are; i) lower net driving pressure,

FIGURE 12.1 Damage of membrane elements resulting from clogging of bundles or spacers: a) Telescoping; b) Squeezed element; and c) Channelling
(Source: Schippers et al. 2019).

which requires a higher pressure to maintain the capacity; ii) damage to elements because of telescoping in spiral-wound; channelling in spiral-wound; squeezing spiral-wound; and breaking fibers as shown in Figure 12.1.

The fouling and scaling could also result from local clogging, which causes uneven flow distribution, resulting in locally high conversion and high concentration polarization; deposition of particles; local precipitation of sparingly soluble compounds; growth and attachment of bacteria.

12.2.2 Increase Hydraulic Resistance of the Membrane

The increase in hydraulic resistance of the membrane mainly results from deposition and/or adsorption of material and/or bacterial growth (biomass) on the membrane surface that results in higher required pressure to maintain capacity, or lower capacity when pressure is not increased.

12.2.3 Decrease in Rejection Owing to Concentration Polarization in the Foul Layer

Because of fouling and scaling, concentration, polarization increases when crossflow velocity close to the membrane decreases, mainly because of uneven flow distribution and foul layers. Consequently, the dissolved salts and organic compounds, colloidal matter, and suspended matter accumulate at the surface of the membrane. The sparingly soluble compounds might precipitate and the rejection of the salts may decrease because of higher concentration at the membrane surface.

To overcome membrane fouling, membrane manufacturers recommend cleaning in place (CIP). The general criteria to apply CIP are when:

- Mass transfer coefficient (MTC) or normalized flux has dropped by 10%.
- Normalized salt passage has increased by 10%.

- Normalized differential pressure (feed pressure – concentrate pressure) has increased by 15%.

In the market, there is a wide range of chemicals that are used for CIP; while using these chemicals, compatibility with the membrane must be secured.

12.3 FOULING TYPES

Fouling can be classified into four types: particulate fouling, organic fouling, biological fouling, and inorganic fouling (scaling)

12.3.1 PARTICULATE FOULING

Particulate/colloidal fouling is caused by the accumulation of particles (> 1μm) and colloids (0.001–1μm) on the membrane surface. These particles and colloids can be inorganic (e.g., clay minerals, colloidal silica, aluminum, iron, and manganese oxides) or organic (e.g., large polysaccharide molecules, fulvic compounds, extracellular polymer substances (EPS), transparent exopolymer particles (TEP), and proteins). Fouling potential of these foulant matters depends on several factors: feed water compositions (e.g., foulant type and concentration, pH, ionic strength), membrane properties (e.g., roughness, charge, hydrophobicity, surface functional group) and hydrodynamic conditions (e.g., flux, cross- flow velocity, recovery, temperature, module, and spacer design) (Tang et al. 2011).

In general, there are four mechanisms of particulate fouling, as shown in Figure 12.2.

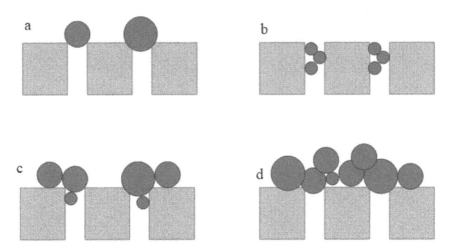

FIGURE 12.2 Mechanisms of particulate fouling: a) complete blocking; b) standard blocking; c) intermediate blocking; d) cake formation/filtration.

Initially, particles/colloids begin to deposit on the membrane surface, blocking the membrane pores. This initial phase is known as pore blocking, which may entail plugging of pores (complete blocking), constriction of pores because of deposition of particles around pores entry (standard blocking), or a combination of the previous two (intermediate blocking). The next stage of particulate fouling includes the development of a cake layer on the membrane surface as additional particles continue to deposit on the initial layer. Once the cake starts to form, the cake layer takes over the role of the membrane performance and controls the transport and removal process (Pearce 2007). Since RO membranes are considered nonporous, unlike MF and UF membranes, the dominant particulate fouling mechanism in RO is the cake formation (Zhu and Elimelech 1997).

Henry et al. (2012) described the mechanisms of particulate fouling as a result of a combination of four elementary phenomena (Figure 12.3), depending on the particle – particle, particle – fluid and particle – surface interactions.

Various methods were developed by researchers to assess the particulate fouling potential. However, silt density index (SDI) and modified fouling index (MFI$_{0.45}$) are the most common methods applied for this purpose; both indices are standard testing methods in the ASTM.

12.3.1.1 Silt Density Index (SDI)

Silt density index (SDI) is commonly applied as a parameter for the fouling potential of feed water for reverse osmosis and nanofiltration plants. Silt density index is determined by measuring the rate of plugging of a 0.45 μm membrane filter at 210 kPa (30 Psi) according to the standard protocol of ASTM (ASTM 2014). SDI measures the decline in filtration rate expressed in percentage flux decline per minute. The following steps need to be taken while measuring the SDI.

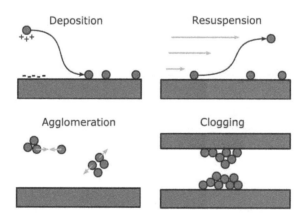

FIGURE 12.3 Mechanism of particulate fouling described by Henry et al. (2012).

Fouling and Scaling in Reverse Osmosis

i. Time t_1 required to filter the first 500 mL is determined.
ii. Fifteen minutes (t_f) after the start of the first volume filtration time t_2 needed to filter, the second 500 mL is determined.
iii. SDI is calculated using equation 12.1

$$SDI = \frac{100\%}{t_f} \left(1 - \frac{t_1}{t_2}\right) = \frac{\%P}{t_f} \tag{12.1}$$

Shorter time (t_f) has to be considered, such as 10, 5, or 3 minutes, if the plugging ratio $(\%P)$ exceeds 75%.

Despite the main advantage of the SDI method represented in its simple performance, the method has some deficiencies: i) it does not take into account the fouling mechanisms; ii) it has no linear correlation with particulate matter concentration; and iii) it has no temperature correction.

12.3.1.2 Modified Fouling Index (MFI$_{0.45}$)

The modified fouling index (MFI$_{0.45}$) was developed by Schippers and Verdouw (1980) to overcome the deficiencies of SDI. MFI is based on the cake filtration mechanism that occurs during a distinct period of the test. At constant pressure, cake filtration can be described by Equation 12.2.

$$\frac{t}{V} = \frac{\eta.R_m}{\Delta P.A} + \frac{\eta.I}{2\Delta P.A^2}.V \tag{12.2}$$

where
V = filtrate volume (L or m^3)
t = filtration time (s)
A = membrane area (m^2)
ΔP = applied pressure (Pa)
η = water viscosity (Pa.s)
R_m = clean membrane resistance (m^{-1})
I = fouling index (m^{-2})

The MFI$_{0.45}$ is calculated from the slope of t/V versus V plot, and is based on the correction standard testing conditions proposed by Schippers and Verdouw (1980), as in Equation 12.3.

$$MFI = \left(\frac{\eta_{20°C}}{\eta}\right)\left(\frac{\Delta P}{\Delta P_o}\right)\left(\frac{A^2}{A_o^2}\right) \times slope \tag{12.3}$$

where $\eta_{20°}$ is the water viscosity at 20°C; ΔP_o represents the standard pressure (200 kPa); and A_o is the standard membrane area (13.8 x 10^{-4} m^2).

12.3.1.3 Modified Fouling Index (MFI – UF)

MFI (0.45) and SDI fail to predict the rate of fouling in RO membranes as they make use of membranes with pore size of 0.45μm, while the particles that are smaller than 0.45 μm are more likely to be responsible for fouling in RO membranes (Schippers et al. 1981). To overcome this, the MFI – UF method has been developed using ultrafiltration membranes to capture the small colloids. MFI – UF was performed initially at constant pressure as the standard $MFI_{0.45}$. It was verified that MFI – UF demonstrates a linear correlation with colloidal matter. However, it was found that the cake formed in the MFI – UF test performed at constant pressure is highly compressible, and consequently leads to overestimated measurements. In addition, the test period is relativity long, possibly several hours in duration (Boerlage et al. 1997, 1998, 2000a, 2002b, 2003a, 2003b). Accordingly, Boerlage et al. (2004) further developed the MFI – UF to operate at constant flux instead of constant pressure.

12.3.1.4 MFI-UF at Constant Flux

MFI-UF is measured at constant flux through UF membranes with pore size reduced to 10 kDa according to the protocol developed by (Boerlage et al. 2004) and modified by (Salinas-Rodríguez et al. 2015). At constant flux (J), cake filtration taking place during MFI – UF measurement can be expressed by the general Equation 12.4.

$$\Delta P = J.\eta.R_m + J^2.\eta.I.t \qquad (12.4)$$

Fouling index (I) can be calculated from the slope of cake filtration region in the plot of P versus t (shown in Figure 12.4) using Equation 12.5. The MFI – UF value can be then calculated by correcting I for the standard testing conditions as in Equation 12.6.

$$I = \tan \alpha \times \frac{1}{J^2.\eta} \qquad (12.5)$$

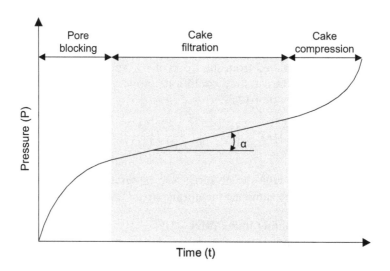

FIGURE 12.4 Fouling mechanisms during MFI – UF at constant flux test.

Fouling and Scaling in Reverse Osmosis

$$MFI - UF = \frac{\eta_{20°C}.I}{2\Delta P_o.A_o^2} \qquad (12.6)$$

It turned out that the MFI - UF constant flux depends, more or less proportional, on flux (Salinas-Rodríguez, et al. 2012). The MFI – UF can currently be measured accurately at the lowest flux of 50 L/m².h; however, the flux in brackish RO is in the range of 20 to 30 L/m².h. and in seawater RO systems in the range of 10–20 L/m².h. Therefore, the MFI–UF needs to be measured at higher flux rates and then extrapolated to simulate the corresponding flux applied in the RO system to be assessed. Particulate fouling rate can be then predicted in terms of the time required for an increase in net driving pressure (*NDP*), using the prediction model shown by Equation 12.7.

$$t = \frac{NDP_t - NDP_0}{J^2.\eta.I.\Omega} \qquad (12.7)$$

where Ω is the particle deposition factor, incorporated in the model to simulate the portion of particles, present in the water passing the membrane, depositing on the RO membrane during the cross-flow filtration. Ω can be calculated based on the operating recovery (R) and the relation between the MFI of concentrate water and the MFI of feedwater, as shown in Equation 12.8. Ω value may vary between 0 and 1, indicating 0% to 100% particle deposition.

$$\Omega = \frac{1}{R} + \frac{MFI_{concentrate}}{MFI_{feed}}.\left(1 - \frac{1}{R}\right) \qquad (12.8)$$

12.3.1.5 Transparent Exopolymer Particles (TEPs)

TEPs are transparent, seasonally abundant organic substances (e.g., algal blooms) in marine and freshwater environments (Passow 2002). Common bloom-forming algae produce algal organic matter (AOM) that are comprised of high-molecular weight biopolymers (polysaccharides and proteins), and often include the sticky transparent exopolymer particles (TEPs) (Villacorte et al. 2013). TEPs can be present in different shapes (e.g., strings, disks, sheets, or fibers) and in different sizes, ranging from a few nanometers in diameter up to hundreds of micrometers in length (Passow 2002, Villacorte et al. 2015). They are sticky and comprise mainly hydrophilic, negatively-charged, acidic polysaccharides. In reverse osmosis membranes, organic fouling often occurs when sticky microbial-derived biopolymers (e.g., transparent exopolymer particles) are abundant in the RO feedwater. The accumulation of such materials may result in a substantial decrease of normalized flux in RO membranes. Strong correlations between TEP produced by several algal spices and MFI-UF 100kDa has been demonstrated showing high particulate fouling potential. (Dhakal et al. 2019; Villacorte et al. 2014). Furthermore, it may further initiate or enhance biological fouling as they can serve as a *conditioning layer*, which is an ideal attachment sites for bacteria – a platform where bacteria can effectively utilize biodegradable nutrients (C, P, N) from the feedwater while excreting more extracellular substances, resulting in rapid buildup of biofilm. Moreover, the accumulated sticky substances may enhance the deposition of other colloids/particles from the feedwater to the membrane/spacers and may further aggravate fouling problems. TEP can be measured according to the protocol described by Villacorte et al. (2015).

12.3.2 Biological Fouling

Biofouling is a special case of fouling whereby microorganisms, present in the reverse osmosis (RO) feedwater, grow in the membrane system by utilizing the biodegradable substances from the water phase and converting them into metabolic products and biomass. Microorganisms can multiply to form a thick layer of slime called a biofilm (Flemming 2011). Biofilm formation on RO membrane surfaces is inevitable (Van Loosdrecht et al. 2012) if the feed water supports bacterial growth because of the presence of dissolved nutrients (Figure 12.5). Moreover, bacteria are always present in RO feedwater, even after ultrafiltration pretreatment (Ferrer et al. 2015). Biofilm formation may cause biofouling in some cases, when biofilm formation is excessive to the extent that operational problems arise (Vrouwenvelder and Van der Kooij 2001).

To date, there is no standard method to monitor biological/organic fouling in RO membrane system. However, several approaches have been followed to monitor biofilm development on RO membrane surface and biofouling potential in RO feed water.

12.3.2.1 Monitoring the Biofilm Development

The most and commonly applied method to monitor biofouling in full-scale RO plants is monitoring the development of pressure drop across the stages of the RO membrane system.

Other methods to monitor biofouling is the direct detection of biofilm formation on RO membranes using online sensors such as the electrical potential measurements (Sung et al. 2003), biosensors (Lee and Kim 2011), ultrasonic

 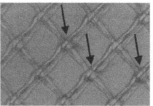

FIGURE 12.5 Biofilm formation on feed spacer taken during autopsy of spiral-wound membrane (Vrouwenvelder et al. 2009).

Fouling and Scaling in Reverse Osmosis

333

time-domain reflectometry (Kujundzic et al. 2007) and online fluorometers (Ho et al. 2004). The major challenges of these methods are the fouling on these sensors over a long period of use, the need for frequent calibration of the sensors, the high pressure applied in RO system, and the use of spiral- wound modules (Nguyen et al. 2012).

Additionally, such tools as the modified biofilm formation rate monitor (mBFR) and the membrane fouling simulator (MFS) are used to simulate biofilm formation on RO membrane surfaces, in which they are connected in parallel to the RO membrane unit (Vrouwenvelder et al. 2006). In mBFR, biofilm formation is measured e.g. as ATP on Teflon/glass rings. While, the Membrane Fouling Simulator makes use of a flat membrane and a spacer. The development of the pressure increase across the spacer is monitored. These two monitors are strong tools in trouble shooting and research and development.

12.3.2.2 Monitoring Biofouling Potential

Another approach is to measure biological/organic fouling potential along pretreatment systems and in the feedwater of RO membrane systems. This approach is very attractive for successful operation and control of RO membrane systems because it can be used as an early warning system by adjusting the operational conditions of the pretreatment processes to meet the required quality of RO feedwater, and consequently, to achieve better control of biofouling in RO systems. In addition measuring biofouling potential facilitates research and development in determining the effectiveness of pre-treatment systems and operation. Biofouling potential methods include assimilable organic carbon (AOC), biodegradable dissolved organic carbon (BDOC), bacterial growth potential (BGP).

12.3.2.2.1 Assimilable Organic Carbon (AOC)

AOC is the fraction (0.1% – 10%) of dissolved organic carbon that can be utilized by heterotrophic bacteria for their growth. AOC takes into account easily assimilable organic compounds and contains many low-molecular-weight organic molecules, namely sugars, organic acids, and amino acids (Hammes and Egli 2005). Two AOC methods have recently been developed in seawater to measure the growth potential in the pretreatment and in the feed of a SWRO membrane system by Weinrich et al. (2011) and Jeong et al. (2013), using a single strain of bacteria (*Vibrio fischeri* and *Vibrio harveyi*, respectively). Hijnen et al. (2009) reported/suggested that the AOC threshold concentration is 1 µg/L for biofilm formation in freshwater RO membranes, which is very low and very difficult to achieve through pretreatment. The rate of biofouling is dependent on the AOC concentration present in the RO feedwater. Weinrich et al. (2009) reported that 10 µg/L of AOC in water results in significant heterotrophic bacterial growth. Badruzzaman et al. (2019) reported a preliminary threshold of 50 µg/L of AOC based on a study of three full-scale seawater RO desalination plants, in which AOC ranged between 10 and 180 µg/L of AOC through pretreatment. Jeong and Vigneswaran (2015) found a very good linear relationship between AOC concentration and low molecular weight neutral (LMW-N) organics concentration.

12.3.2.2.2 Biodegradable Dissolved Organic Carbon (BDOC)

BDOC is the total amount of organic matter that is biodegraded by an inoculum of suspended or fixed bacteria over a period of time (Escobar and Randall 2001). The main difference between AOC and BDOC assays is that BDOC assays assess the concentration of DOC removed through biofilm-related microbial growth, while AOC assays assess the amount of cells produced through utilization of bioavailable carbon (Hammes 2008). The accuracy of the BDOC method is problematic because of the limited accuracy and reproducibility of DOC measurements. Escobar and Randall (2001) proposed that both AOC and BDOC should be used as complementary measurements of bacterial regrowth potential.

12.3.2.2.3 Bacterial Growth Potential (BGP)

The BGP method measures the ability of indigenous microbial consortia (mixed cultures) to grow in water samples because of easy biodegradable nutrients present in the sample. The higher growth potential, the higher the possibility of biofouling is. The BGP method is similar to AOC method, but in the BGP method, the inoculum consists of an indigenous microbial consortium with constant concentration, rather than a single pure culture. The use of an indigenous microbial consortium further broadens and diversifies the substrate utilization range when compared to a single pure culture. Ross et al. (2013) demonstrated that bacterial growth in indigenous microbial consortia was higher (> 20 %) than bacterial growth in pure strains and provides a more realistic interpretation of growth potential in water.

BGP measurement comprises four steps that include bacterial inactivation, bacterial inoculation, incubation, and bacterial enumeration. In this method, bacterial inactivation can be achieved by either pasteurisation (70°C for 30 min) or by filtration (0.22 µm) and inoculation (with 10,000 intact cell/mL of an indigenous microbial consortium, measured by flow cytometry). Microbial inactivation allows the standardization of the initial microbial population by adding a constant inoculum concentration. The sample is then incubated at 30°C, and is monitored over time until the maximum bacterial growth is reached. Several methods have been recently developed to monitor the bacterial growth of an indigenous microbial consortium in seawater, including microbial adenosine triphosphate (ATP) (Abushaban et al. 2018, 2019b), total cell counts measured by flow cytometry (Dhakal 2017, Farhat et al. 2018), turbidity (Dixon et al. 2012), and microbial electrolysis cell biosensors (Quek et al. 2015). Abushaban et al. 2019a used an ATP-based BGP method that employed an indigenous microbial consortium to monitor BGP along the pretreatment of three full-scale SWRO desalination plants, and reported high removal (> 50%) of BGP in dual media filtration (DMF) in two SWRO plants. For the three SWRO plants investigated, a preliminary correlation was observed between BGP in SWRO feedwater and chemical cleaning frequency. However, to establish a real correlation, more data needs to be collected and many more SWRO plants need to be monitored for longer periods of time with different operating conditions.

Fouling and Scaling in Reverse Osmosis   335

12.3.3 Performance Restoration

Biofouling of RO membranes can be expected when the RO feed water has high biofouling potential. The performance of RO membrane systems can be restored when there is significant drop of differential pressure and/or high flux decline by executing chemical cleaning (Vrouwenvelder et al. 2011a). The concept of cleaning is to remove and the accumulated biomass from the membrane surface so that the initial flux can be recovered. Cleaning can be performed both as physical cleaning and chemical cleaning (Nguyen et al. 2012).

Generally, physical cleaning is applied prior to chemical cleaning (Cornelissen et al. 2009). Physical cleaning includes air/water flushing cleaning. Physical cleaning removes mostly nonadhesive fouling. Cornelissen et al. (2009) reported that air/water cleaning is an efficient way to control biofouling in spiral-wound membranes.

Usually membrane cleaning is achieved with chemical solutions, which is done in place (cleaning in place – CIP). Several types of chemical cleaning agents (e.g., alkaline, acids, biocides, detergents, and enzymes) are recommended by membrane manufacturers. Chemical cleaning is efficient in killing or inactivating microorganisms, but it does not remove the accumulated biomass adequately (Flemming et al. 1997, Vrouwenvelder et al. 1998). As a consequence, the remaining inactivated biomass is available as food by surviving bacteria, causing rapid regrowth of bacteria (Vrouwenvelder et al. 2010). Thus, membrane cleaning can partially remove the biomass, but regular CIP is needed.

Frequently pre-chlorination is applied to surface water intakes to control marine growth. Chlorine is very efficient in killing bacteria, but is does not control biofouling at all. This is because chlorination produces biodegradable organic compounds by breaking down natural organic matter including humic compounds. In addition chlorination is a permanent risk to damage Thin Film Composite membranes , which are easily attacked by chlorine. Monochloramine is also used to control biofouling in some treatment plants, primarily in waste water applications.

12.3.4 Organic fouling

Organic fouling refers to fouling in RO due to oragnics presence in water. Surface water (lake, river) typically contains higher natural organic matter (NOM) compared to ground water. It has been believed that for such type of water with high NOM, flux decline in membrane systems is mainly due to organic fouling. NOM can be classified as hydrophilic, hydrophobic and transphilic. Dissolved organic carbon (DOC) is considered as an indicator for organic fouling. Moreover, measuring only DOC is neither proper nor adequate. NOM can be in general characterized by size (expressed as molecular weight), structure and functionality (charge density). In general, following parameters can be measured to monitor organic fouling in RO; i) Liquid chromatography organic carbon detection (LC-OCD) based in size exclusion and categorized organic carbon concentrations into biopolymers, humic substances, building blocks, low molecular weight (LMW)

Membrane Desalination

acids, and neutrals; ii) Total organic/dissolved organic carbon (TOC/DOC); iii) UV_{254}; or SUVA as an index of aromaticity (hydrophobicity) of water calculated as SUVA = UV_{254}/DOC iv) Fluorescence excitation and emission matrix (FEEM); v) Fourier transform infrared spectroscopy (FTIR).

12.3.5 SCALING IN REVERSE OSMOSIS SYSTEMS

Scaling in reverse osmosis refers to the precipitation of sparingly soluble inorganic compounds present in the feedwater onto the membrane surface and feed spacer.

Precipitation usually occurs when the ionic product of a certain salts exceeds the solubility product (Antony et al. 2011). Therefore, scaling is directly related to the concentrations of inorganic ions in the feedwater and to the recovery (ratio of the permeate water to the feedwater) of the RO system. At higher recovery, the concentrate water in the last stage becomes more concentrated and, therefore, in many cases, exceeds the solubility limit for several types of salts, resulting in scaling (Kucera 2015). Therefore, the concentration factor is calculated as in Equation 12.9:

$$CF = \frac{C_c}{C_f} \qquad (12.9)$$

where, C_c = concentration in concentrate (brine); and C_f = concentration in feed water

Concentration factors depend upon the recovery of the systems as in Equation 12.10

$$CF = \frac{1 - R(1 - f)}{1 - R} \qquad (12.10)$$

where R = system recovery; and f = rejection when f= 100%

$$CF = \frac{1}{1 - R} \qquad (12.11)$$

Scaling is considered as one of the major challenges of RO application, and it is one of the main factors that limit system recovery in RO. The consequences of scaling are permeability decline of the membrane, increase in pressure drop, increase salt passage, and shorter lifetime of membranes resulting from frequent cleanings. For a scaled membrane, higher than normal operating pressure is required to produce permeate water, which results in higher operational costs (Kucera 2015).

12.3.5.1 Scaling Compounds on the RO Membranes

Several types of scaling such as calcium carbonate, calcium sulphate, barium sulphate, calcium phosphate, calcium fluoride, strontium sulphate, and silica may occur on the membrane surface.

Fouling and Scaling in Reverse Osmosis
337

12.3.4.1.1 Calcium Carbonate Scaling

Calcium carbonate is one of the most common scales encountered in RO applications (Kucera 2015). The formation and degree of $CaCO_3$ scaling mainly depends on the concentrations of calcium, bicarbonate and pH in the feedwater, as well as the recovery of the RO (Tzotzi et al. 2007, Antony et al. 2011). Other factors which affect the precipitation of $CaCO_3$ are temperature, TDS and the presence of inorganic ions and organic substances.

In the literature, six forms of $CaCO_3$ scale deposits are reported to exist, depending on the experimental conditions and presence of foreign substances (impurities), such as three anhydrous forms (calcite, aragonite, and vaterite), two hydrated forms (calcium carbonate monohydrate and calcium carbonate hexahydrate) and one amorphous calcium carbonate (Brečević and Nielsen 1989, Chakraborty et al. 1994, Elfil and Roques 2001, Coleyshaw et al. 2003).

12.3.5.1.2 Calcium Sulphate Scaling

Calcium sulphate is one of the common specie of the nonalkaline scales encountered on the RO membrane surface (Antony et al. 2011). The $CaSO_4$ precipitation can occur when the ionic product of the Ca^{2+} and SO_4^{2-} ions exceeds solubility product (K_{sp}) according to the following reaction Equation 12.12:

$$Ca^{2+} + SO_4^{2-} + xH_2O \rightarrow CaSO_4 \cdot xH_2O \downarrow \qquad (12.12)$$

where x can be 0, ½, or 2, based on different forms of calcium sulphate.

The calcium sulphate scale can occur in three different forms: gypsum ($CaSO_4 \cdot 2H_2O$), hemihydrate ($CaSO_4 \cdot 1/2H_2O$) and anhydrite ($CaSO_4$), where gypsum is the commonly encountered form at ambient temperatures of 20°C (Lee and Lee 2000, Antony et al. 2011).

12.3.5.1.3 Barium Sulphate Scaling

Barium sulphate precipitation results in very hard deposits on the membrane surface (Boerlage et al. 2000b). The solubility of the barium sulphate is very low (1×10^{-5} mol/L or 2.33 mg/L in pure water) (Van der Leeden 1991). Therefore, concentrate water at very low recoveries can be supersaturated with respect to barium sulphate. It is worth mentioning that precipitation is not only governed by the supersaturation, but also depends on the precipitation kinetics that involve the formation of nuclei and further crystal growth. Boerlage et al. (2002a) reported that $BaSO_4$ has a long stable phase prior to nucleation in the supersaturated state. Consequently substantial supersaturation is allowable.

12.3.5.1.4 Calcium Phosphate Scaling

Calcium phosphate scaling can occur on the membrane surface when high concentrations of calcium and orthophosphate ions are present in the feed water (Greenberg et al. 2005, Chesters 2009). Calcium phosphate can precipitate in various forms including but not limited to such as amorphous calcium phosphate, dicalcium phosphate dihydrate (CaHPO$_4$·2H$_2$O), dicalcium phosphate anhydrous (CaHPO$_4$), octacalcium phosphate (Ca$_8$H$_2$(PO$_4$)$_6$·5H$_2$O), tricalcium-phosphate (Ca$_3$(PO$_4$)$_2$), and hydroxyapatite (Ca$_5$(PO$_4$)$_3$OH) (Dorozhkin 2017). Calcium phosphate scaling is, in particular, important in waste water treatment due to the rather high phosphate concentrations.

12.3.5.2 Mechanism of Scaling in Reverse Osmosis
Scale formation is a complex process in which both crystallization and hydrodynamic transport mechanisms are involved (Oh et al. 2009). There are two types of crystallization identified as heterogeneous (also called surface crystallization) and homogeneous (bulk crystallization). Surface crystallization refers to the formation of crystals on the surface of the RO membrane, while in the case of the bulk crystallization process, crystals form in the bulk solution and then precipitate on the membrane surface. The mechanism of crystallization is illustrated in Figure 12.6.

12.3.5.3 Prediction of Scaling
The commonly applied method to determine the scaling potential are briefly described in the following sections.

12.3.5.3.1 Scaling Indices
There are a number of indices available to measure the scaling tendency of the sparingly soluble salts in a water solution. The most commonly used in RO applications are:

Saturation index (SI): Saturation index is calculated by subtracting the logarithm of the thermodynamic solubility product (Log K_{sp}) from the logarithm of the ion activity product (Log IAP) as in Equation 12.13.

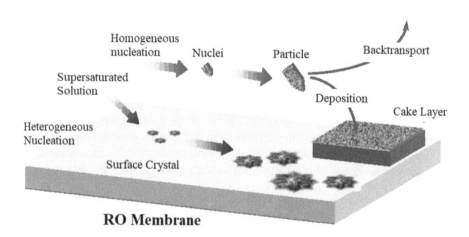

FIGURE 12.6 Scale formation mechanism in RO system. (Source: Oh et al. 2009).

Fouling and Scaling in Reverse Osmosis

$$SI = \log \frac{IAP}{K_{sp}} \qquad (12.13)$$

when
    SI = 0, the concentrate is in equilibrium
    SI > 0, the concentrate is supersaturated
    SI < 0, the concentrate is under saturated

Supersaturation ratio (S_r): Supersaturation ratio can be calculated by dividing the square root of ion activity product (IAP) by the square root of the thermodynamic solubility product (K_{sp}) as in Equation 12.14

$$S_r = \sqrt{\frac{IAP}{K_{sp}}} \qquad (12.14)$$

when
    S_r = 1, the concentrate is in equilibrium
    S_r > 1, the concentrate is supersaturated
    S_r < 1, the concentrate is under saturated

12.3.5.3.2 Scaling Prediction with Computer Software

A number of commercial programs are available which can be used to predict the scaling potential in RO. Most of these programs are developed by antiscalant suppliers and membrane manufacturers. The programs include but not limited to:

- Genesys Membrane Master (MM4) – Genesys International
- Sokalant RO-Xpert-BASF
- Hyd-RO-dose – French Creek Software
- Argo Analyzer-Suez
- Avista Advisor – Avista Technologies, Inc.
- Proton-American Water Chemicals
- WAVE-DuPont membrane projection software
- IMSDesign– Hydranautics membrane projection software

12.3.5.4 Techniques to Control Scaling

To prevent scaling in RO applications, various chemical, physical, and mechanical approaches have been proposed, which can be summarized into three groups: optimization of operating parameters and system design; altering feedwater characteristics; and addition of scale inhibitors (Antony et al. 2011).

12.3.5.4.1 Addition of Scale Inhibitors

Scale inhibitor (antiscalant) additions to feedwater is one of the most widely used and an effective techniques to prevent scaling in RO applications (Pervov 1991, Lee et al. 1999, Greenlee et al. 2010). A number of commercial antiscalants are available that are designed for specific types of scale. The commonly used antiscalants in RO applications can be categorized in three different groups based on their compositions and properties, and include: polyphosphates, phosphonates/organophosphates, and poly acrylates/polymaletes (Antony et al. 2011, Van Engelen and Nolles 2013). A factor contributing to the attractiveness of antiscalants is the low-dose requirement to overcome scaling problems (Antony et al. 2011).

12.3.5.4.2 Operate at Low Recovery

Lowering RO recovery is also a scaling control technique, though it is not a desirable one as this technique leads to a high concentrate production and specific energy consumption (kWh/m^3). In this method, the recovery of the RO is decreased to an extent where the sparingly soluble salts remain undersaturated. In addition, the effect of concentration polarization is also diminished with this approach.

12.3.5.4.3 Acid addition

Addition of acid is one of the earliest techniques to prevent the precipitation of calcium carbonate. The solubility of calcium carbonate is increased when pH of the water is lowered. Acidification is also not an attractive approach since massive amount of acid is needed to lower the feedwater pH when the concentration of bicarbonate is high.

REFERENCES

Abushaban, A., M. N. Mangal, S. G. Salinas-Rodriguez, C. Nnebuo, S. Mondal, S. A. Goueli, J. C. Schippers, and M. D. Kennedy. 2018. Direct measurement of ATP in seawater and application of ATP to monitor bacterial growth potential in SWRO pre-treatment systems. *Desalination and Water Treatment* **99**:91–101.

Abushaban, A., S. G. Salinas-Rodriguez, N. Dhakal, J. C. Schippers, and M. D. Kennedy. 2019a. Assessing pretreatment and seawater reverse osmosis performance using an ATP- based bacterial growth potential method. *Desalination* **467**:210–218.

Abushaban, A., S. G. Salinas-Rodriguez, M. N. Mangal, S. Mondal, S. A. Goueli, A. Knezev, J. S. Vrouwenvelder, J. C. Schippers, and M. D. Kennedy. 2019b. ATP measurement in seawater reverse osmosis systems: Eliminating seawater matrix effects using a filtration-based method. *Desalination* **453**:1–9.

Antony, A., J. H. Low, S. Gray, A. E. Childress, P. Le-Clech, and G. Leslie. 2011. Scale formation and control in high pressure membrane water treatment systems: A review. *Journal of Membrane Science* **383**:1–16.

ASTM. 2014. *ASTM D4189-07, Standard Test Method for Silt Density Index (SDI) of Water.* West Conshohocken, PA.

Badruzzaman, M., N. Voutchkov, L. Weinrich, and J. G. Jacangelo. 2019. Selection of pre-treatment technologies for seawater reverse osmosis plants: A review. *Desalination* **449**:78–91.

Boerlage, S. F. E., M. Kennedy, M. P. Aniye, and J. C. Schippers. 2003a. Applications of the MFI-UF to measure and predict particulate fouling in RO systems. *Journal of Membrane Science* **220**:97–116.

Boerlage, S. F. E., M. Kennedy, Z. Tarawneh, R. De Faber, and J. C. Schippers. 2004. Development of the MFI-UF in constant flux filtration. *Desalination* **161**:103–113.

Boerlage, S. F. E., M. D. Kennedy, M. P. Aniye, E. Abogrean, Z. S. Tarawneh, and J. C. Schippers. 2003b. The MFI-UF as a water quality test and monitor. *Journal of Membrane Science* **211**:271–289.

Boerlage, Ś. F. E., M. D. Kennedy, M. P. Aniye, E. M. Abogrean, D. E. Y. El-Hodali, Z. S. Tarawneh, and J. C. Schippers. 2000a. Modified fouling index ultrafiltration to compare pretreatment processes of reverse osmosis feedwater. *Desalination* **131**:201–214.

Boerlage, S. F. E., M. D. Kennedy, M. P. Aniye, E. M. Abogrean, G. Galjaard, and J. C. Schippers. 1998. Monitoring particulate fouling in membrane systems. *Desalination* **118**:131–142.

Boerlage, S. F. E., M. D. Kennedy, P. A. C. Bonne, G. Galjaard, and J. C. Schippers. 1997. Prediction of flux decline in membrane systems due to particulate fouling. *Desalination* **113**:231–233.

Boerlage, Ś. F. E., M. D. Kennedy, I. Bremere, G. J. Witkamp, J. P. Van der Hoek, and J. C. Schippers. 2000b. Stable barium sulphate supersaturation in reverse osmosis. *Journal of Membrane Science* **179**:53–68.

Boerlage, S. F. E., M. D. Kennedy, I. Bremere, G. J. Witkamp, J. P. Van Der Hoek, and J. C. Schippers. 2002a. The scaling potential of barium sulphate in reverse osmosis systems. *Journal of Membrane Science* **197**:251–268.

Boerlage, S. F. E., M. D. Kennedy, M. R. Dickson, D. E. Y. El-Hodali, and J. C. Schippers. 2002b. The modified fouling index using ultrafiltration membranes (MFI-UF): Characterisation, filtration mechanisms and proposed reference membrane. *Journal of Membrane Science* **197**:1–21.

Brečević, L., and A. E. Nielsen. 1989. Solubility of amorphous calcium carbonate. *Journal of Crystal Growth* **98**:504–510.

Chakraborty, D., V. K. Agarwal, S. K. Bhatia, and J. Bellare. 1994. Steady-state transitions and polymorph transformations in continuous precipitation of calcium carbonate. *Industrial & Engineering Chemistry Research* **33**:2187–2197.

Coleyshaw, E. E., G. Crump, and W. P. Griffith. 2003. Vibrational spectra of the hydrated carbonate minerals ikaite, monohydrocalcite, lansfordite and nesquehonite. *Spectrochimica Acta Part A: Molecular and Biomolecular Spectroscopy* **59**:2231–2239.

Cornelissen, E. R., L. Rebour, D. Van der Kooij, and L. P. Wessels. 2009. Optimization of air/water cleaning (AWC) in spiral wound elements. *Desalination* **236**:266–272.

Dhakal, N. 2017. *Controlling Biofouling in Seawater Reverse Osmosis Membrane Systems.* Delft University of Technology, Taylor & Francis Group.

Dhakal, N., S. G. S., Salinas-Rodríguez, J.C. Schippers, M.D. Kennedy. 2018. "Fouling of ultrafiltration membranes by organic matter generated by marine algal species." *Journal of Membrane Science* **555**: 418–428.

Dixon, M. B., T. Qiu, M. Blaikie, and C. Pelekani. 2012. The application of the bacterial regrowth potential method and flow cytometry for biofouling detection at the Penneshaw desalination plant in South Australia. *Desalination* **284**:245–252.

Dorozhkin, S. V. 2017. Calcium orthophosphates (CaPO4): Occurrence and properties. *Morphologie* **101**: 125–142.

Elfil, H., and H. Roques. 2001. Role of hydrate phases of calcium carbonate on the scaling phenomenon. *Desalination* **137**:177–186.

Escobar, I. C., and A. A. Randall. 2001. Assimilable organic carbon (AOC) and biodegradable dissolved organic carbon (BDOC): Complementary measurements. *Water Research* **35**:4444–4454.

Farhat, N., F. Hammes, E. Prest, and J. Vrouwenvelder. 2018. A uniform bacterial growth potential assay for different water types. *Water Research* **142**:227–235.

Ferrer, O., S. Casas, C. Galvañ, F. Lucena, A. Bosch, B. Galofré, J. Mesa, J. Jofre, and X. Bernat. 2015. Direct ultrafiltration performance and membrane integrity monitoring by microbiological analysis. *Water Research* **83**:121–131.

Flemming, H.-C. 2011. Microbial biofouling: Unsolved problems, insufficient approaches, and possible solutions. In *Biofilm Highlights* (pp. 81–109). Springer.

Flemming, H. C., G. Schaule, T. Griebe, J. Schmitt, and A. Tamachkiarowa. 1997. Biofouling – The Achilles heel of membrane processes. *Desalination* 113:215–225.

Greenberg, G., D. Hasson, and R. Semiat. 2005. Limits of RO recovery imposed by calcium phosphate precipitation. *Desalination* 183:273–288.

Greenlee, L. F., F. Testa, D. F. Lawler, B. D. Freeman, and P. Moulin. 2010. The effect of antiscalant addition on calcium carbonate precipitation for a simplified synthetic brackish water reverse osmosis concentrate. *Water Research* 44:2957–2969.

Guo, W., H. H. Ngo, and J. Li. 2012. A mini-review on membrane fouling. *Bioresource Technology* 122: 27–34.

Hammes, F. 2008. A comparison of AOC methods used by different TECHNEAU partners. TECHNEAU 06. Deliverable 3.310.

Hammes, F. A., and T. Egli. 2005. New method for assimilable organic carbon determination using flow-cytometric enumeration and a natural microbial consortium as inoculum. *Environmental Science & Technology* 39:3289–3294.

Henry, C., J.-P. Minier, and G. Lefèvre. 2012. Towards a description of particulate fouling: From single particle deposition to clogging. *Advances in Colloid and Interface Science* 185–186:34–76.

Hijnen, W. A. M., D. Biraud, E. R. Cornelissen, and D. Van Der Kooij. 2009. Threshold concentration of easily assimilable organic carbon in feedwater for biofouling of spiral-wound membranes. *Environmental Science and Technology* 43:4890–4895.

Ho, B. P., M. W. Wu, E. K. Zeiher, and M. Chattoraj. 2004. *Method of Monitoring Biofouling in Membrane Separation Systems.* Google Patents.

Huber, S. A., A. Balz, M. Abert, and W. Pronk. 2011. Characterisation of aquatic humic and non-humic matter with size-exclusion chromatography – Organic carbon detection – Organic nitrogen detection (LC-OCD-OND). *Water Research* 45:879–885.

Ito, Y., Y. Takahashi, S. Hanada, H. X. Chiura, M. Ijichi, W. Iwasaki, A. Machiyama, T. Kitade, Y. Tanaka, and M. K. A. K. Kogure. 2013. Impact of chemical addition on the establishment of mega-ton per day sized swro desalination plant. In YT; M. Kurihara and K. Kogure.

Jeong, S. 2013. *Novel Membrane Hybrid Systems as Pretreatment to Seawater Reverse Osmosis.* University of Technology, Sydney.

Jeong, S., G. Naidu, S. Vigneswaran, C. H. Ma, and S. A. Rice. 2013. A rapid bioluminescence-based test of assimilable organic carbon for seawater. *Desalination* 317:160–165.

Jeong, S., and S. Vigneswaran. 2015. Practical use of standard pore blocking index as an indicator of biofouling potential in seawater desalination. *Desalination* 365:8–14.

Kucera, J. 2015. *Reverse Osmosis: Industrial Processes and Applications.* Salem, Massachusetts, John Wiley & Sons, Inc. Hoboken, and Scrivener Publishing LLC.

Kujundzic, E., A. C. Fonseca, E. A. Evans, M. Peterson, A. R. Greenberg, and M. Hernandez. 2007. Ultrasonic monitoring of early stage biofilm growth on polymeric surfaces. *Journal of Microbiological Methods* 68:458–467.

Lee, J., and I. S. Kim. 2011. Microbial community in seawater reverse osmosis and rapid diagnosis of membrane biofouling. *Desalination* 273:118–126.

Lee, S., J. Kim, and C.-H. Lee. 1999. Analysis of CaSO4 scale formation mechanism in various nanofiltration modules. *Journal of Membrane Science* 163:63–74.

Lee, S., and C.-H. Lee. 2000. Effect of operating conditions on CaSO4 scale formation mechanism in nanofiltration for water softening. *Water Research* 34:3854–3866.

Matin, A., Z. Khan, S. M. J. Zaidi, and M. C. Boyce. 2011. Biofouling in reverse osmosis membranes for seawater desalination: Phenomena and prevention. *Desalination* 281:1–16.

Nguyen, T., F. A. Roddick, and L. Fan. 2012. Biofouling of water treatment membranes: A review of the underlying causes, monitoring techniques and control measures. *Membranes (Basel)* **2**:804–840.

Oh, H.-J., Y.-K. Choung, S. Lee, J.-S. Choi, T.-M. Hwang, and J. H. Kim. 2009. Scale formation in reverse osmosis desalination: Model development. *Desalination* **238**:333–346.

Passow, U. 2002. Transparent exopolymer particles (TEP) in aquatic environments. *Progress in Oceanography* **55**:287–333.

Pearce, G. 2007. Introduction to membranes: Fouling control. *Filtration & Separation* **44**:30–32.

Pervov, A. G. 1991. Scale formation prognosis and cleaning procedure schedules in reverse osmosis systems operation. *Desalination* **83**:77–118.

Quek, S.-B., L. Cheng, and R. Cord-Ruwisch. 2015. Detection of low concentration of assimilable organic carbon in seawater prior to reverse osmosis membrane using microbial electrolysis cell biosensor. *Desalination and Water Treatment* **55**:2885–2890.

Ross, P. S., F. Hammes, M. Dignum, A. Magic-Knezev, B. Hambsch, and L. C. Rietveld. 2013. A comparative study of three different assimilable organic carbon (AOC) methods: Results of a round-robin test. *Water Science and Technology: Water Supply* **13**:1024–1033.

Salinas-Rodríguez, S. G. S., M.D. Kennedy, G.L.Amy, J.C. Schippers. 2012. "Flux dependency of particulate/colloidal fouling in seawater reverse osmosis systems." *Desalination and Water Treatment* **42**(1–3): 155–162.

Salinas-Rodríguez, S. G., G. L. Amy, J. C. Schippers, and M. D. Kennedy. 2015. The modified fouling index ultrafiltration constant flux for assessing particulate/colloidal fouling of RO systems. *Desalination* **365**:79–91.

Schausberger, P., G. M. Mustafa, G. Leslie, and A. Friedl. 2009. Scaling prediction based on thermodynamic equilibrium calculation – Scopes and limitations. *Desalination* **244**:31–47.

Schippers, J. C., J. H. Hanemaayer, C. A. Smolders, and A. Kostense. 1981. Predicting flux decline of reverse osmosis membranes. *Desalination* **38**:339–348.

Schippers J. C., M. D. Kennedy, and S. G. Salinas-Rodriguez. 2019. *Desalination and Membrane Technology.* Lecture Note, IHE Delft, the Netherlands.

Schippers, J. C., and J. Verdouw. 1980. The modified fouling index, a method of determining the fouling characteristics of water. *Desalination* **32**:137–148.

Sung, J. H., M.-S. Chun, and H. J. Choi. 2003. On the behavior of electrokinetic streaming potential during protein filtration with fully and partially retentive nanopores. *Journal of Colloid and Interface Science* **264**:195–202.

Tang, C. Y., T. H. Chong, and A. G. Fane. 2011. Colloidal interactions and fouling of NF and RO membranes: A review. *Advances in Colloid and Interface Science* **164**:126–143.

Tzotzi, C., T. Pahiadaki, S. G. Yiantsios, A. J. Karabelas, and N. Andritsos. 2007. A study of CaCO3 scale formation and inhibition in RO and NF membrane processes. *Journal of Membrane Science* **296**:171–184.

Van der Leeden, M. C. 1991. The role of polyelectrolytes in barium sulphate precipitation.

Van Engelen, G., and R. Nolles. 2013. A sustainable antiscalant for RO processes. *Desalination and Water Treatment* **51**:921–923.

Van Loosdrecht, M., L. Bereschenko, A. Radu, J. C. Kruithof, C. Picioreanu, M. L. Johns, and H. S. Vrouwenvelder. 2012. New approaches to characterizing and understanding biofouling of spiral-wound membrane systems. *Water Science and Technology* **66**:88–94.

Villacorte, L. O., Y. Ekowati, H. N. Calix-Ponce, J. C. Schippers, G. L. Amy, and M. D. Kennedy. 2015. Improved method for measuring transparent exopolymer particles (TEP) and their precursors in fresh and saline water. *Water Research* **70**:300–312.

Villacorte, L. O., Y. Ekowati, H. Winters, G. Amy, J. C. Schippers, and M. D. Kennedy. 2013. Characterisation of transparent exopolymer particles (TEP) produced during al gal bloom: A membrane treatment perspective. *Desalination & Water Treatment* **51**(4–6): 1021–1033.

Vrouwenvelder, J., J. Van Paassen, L. Wessels, A. Van Dam, and S. Bakker. 2006. The membrane fouling simulator: A practical tool for fouling prediction and control. *Journal of Membrane Science* **281**:316–324.

Vrouwenvelder, J. S., D. A. Graf von der Schulenburg, J. C. Kruithof, M. L. Johns, and M. C. M. van Loosdrecht. 2009. Biofouling of spiral-wound nanofiltration and reverse osmosis membranes: A feed spacer problem. *Water Research* **43**:583–594.

Vrouwenvelder, J. S., J. Kruithof, and M. C. Van Loosdrecht. 2011a. *Biofouling of Spiral Wound Membrane Systems.* Iwa Publishing.

Vrouwenvelder, J. S., J. Kruithof, and M. C. M. Van Loosdrecht. 2011b. *Biofouling Studies in NF and RO Installations Page 360 Biofouling of Spiral Wound Membrane Systems.* IWA Publishing, London, UK.

Vrouwenvelder, J. S., J. C. Kruithof, and M. C. Van Loosdrecht. 2010. Integrated approach for biofouling control. *Water Science and Technology* **62**:2477–2490.

Vrouwenvelder, J. S., and D. Van der Kooij. 2001. Diagnosis, prediction and prevention of biofouling of NF and RO membranes. *Desalination* **139**:65–71.

Vrouwenvelder, J. S., J. A. M. Van Paassen, H. C. Folmer, J. A. M. H. Hofman, M. M. Nederlof, and D. van der Kooij. 1998. Biofouling of membranes for drinking water production. *Desalination* **118**:157–166.

Weinrich, L. A., E. Giraldo, and M. W. LeChevallier. 2009. Development and application of a bioluminescence-based test for assimilable organic carbon in reclaimed waters. *Applied and Environmental Microbiology* **75**:7385–7390.

Weinrich, L. A., O. D. Schneider, and M. W. LeChevallier. 2011. Bioluminescence-based method for measuring assimilable organic carbon in pretreatment water for reverse osmosis membrane desalination. *Applied and Environmental Microbiology* **77**:1148–1150.

Zhu, X., and M. Elimelech. 1997. Colloidal fouling of reverse osmosis membranes: Measurements and fouling mechanisms. *Environmental Science & Technology* **31**:3654–3662.

13 Sustainable Development and Future Trends in Desalination Technology

Mattheus Goosen
Office of Research & Graduate Studies
Alfaisal University
Riyadh, Saudi Arabia

Hacene Mahmoudi
Faculty of Technology
Hassiba Ben Bouali University
Chlef, Algeria

13.1 INTRODUCTION

The high cost of energy is one of the main challenges facing the economic sustainability of desalination plants. Energy generation is also associated with air pollution and environmental degradation (Fletcher et al., 2019; Margulis et al., 2010). Furthermore, the scientific community and much of the public, now recognize that climate change is likewise a major factor in sustainable economic development. This change is not only influenced by human activity but it is also accelerating, thus becoming a foremost hazard to the world's economic growth and to environmental stability (Cox et al., 2019). Consequently, water resource planners must consider that recent climate history may not be an adequate predictor of the future. Fluctuating climate necessitates the need for us to understand how these changes occur to be able to plan appropriately. Adaptive infrastructure planning, for example, may be employed to help address the problems resulting from climate change uncertainty. Effective planning models play a critical role in targeting resources, thus saving financial resources (Fletcher et al., 2019). A United Nations report estimated that the cost of climate change adaptation investments in the developing world may reach $500 billion per year by 2050. It

346 Membrane Desalination

is therefore essential to target infrastructure investments efficiently to reach the widest number of vulnerable communities (Puig et al., 2016).

Pramanik et al. (2017) reported that hypersaline brines are of growing environmental concern. Prominent examples of such high-salinity brines include water produced from the oil and gas industry, waste streams of minimum/zero liquid discharge operations, inland desalination concentrate, landfill leachate and flue gas desulfurization wastewater. Very high total dissolved solids (TDS) >60,000 ppm pose considerable technical challenges in treatment. While reverse osmosis (RO) is the most energy-efficient and cost-effective technique for desalinating seawater, exceedingly high operating pressures are needed to overcome the osmotic pressure of hypersaline brines, which prohibits the application of RO.

The influence of continuous discharge of high-salinity brine from desalination plants, both membrane and thermal, into coastal environments are also a matter of grave public concern (Petersen et al., 2019). The results of Petersen et al. (2019) indicated that to minimize environmental impacts, discharge should target waters where a long history of human activity has already compromised the natural setting. Aside from concerns over brine discharge, effective metering of potable water also plays a role in the sustainable management of drinking water distribution systems (Maiolo et al., 2019). In this case, information is required on the operating status of system components to identify the best operational management measures. Fantini et al. (2016) found that smart metering networks in the energy sector allow operators and companies to improve production efficiency and offer customers enhanced service.

The aim of this chapter is to provide a critical review and analysis of sustainable developments and future trends as they relate to desalination technology. Emphasis is placed on infrastructure planning and management, the problem of brine discharge, energy efficiency and costs, membrane desalination and applications of renewable energy technology.

13.2 FLEXIBLE INFRASTRUCTURE PLANNING AND SUSTAINABLE DEVELOPMENT

Uncertainty in climate change projections poses a challenge to infrastructure planning, owing to the high cost of investments for any future developments such as increased desalination capacity (Fletcher et al., 2019; Margulis et al., 2010). Effective planning models thus play a critical role in targeting capital, consequently saving financial resources. Getting ready for climate change by adding extra desalination capacity, for example, incurs a high risk of expensive overbuilding in resource-scarce areas. However, enabling flexibility often requires substantial proactive planning or upfront investment. In the case of water resources, it is difficult for planners to know if and when to take action. Problems also arise if a water resources infrastructure, such as a water distribution network, cannot be adapted quickly. Fletcher et al. (2019) argued that techniques or models are needed to weigh the risks and benefits of static, as opposed to flexible, infrastructure approaches in responding to climate change uncertainty.

Sustainable Development and Future Trends

A technique called adaptive infrastructure planning may be employed to help address the problems resulting from climate change uncertainty. For instance, robust decision making uses iterative scenario development to minimize remorse from both overbuilding excessive infrastructure and being unprepared (Lempert et al., 2006). Furthermore, adaptive management requires ability to learn over time as more information is collected (Pahl-Wostl, 2007). For example, suppose current regional projections estimate a range between 0.5 and 1.5°C of change over the next 20 years. If after two decades, a 1.5°C of change is observed, then it can be argued that the climate is warming in this region more rapidly than expected. The temperature projections can then be shifted upward for the subsequent two decades. Fletcher et al. (2019) reported on a planning framework that explicitly models the potential to learn about climate uncertainty over time. They used probable learning to develop and evaluate flexible planning strategies in comparison to static approaches. A comprehensive set of virtual climate observations were developed that reflected many possible future regional climates, some of which were drier and some of which were wetter. Updated estimates reflected what would be learned if the virtual observation came to pass. Fletcher et al. (2019) employed this framework model to evaluate flexible infrastructure planning approaches and compared them to static approaches.

The *2016 Adaptation Finance Gap Report* from the United Nations Environment Program evaluated the expenses of meeting adaptation requirements and assessed the funding that was available for doing so (Puig et al., 2016). The report suggested that although international public funding for adaptation has increased in recent years, previous assessments of the costs of adaptation have been significantly underestimated. This left a gap, the adaptation finance gap, which needs to be filled if societies (i.e., nations) are to meet the goals of the Paris Agreement. The United Nations report estimated that the cost of climate change adaptation investments in the developing world may reach $500 billion per year by 2050. It is therefore essential to target infrastructure investments efficiently so that they reach the widest number of susceptible people.

Flexible planning strategies can substantially reduce the cost of infrastructure investments. Fletcher et al. (2019) claimed that theirs was the first framework that values the ability of flexible approaches to respond to climate learning. Their results showed that climate change uncertainty can be reduced over the lifetime of an infrastructure project across different climate change trajectories. Flexibility was effective in preventing unnecessary infrastructure additions while maintaining reliability.

In a related study by Margulis et al. (2010) for the World Bank, the authors concluded that as developing countries weigh how best to revitalize their economies using a sustainable development path, they will have to factor in the reality that the global annual average temperature is expected to be 2°C above preindustrial levels by 2050. A 2°C warmer world will experience more intense rainfall, as well as more frequent and more intense droughts, floods, heat waves and other extreme weather events. This will have a significant effect on how nations manage their economies, care for their people, and design their development paths. It was argued that countries will need to adopt measures to adapt to climate change, similar to what was recommended by Fletcher et al. (2019). Hopefully, this strategy will offer a way to make the effects of climate change less disruptive and spare the poor and the vulnerable from shouldering an unduly high burden.

Margulis et al. (2010) further noted that given the uncertainty surrounding both climate outcomes and longer-term projections of social and economic development, countries should try to delay adaptation decisions as much as possible and focus on low-regret actions. They should also enhance the resilience of vulnerable sectors. In agriculture, for example, this would mean better management of water resources by giving policymakers greater flexibility in handling either droughts or waterlogging caused by floods.

The instinctive approach to cost adaptation related to climate change was also described by Margulis et al. (2010). This methodological approach consists of two tracks, comparing a future world without climate change with a future world with climate change. The differences between the two biospheres necessitate a series of actions to adapt to the new world conditions. The costs of these additional actions are the costs of adapting to climate change. Figure 13.1 summarizes the methodological approaches of the two tracks. Margulis et al. (2010)

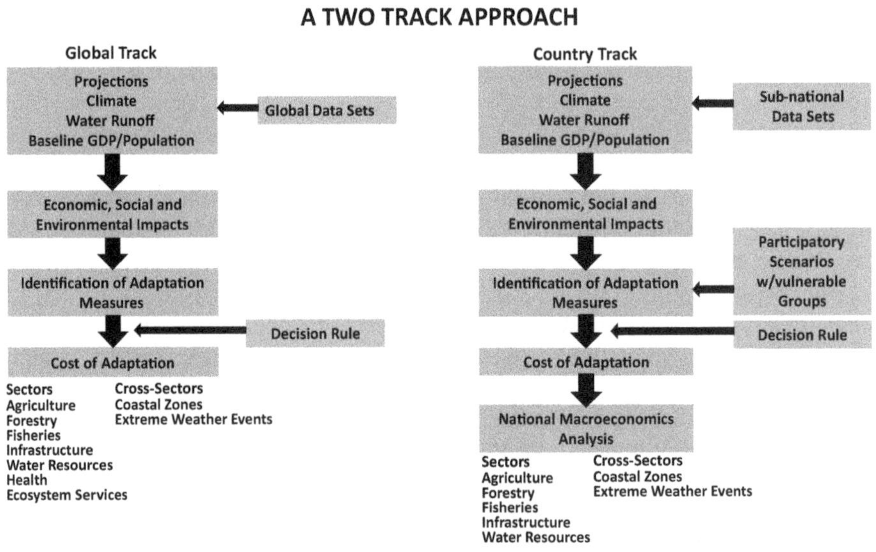

FIGURE 13.1 Economics of adaptation to climate change. Summary of methodological approaches of two tracks: global and country. The intuitive approach to costing adaptation involves comparing a future world without climate change with a future world with climate change. The difference between the two worlds entails a series of actions to adapt to the new world conditions, and the costs of these additional actions are the costs of adapting to climate change (Margulis et al., 2010).

Sustainable Development and Future Trends   **349**

explained that overall, the global study estimated that the cost between 2010 and 2050 of adapting to an approximately 2°C warmer world by 2050 is in the range of $70 billion to $100 billion per year. Similar results were reported by Puig et al. (2016) for the United Nations were the estimated cost of climate change adaptation investments in the developing world was predicted as $500 billion per year by 2050. However, this was more than five times greater than that estimated by Margulis et al. (2010).

Cox et al. (2019) reported that global climate model (GCM) projections are generally considered the best source of information for predicting future weather and hydrologic conditions in the face of a changing environment. Understanding and interpreting GCM projections is therefore critical for water resources planning. Regrettably, this can be a challenging task as climate model data, particularly precipitation data, have a large scatter and often lack apparent trends, as likewise observed by Fletcher et al. (2019) and Margulis et al. (2010). The paper by Cox et al. (2019) demonstrated a simple, practical method for synthesizing climate model data into more informative metrics using case studies. Their results identified significant increasing trends in expected 21st century temperatures for most GCM projections. Significant trends were also identified for probable future monthly and 24-hour maximum precipitation and drought severity. Implications of this work for water resources planning were discussed.

Scientific communities, and much of the public, recognize that climate change is not only influenced by human activity, but that it is also accelerating Cox et al. (2019). Consequently, water resources planners must consider that recent climate history is not an adequate predictor of the future. Fluctuating climate necessitates the need for mankind to understand how these changes occur to be able to plan properly. As Fletcher et al. (2019) and Margulis et al. (2010) concluded, the variability, nonstationarity and trends in climate data must be assessed and then incorporated into resource planning models so that future trends may be more reliably predicted. This conclusion is correct whether planning emphasis is on typical conditions such as available water resources, or on extreme conditions such as floods and droughts. A wealth of data now exists, including 20th century observations and 21st century climate model projections, to help scientists comprehend these dynamic forces. Cox et al. (2019) concluded that the methodology presented in their study could serve as a useful, low-cost, initial step in long-term planning.

13.3 MANAGEMENT AND TREATMENT OF HYPERSALINE BRINE SOLUTIONS THROUGH USE OF LOW GRADE HEAT AND SOLVENT EXTRACTION

Concentrated brine streams coming out of desalination plants pose a serious environmental risk to many communities. Pramanik et al. (2017) critically examined conventional technologies aimed at the management and treatment of brine solutions to reduce their environmental impact. Their study critically elucidated the challenges for membrane processes for producing clean water and resource recovery. The researchers noted that sustainable management of the concentrate streams because of the toxic nature of some of the compounds in the solution was a serious concern. Pramanik et al. (2017) argued that conventional processes are probably the best treatments because of their efficient removal and detoxification of organics. However, high energy consumption for the oxidation process and sludge production for the coagulation process is a substantial barrier to the use of these treatments. Furthermore, the formation of by-product compounds was another concern associated with oxidative treatments. The authors went on to explain that the oxidative process would be a suitable option to improve the biodegradability of brine solutions and make downstream biological treatment feasible because it generates biodegradable organic compounds.

As also noted by Pramanik et al. (2017), prominent examples of high-salinity brines include produced water from the oil and gas industry, waste streams of minimum/zero liquid discharge operations, inland desalination concentrate, landfill leachate and flue gas desulfurization wastewater. Very high total dissolved solids (TDS) >60,000 ppm poses considerable technical challenges in treatment. While reverse osmosis (RO) is the most energy-efficient and cost-effective technique for desalinating seawater, exceedingly high operating pressures are needed to overcome the osmotic pressure of hypersaline brines, which prohibits the application of RO. Evaporation-based thermal methods, e.g., multiple effect distillation, thermal brine concentrator and crystallizer, are the prevailing processes to desalinate or dewater highly concentrated brines. However, because the enthalpy of vaporization for water is enormous ($\approx$630 kWh/m^3) and the energy efficiency of evaporative phase-change methods is thermodynamically constrained, (Brogioli et al., 2019) these processes inherently require intensive thermal energy input, even though the quality of energy is lower (heat as opposed to electricity for RO). Therefore, there is a pressing need to develop energy-efficient technologies for the more sustainable desalination of environmentally-relevant hypersaline streams.

In an effort to develop more energy-efficient technologies it is important to remember that low-temperature heat sources (i.e., 80°C –120°C), including low-concentration solar, shallow-well geothermal, household cogeneration, and industrial waste heat, are widely abundant and have the potential to be recovered as electric energy. As an example, the integration of distillation with salinity gradient power technologies has been proposed by Brogioli et al. 2019 as an alternative to traditional technologies for harnessing such low-grade heat. In this method, heat is used to distill a salt solution, thus producing a concentrated solution and pure solvent. Controlled mixing of the two solutions at different concentrations in a salinity gradient power process converts the mixing Gibbs free energy (i.e., heat) to useful work.

As an example of another technology, Boo et al. (2019) described the membraneless and nonevaporative desalination of hypersaline brines by a temperature swing solvent extraction (TSSE) technique. Solvent extraction is a separation method that is widely employed for chemical engineering processes (Figure 13.2). Application of solvent extraction for desalination was first explored using amine

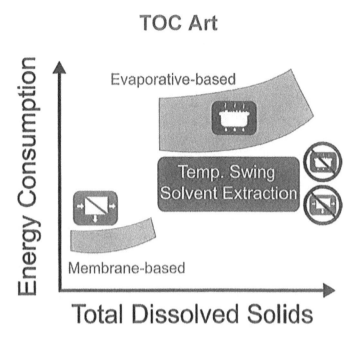

FIGURE 13.2 Schematic of membraneless and nonevaporative desalination of hypersaline brines by temperature swing solvent extraction (Boo et al., 2019).

solvents in the 1950s, but the effort was limited to desalting brackish water of relatively low salinity (<10,000 ppm TDS). More recently, the technique was investigated for desalination of seawater simulated by 3.5 w/w% NaCl solution with decanoic acid as the solvent. In the study by Boo et al. (2019), temperature swing solvent extraction (TSSE) desalination of high-salinity brines using three amine solvents was demonstrated. The performance metrics of water extraction, salt removal, product water quality and osmotic pressure reduction were evaluated, and the most suitable solvents for specific performance objectives were identified. Attainable water recovery for desalination of high-salinity brines was further assessed in semi-batch TSSE experiments with multiple extraction cycles. The implications of TSSE as an alternative membraneless and nonevaporative technique for hypersaline desalination was discussed.

13.4 EFFECTS OF BRINE DISCHARGE AND METERING TRENDS ON DESALINATION PLANT CONSTRUCTION

The influence of continuous discharge of high-salinity brine from desalination plants, both membrane and thermal, into coastal environments are a matter of grave public concern (Petersen et al., 2019). In a case study from Carlsbad, California by Petersen et al. (2019), in situ measurements of water chemistry and biological indicators in coastal waters (i.e., up to ~2 km from shore) were reported before and after a newly constructed desalination plant began operation. A bottom water salinity irregularity indicated that the three-dimensional footprint of the brine discharge plume extended 600 m offshore with salinity up to 2.7 units above ambient temperature. This exceeded the maximum salinity permitted for this location based on the local government regulations (i.e., California Ocean Plan; 2015 Amendment to Water Quality Control Plan). The coastal wave energy and mixing potential at the case study site (i.e., Carlsbad Beach, California) was high compared to other locations along the southern and central California coastline. The lack of sufficient mixing of the brine at Carlsbad Beach raised concern regarding the efficiency of brine mixing at other proposed sites for desalination plants along the coast of California. As a positive observation, no significant changes in the assessed biological indicators were observed at the discharge site.

A model of mean ocean wave potential has been used as an indicator of coastal mixing where desalination facilities are proposed. The results of Petersen et al. 2019 indicated that to minimize environmental impacts, discharge should target waters where a long history of human activity has already compromised the natural setting. The authors recommended that to ensure adequate mixing of the discharge brine, desalination plants should be constructed at high-energy sites with sandy substrates, and should be discharged through diffusor systems. Furthermore, where possible, plants should be colocated with other water-discharging facilities such as power plant or wastewater treatment plants to restrict potential impacts to areas that are already disturbed from man-made activities. Finally, it was stressed that there is a need for long-term monitoring of coastal ecosystems in areas surrounding proposed and current desalination facilities to fully understand the ecological response to long-term exposure to salinity increases.

Aside from concerns over brine discharge, effective metering of potable water also plays a role in the sustainable management of drinking water distribution systems (Maiolo et al., 2019). In this case, information is required on the operating status of system components to identify the best operational management measures. The ability to acquire information on tank levels, pipeline flow, and real-time pressure provides for efficient and cost-effective management and enables wider monitoring. The technology of measuring instruments for hydrodynamic variables used to monitor potable water systems differs with respect to independence from electronic data acquisition components and ability to connect to remote data communication systems. Advanced water measurement infrastructure is characterized by the ability to capture data with measurable errors from anywhere in the system, without restrictions on communication type. Maiolo et al. (2019) assessed hydrodynamic parameters and proposed a water meter classification system. Their study included analysis of the main water meter and data teleacquisition infrastructure. Several selection criteria were evaluated to determine their ability to support mathematical hydraulic models and expert systems for water distribution system management.

In the construction of desalination plants, the ability to set up a sustainable development model is linked to resource monitoring, quantifying sustainable use, and identifying management operations (Maiolo et al., 2019). Measuring instruments, real-time network monitoring, and control equipment must be included in any water distribution system to enable real-time intervention for malfunctions or hydrodynamic imbalances. Such control functions are linked to allow for active management of a water delivery arrangement. Real-time measurement of hydrodynamic parameters is very important since it enables control (e.g., operating corrections), both in desalination plants, as well as in the water distribution system coming out of the plant.

A wide variety of wireless communication technologies is available, matching the range of services and requirements that might use them. Every application has its own communication requirements, and it is virtually impossible to identify a single technology that meets the requirements of each application (Maiolo et al., 2019). Figure 13.3 shows a possible radio technology classification based on coverage offered and transmission speed. Short-range devices usually operate on unlicensed bands. Application-specific technologies have been established for metering, with others, such as low-power Bluetooth for wearable equipment and healthcare, and NFC (Near-Field Communication) for payment (Fantini et al., 2016). Fantini et al. (2016) concluded that smart metering networks in the energy sector allow operators and companies to improve production efficiency and offer customers an enhanced service. The ability to acquire real-time information on hydrodynamic variables in water systems improves sustainability and efficiency. Current technology for remote measurement and monitoring in water supply systems enables the use of mathematical hydraulic modeling capable of reproducing realistic operating conditions. Fantini et al. (2016) went on to explain that the availability of real-time data, with errors that can be measured precisely and from any point in the system, opens the viewpoint of management operation reinforced by artificial intelligence.

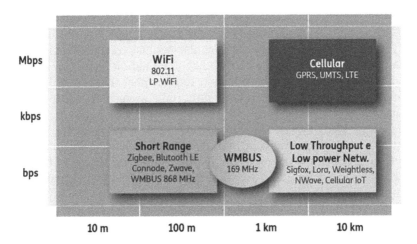

FIGURE 13.3 Range and bit rate of some communication technologies for the Internet of Things (IoT) (Fantini et al., 2016).

13.5 ENERGY EFFICIENCY, COST ANALYSIS, AND ROAD MAP TO SUSTAINABLE DESALINATION

The high cost of energy is one of the main challenges facing the economic sustainability of desalination plants. Energy generation is also associated with air pollution and environmental appropriateness (Fletcher et al., 2019; Margulis et al., 2010). Levelized energy cost analysis is normally employed in assessing the average total cost to build and operate a power-generating plant. This allows for comparison of different methods of electricity generation on a consistent basis. In other words, it is an economic assessment of the average total cost to build and operate a power-generating asset over its lifetime, divided by the total energy output of the asset (i.e., the plant) over that lifetime. Kaya et al. (2019) performed a levelized energy cost analysis for a solar desalination plant using the Emirate of Abu Dhabi as a case study. The aim was to understand how a sustainable desalination scheme could be implemented by looking at recent developments in both the desalination and the energy sectors.

Kaya et al. (2019) noted that conventional desalination technologies such as multistage flash (MSF) and multiple effect distillation (MED) require thermal heat, which can be acquired either from the burning of fossil fuels or through concentrated solar or geothermal energy as more sustainable options. Furthermore, the third technology for desalination, reverse osmosis (RO), has seen an exponential increase in use as a result of decreasing cost and increasing effectiveness. In most parts of the world, RO has become the main choice for desalination technology. RO systems require direct electricity as the energy form to operate during the desalination process as opposed to MSF and MED. Kaya et al. (2019) went on to argue that the cost reduction in direct-electricity-producing renewable energy technologies such as wind and solar photovoltaic (PV) has been much sharper compared to the thermal-heat-producing renewables such as concentrated solar and geothermal. One of their main conclusions, which was supported by the work of Caldera et al. (2018), was that because of rapid technological developments in both RO desalination and solar PV, a combination of these two technologies may result in the cheapest and cleanest commercial process for large-scale desalination of seawater. As scientists and decision makers, it is important to remember that cost reduction and environmental protection are equally important when it comes to long term sustainability.

In the case study reported by Kaya et al. (2019), it can be seen from Figure 13.4 that Abu Dhabi's current practice of seawater desalination is not very sustainable, considering the production of harmful emissions such as CO_2, NO_x and SO_2. Furthermore, Figure 13.5 depicts that while desalination will play a growing role in the future for Abu Dhabi to meet its water demand, the current practice of using MSF and MED technologies for desalination is not a good option regarding the sustainability criteria from environment and health viewpoints. Kaya et al. (2019) concluded that meeting the increasing demand for freshwater through seawater desalination in a nonpolluting, inexpensive and sustainable way is essential for cities in the Persian Gulf.

Sustainable Development and Future Trends

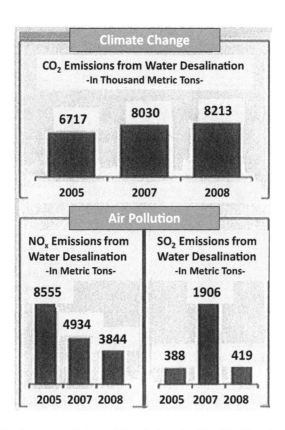

FIGURE 13.4 Environmentally harmful emissions in Abu Dhabi owing to desalination (Kaya et al. 2019).

To improve energy efficiency in industry, low-grade heat recovery technologies have been progressing unceasingly. Ling-Chin et al. (2019), for example, reported on state-of-the-art technologies for low-grade heat recovery and utilization in industry. The technologies assessed included adsorption, absorption, liquid desiccant, organic Rankine cycles (ORC) and Kalina cycles. To achieve a low-carbon future, it was envisaged that R&D for low-grade heat utilization and recovery technologies will continue. Worldwide commercial applications that utilize or recover industrial low-grade heat, which is otherwise discharged to the environment, are possible for useful work by taking advantage of the state-of-the-art technologies discussed by Ling-Chin et al. (2019). The authors noted however that the number of commercial applications is still very limited, even though industrial waste heat is abundantly available and the concept of utilization or recovery is not new. More effective use of waste heat is another area in which significant sustainable development contributions can be made by both scientists and decision makers.

From a business and organizational perspective, Ling-Chin et al. (2019) concluded that capital cost and payback period are the key factors that should be

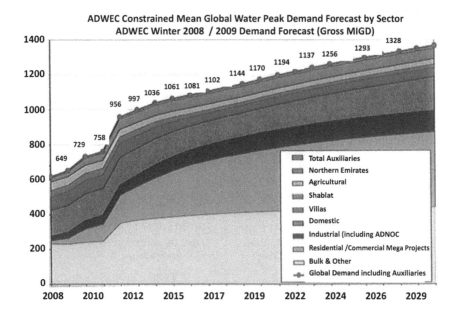

FIGURE 13.5 Abu Dhabi's water demand projections (in millions of imperial gallons per day (MIGD)) (Kaya et al. (2019).

considered by the management boards of industrial organizations. Currently, however, commercial applications are hindered by regulatory barriers such as the absence of financial incentives, tax breaks, strong policies, and legislation. Nonetheless, successful accounts of existing commercial applications by Ling-Chin et al. (2019) indicate additional opportunities. Certainly, these accomplishments have set important precedents for feasible uptake of the technologies in various industry sectors. The authors concluded that prior to commercial applications, systems must be designed with precaution and assessed thoroughly using, a whole-system approach, taking into account technical, economic, legislative, social, and environmental concerns.

In a related sustainability case study from Japan by Saito et al. (2019), it was reasoned that design of future possible scenarios is crucial for helping decision makers to identify the potential impact of different policy options. It was noted that there is a lack of reported scenario approach studies in Asia. A new five-year research project called PANCES was developed for predicting and assessing the natural capital and ecosystem services in Japan using an integrated social-ecological system approach through the participation of 15 research institutions and more than 100 researchers. PANCES conducts the development of national-scale future scenarios for exploring potential changes in natural capital and ecosystem services, as well as human well-being until 2050, using key direct and indirect drivers including climate change, depopulation and superaging, as well as globalization and technological innovation.

Four future scenarios were created: "Natural capital-based compact society," "Natural capital-based dispersed society," "Produced capital-based compact society," and "Produced capital-based dispersed society," respectively, in addition to the business-as-usual scenario. Saito et al. (2019) explained that the strength of this sustainability case study was that it described possible national-scale future scenarios for Japan which would help decision makers to correctly predict and assess natural capital and ecosystem services.

For future sustainable seawater desalination, Shahzad et al. (2019a) maintained that the importance of achieving better energy efficiency of the existing 19,500 commercial-scale desalination plants cannot be overemphasized. They reported that a major concern of the desalination industry is the inadequate approach to energy efficiency evaluation. For example, while many seawater desalination processes have been reported, the grade of the energy supplied has not been adequately described. Effectiveness comparison should only be conducted for the same energy input processes. The authors went on to explain that the misconception of considering all derived energies as equivalent in the desalination industry has severe economic and environmental consequences. For example, energy and desalination system planners may make serious judgmental errors in the process selection of installations, without realizing it, if data is either flawed or inaccurate. The authors found inferior efficiency technologies that were being implemented as a result of poor decisions in many water-stressed countries. This can burden a country's economy with higher energy costs as well as causing undesirable environmental effects on the surroundings. In their article, Shahzad et al. (2019a) presented a standard primary energy-based thermodynamic framework that addressed energy efficiency equitably and accurately. They showed unmistakably that a thermally driven process consumes 2.5%–3% of standard primary energy (SPE) when combined with power plants. The authors proposed a standard universal performance ratio –based evaluation method that revealed all desalination processes performance, varying from 10%–14% of the thermodynamic limit. They concluded that to achieve 2030 sustainability goals, innovative processes are required to meet the 25%–30% of the thermodynamic limit.

In a related study, Shahzad et al. (2019b) confirmed that considering different grades of energy as equivalent in the desalination industry could have negative economic and environmental consequences. While this approach will suffice for the comparison of same energy input processes, omitting the grade of energy when comparing diverse technologies may lead to incorrect conclusions and, as a consequence, inefficient installations. A standard primary energy-based thermodynamic framework was presented, as described previously, (Shahzad et al., 2019a) that addressed the energy effectiveness of assorted desalination processes. A roadmap was described to increase the efficiency level up to 25%–30% of the thermodynamic limit. However, this will require a technological shift in the capability of dissolved salt separation processes, such as membrane desalination.

The water, energy and environment relationship or association is important for future sustainability. In Gulf Cooperation Council (GCC) countries, the water demand is expected to grow up to 46 BCM (billion cubic meters) by

2030, compared to 28 BCM in 2000, as shown in Figure 13.6A (Shahzad et al., 2019a). The current fresh water sources are not sufficient to fulfill the increasing water demand. The projected demand of potable water can only be supplied by seawater desalination processes. It has been reported that the world's desalination capacities may be doubled by 2030 with the current cumulative annual growth rates shown in Figure 13.6B.

13.6 COMPARISON OF ENERGY CONSUMPTION IN DESALINATION BY CAPACITIVE DEIONIZATION AND REVERSE OSMOSIS

It is important to consider novel desalination processes, even if they do not currently meet the economic feasibility of conventional processes. Ramachandran et al. (2019) remarked on energy consumption in desalination by comparing

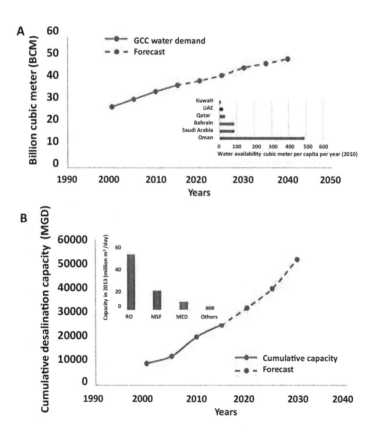

FIGURE 13.6 A Water demand trend from 1990 to 2040 in the Gulf Cooperation Council countries; and B Global cumulative desalination trend and forecast up to 2030 (Shahzad et al. 2019b).

capacitive deionization and reverse osmosis. Capacitive deionization (CDI) is an emerging brackish water desalination technology in which ions are removed from water by electrostatically adsorbing them onto porous electrodes (Figure 13.7). Qin et al. (2019) estimated very high values of energy consumption in CDI for brackish water desalination (e.g., ~1 to 2 g/L salt solution). These values and trends did not agree with published experiments for CDI by Wang et al. (2019). The differences were attributed to important scaling errors and incorrect values in their model resistance parameters. This was a similar observation and conclusion to that reported by Shahzad et al (2019a, 2019b), who maintained that energy and desalination system planners may make serious judgmental errors in the process selection of installations, without realizing it, if data is either flawed or inaccurate. Ramachandran et al. (2019) concluded that results from experiments and experimentally validated models suggest that brackish water desalination with reasonably high salt rejection (±70%) and water recovery (± 80%) can be achieved by capacitive deionization in an energy-efficient manner. While capacitive deionization (CDI) is a significantly less mature technology than reverse osmosis, experiments and experimentally validated models show that it has strong potential to become a competitor to other brackish water desalination technologies.

Understanding the key drivers of energy consumption in capacitive deionization (CDI) and benchmarking CDI with RO, the current state-of-the-art for brackish and seawater desalination, is crucial to guide the future expansion of desalination technologies. Qin et al. (2019) developed system-scale models to analyze the energy consumption and energy efficiency of CDI and RO over a wide range of material properties and operating conditions. Their analysis showed that RO is significantly more energy efficient than CDI, particularly

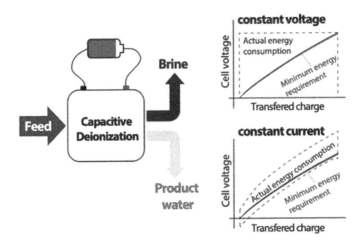

FIGURE 13.7 Energy efficiency of capacitive deionization. Ions are removed from water by electrostatically adsorbing them onto porous electrodes (Wang et al., 2019).

when targeting higher salinity feed streams and higher salt rejection values. For brackish water with a salt concentration of 2000 mg L^{-1}, achieving 50% water recovery and 75% salt rejection with an average water flux of 10 $Lm^{-2} h^{-1}$, using CDI requires a specific energy consumption of 0.85 $kWhm^{-3}$, more than eight times that of RO (0.09 $kWhm^{-3}$). Significantly, the results of Qin et al. (2019) also indicated that current efforts to improve electrode materials can only slightly reduce the energy consumption of CDI.

The modeling study by Qin et al. (2019) focused on a multicell CDI stack. The cells were connected in parallel with respect to both electrical and fluid flow. Each cell consisted of a pair of current collectors and porous electrodes. The positively charged electrode was equipped with an anion-exchange membrane (AEM), whereas the negatively charged electrode used a cation-exchange membrane (CEM). This CDI configuration is often referred to as membrane capacitive deionization (MCDI) and is employed to improve charge efficiency. Sodium chloride (NaCl) solution, in concentrations representative of brackish water, was used as the feed. When NaCl solution was fed between the electrodes and an electric potential applied, sodium ions were transported through the CEM and adsorbed onto the negative electrode. On the other hand, chloride ions are transported through the AEM and adsorbed onto the positive electrode (Figure 13.8A). A constant current was applied via a power supply unit and the voltage throughout the charging cycle was prevented from exceeding 1.5 V to ensure that Faradaic reactions (i.e., charge-transfer reactions occurring at the surface of the electrode) did not occur. During the discharge phase, the electrodes were short-circuited while feed solution continued to flow through the channel. The ions accumulated in the electrodes during the charging step were released into the flow spacer channel, forming the brine and regenerating the porous electrodes for the next cycle of salt adsorption (Figure 13.8B). To alter the water recovery ratio, the duration of the discharging step was varied, while holding the flow rate of both steps constant.

The capital costs of CDI are generally controlled by the price of the electrodes and ion-exchange membranes. Qin et al. (2019) has shown that the capital cost of CDI is currently extremely high, compared to brackish water RO. This is demonstrated by the price per unit area of frequently used ion-exchange membranes in CDI, which is more than one order of magnitude higher than a typical commercial brackish water RO membrane. This high cost hinders the scale-up of this technology. In addition, the high cost and low stability of electrode materials results in high life-cycle cost, further detracting from the feasibility of large-scale CDI. Qin et al. (2019) concluded that their rigorous modeling of both CDI and RO revealed that RO is currently the more energy efficient technology for desalination of brackish waters. Though CDI excels energetically for very low salt rejection (< 25%) and high water recovery, such conditions are not of practical interest. RO has also demonstrated superior long-term performance, lower capital cost, and has the potential to achieve greater solute rejection, including the removal of neutral solutes, further reinforcing its supremacy for low salinity desalination.

Sustainable Development and Future Trends

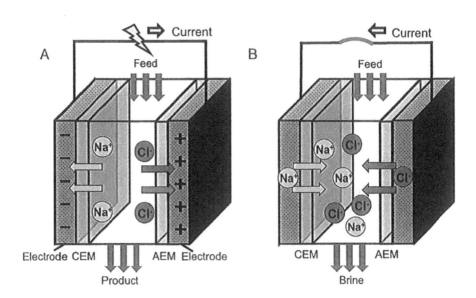

FIGURE 13.8 Schematics of membrane capacitive deionization (MCDI) showing: (A) the charging step; and (B) the discharging step. When brackish water is fed into the cell and an external voltage is applied, cations migrate through the cation-exchange membrane (CEM) and adsorb on the negative electrode, while anions migrate through the anion-exchange membrane (AEM) and adsorb on the positive electrode. During the discharging step, the power supply is disengaged and the electrodes are short-circuited. The ions stored in the electrodes are released into the flow channel, forming a brine and regenerating the porous electrodes for the next cycle of salt adsorption (Qin et al. (2019).

13.7 LATEST DEVELOPMENTS AND FUTURE DIRECTIONS IN MEMBRANE DESALINATION

Reverse osmosis (RO) based on polyamide (PA) thin-film composite (TFC) membranes have dominated the desalination industry (Xu et al., 2019). Two-dimensional (2D) nanoporous membranes with subnanopores, typically represented by graphene and its derivatives accompanied by some analogues such as molybdenum disulfide (MoS2), have displayed great potential for significant improvements in designing new kinds of desalination membranes. In a review paper Xu et al. (2019) discussed the development of 2D nanoporous membranes with subnanopores on desalination, with emphasis on simulations coupled with experimental studies. The authors anticipated that their studies could provide clues and insights for the future exploration of novel desalination membranes, and further contribute to the advance of desalination membranes synthesis and large-scale production. Xu et al. (2019) concluded that for the commercialization of 2D nanoporous membranes, the most important issue is to discover the ideal configuration. Given the ultrathin thickness (~1 nm) of 2D nanoporous membranes, it is most feasible to construct multilayer membranes based on suitable supports. It was recommended that desalination performance of multilayer graphene nanoporous membranes be further investigated.

Albeirutty et al. (2019) noted that the development of innovative materials, such as nanocomposite membranes, as well as the study of advanced membrane processes, for example membrane distillation (MD), are the focus of many universities and research centers. The driving force in many cases is to be seen as a leader in the development and application of sustainable desalination technology. In the situation of arid countries such as Saudi Arabia, Albeirutty et al. (2019) argued that such activities must be matched with augmentation of the water supply in urban areas through desalination and water treatment, capacity building, technology transfer and localization and strengthening of international collaboration. Their current research focused on development of blended fouling resistive polymeric membranes through the incorporation of nanoparticles in the membrane matrix, membrane modules design, membrane distillation applications, and renewable energy desalination. The importance of establishing strong international collaborations to accelerate knowledge transfer and improve quality of the research was also emphasized.

Thorough characterization of seawater RO fouled membranes from large scale commercial desalination plants is critical to any significant improvements in performance. Oussama et al. (2019), in a case study from Algeria, assessed used RO membranes taken from a commercial water company seawater desalination unit. Such tests help make it possible to understand the links between performance degradation observed at the macroscopic scale and at the scale at which aging takes place. External and internal visual observations allow seeing the state of degradation. Microscopic analysis of the used membranes surface showed the importance of fouling. In addition, quantification and identification analyses determined a high fouling rate in the used membrane with foulants that were inorganic and organic in nature. From the results of the study by Oussama et al. (2019), the following conclusions could be drawn: the extent of fouling was uneven on the membrane surface with the areas below or near the feed spacer being most affected; inorganic elements were found with a high percentage of Fe and Al; the use of iron chlorides as coagulants and maleic acid – based antiscalants can contribute to the increase in the percentage of these elements in the fouling layer; and the adhesive complex (i.e., fouling layer) was composed of particulate colloidal matter (Al, Fe and Si) and organic substances.

13.8 APPLICATION OF RENEWABLE ENERGY TECHNOLOGY FOR WATER DESALINATION

Conventional technologies for desalination of salt and brackish water are still limited since they have a high demand for energy which is mostly provided through expensive fossil fuels, whereas less than 1% of the desalination capacity depends on renewables (Bundschuh et al., 2018). However, the upsurge in desalination capacity and the proportional energy rise in demand make the further use of fossil fuels, increasingly economically and environmentally unsustainable.

Sustainable Development and Future Trends

A massive shift from fossil fuel – powered desalination to renewable energy (RE) powered technologies will be essential to meet the growing demand for freshwater production by desalination. Decoupling freshwater production costs from the ever increasing prices of fossil fuels is vital if freshwater is to be provided for agricultural purposes such as irrigation in which water is demanded in large quantities but at lowest possible cost.

Significant efforts have been made in the last two decades for adapting conventional desalination technologies so that they can be powered by renewable energy sources; however, upscaling to larger plants has been hampered by technological, economic, and political/regulatory (e.g., subsidies for fossil fuels) challenges. In the last few years, medium-scale RE-driven desalination plants have been installed worldwide. However, most of them are powered by electricity produced from solar PV and wind commercial units which are available on the market (Bundschuh et al., 2018), rather than by solar or geothermal heat directly, which can provide more sustainable desalination processes. Solar thermal desalination which combines solar heat with desalination technologies such as MSF, MED, and VC (vapor compression) are most suited for areas with high solar irradiation (Ghaffour et al., 2014; Zaragoza et al., 2014).

Pressure-driven membrane desalination processes necessitate electrical energy to supply the mechanical energy for membrane separation and pretreatment, as well as pumping inside the plant, and distribution outside of the plant (Gude, 2016). In his review, Gude (2016) presented the specific energy consumption for thermal and membrane desalination processes in terms of kJ of energy required for producing one unit of freshwater in kilograms (Figure 13.9).

Geothermal energy can be an attractive alternative option or an additional energy source if low-cost, low-enthalpy geothermal sources are available. Such sources include geothermal resources at shallow depth, water coproduced from onshore and offshore hydrocarbon wells or from already existing deep wells,

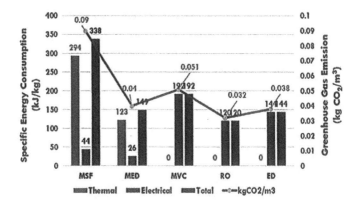

FIGURE 13.9 Specific energy consumption for thermal and membrane desalination processes and greenhouse gas emissions for unit freshwater production (after Bundschuh et al., 2018; Gude, 2016).

364   Membrane Desalination

and residual heat from geothermal power plants (Bundschuh et al., 2018). Geothermal energy is accessible day and night every day of the year and can thus serve as an add-on to energy sources which are only available intermittently. The application of geothermal energy in desalination is still a relatively unexplored technical concept, however.

Freshwater production by RE-powered desalination is a technologically sound option at small- or medium-scale, and is economically viable for water supply in remote areas. However, upscaling to large size is still hindered, owing to the intermittent availability of, for example, wind and solar energy; this is a disadvantage that geothermal energy does not have. This also suggests that the implementation of a combined-cycle solar- and geothermal-powered desalination process without the need for energy storage has the potential for stable energy production. Likewise, the development and improvement of innovative desalination technologies that do not need continuous operation, such as adsorption distillation (AD) and membrane distillation (MD) and are consequently more suitable for RE use, can be utilized to overcome the energy intermittency limitation.

Ullah and Rasul (2019) highlighted the environmental impacts of desalination technologies along with an economic analysis and cost comparison of conventional desalination methods with different solar energy – based technologies. Their review was part of an investigation into integration of solar thermal desalination into existing grid infrastructure in the Australian context. The authors concluded that large-scale desalination projects can have major environmental impacts during construction and operation similar to any major project. Desalination may produce high-quality water, but may also end up introducing harmful biological or chemical contaminants into the water supply. All water from desalination plants must be monitored and regulated to ensure safe public and marine health. Further research into identifying all contaminants in desalination brines and the mitigation of impacts of brine discharge is needed. There should be no brine disposal in underground aquifers until in-depth and expert groundwater surveys are carried out, and there is no potential risk of brine plumes appearing in fresh water wells.

13.9 CHALLENGES SURROUNDING THE USE OF NUCLEAR TECHNOLOGY AS AN ENERGY SOURCE

Goosen et al. (2014) in a critical review discussed the controversy surrounding conventional energy production using nuclear technology. The feasibility of integrated nuclear desalination plants has been proven with over 150 reactor-years of experience, chiefly in Kazakhstan, India, and Japan. However, it can be reasoned that nuclear energy should not be considered as a long-term global energy source. The growth of nuclear energy has historically increased the ability of nations to obtain or enrich uranium for nuclear weapons and a large-scale worldwide increase in nuclear energy facilities would aggravate this dilemma, putting the world at greater risk of a nuclear incident. In Japan, some ten desalination

Sustainable Development and Future Trends   **365**

facilities linked to pressurized water reactors operating for electricity production have yielded 1000–3000 m^3/day each of potable water, and over 100 reactor-years of experience have accrued. On the other hand, Goosen et al. (2014) argued that the earthquake and tsunami catastrophe that occurred in Japan in 2011 and its effects on Japan's nuclear reactors have shown the susceptibility of this technology and the need for better safeguards.

ElBaradei (2020) in a recent address at the *Japan Atomic Industrial Forum* stated that three decades ago, nuclear energy was hailed as the energy of the future. He went on to explain that today, its growth is stagnant in many parts of the world and it is not absolutely assured that global environmental considerations will, by themselves, result in new investment in nuclear power generation. The extent to which impartial attention will be given to the positive role of nuclear energy to human well-being and sustainable growth depends on ensuring its peaceful and safe use and its competitiveness in the marketplace. ElBaradei (2020) stressed that these are not competing objectives. All must be pursued equally. He concluded that meeting the challenges to the future of nuclear power consequently necessitates action on two fronts: the restoration of public confidence in the safe and wholly peaceful use of nuclear energy, and the demonstration of economic competitiveness of nuclear power in comparison with other options.

13.10   CONCLUDING REMARKS

One of the main challenges facing the economic sustainability of desalination plants is the high cost of energy. Good infrastructure planning and management, minimizing the problem of brine discharge, enhancing energy efficiency and reducing costs, improved membrane desalination techniques, and applying renewable energy and environmentally friendly technology are all key factors in the successful sustainable commercialization of desalination technology. Energy generation is also associated with air pollution and environmental degradation. Furthermore, the scientific community, as well as the public, now recognize that climate change is likewise a major factor in sustainable economic development. Consequently, water resources planners must consider that recent climate history may not be an adequate predictor of the future.

Although many seawater desalination processes have been reported, the grade of the energy supplied has often not been adequately described. Effectiveness comparison should only be conducted for the same energy input processes. The misunderstanding that considers all derived energies as equal in the desalination industry can have severe economic and environmental consequences. Energy and desalination system planners may make serious judgmental errors in the process selection of installations, without realizing it, if data is either flawed or inaccurate. Substandard efficiency technologies may be implemented as a result of poor decisions in many water-stressed countries. This can burden a country's economy with higher energy costs, and cause undesirable environmental effects. Furthermore, the potential of nuclear energy to effectively

366   Membrane Desalination

contribute to human well-being and sustainable growth is still an open question that will depend on ensuring its peaceful and safe use and its competitiveness in the market place.

It is essential to consider novel desalination processes such as capacitive deionization, even if they do not currently meet the economic feasibility of conventional processes such as reverse osmosis. One can argue that the development of reverse osmosis and photovoltaic panels are excellent examples of technologies that have become very commercially successful, even after a rather slow start many decades ago. Understanding the key drivers of energy consumption in new technologies is crucial to guide any future expansion. Establishing strong international collaborations to accelerate knowledge transfer and improve quality of the research is vital.

Finally, desalination may produce high-quality water, but it may also end up introducing harmful biological or chemical contaminants into the water supply. All water from desalination plants must be monitored and regulated to ensure safe public and marine health. Further research into identifying all contaminants in desalination brines and the mitigation of impacts of brine discharge is needed. There should be no brine disposal in underground aquifers until in-depth and expert groundwater surveys are carried out, and there is no potential risk of brine plumes appearing in fresh water wells.

In conclusion, as scientists and decision makers, it is crucial to remember that cost reduction and environmental protection are equally important in terms of sustainable development of desalination technology.

REFERENCES

Albeirutty, M., Bamaga, O.A., Figoli, A., and Drioli, E., 2019. Desalination research and development in Saudi Arabia: Experience of the center of excellence in desalination technology at King Abdulaziz University. *Journal of Membrane Science and Research*, 5(1), pp.76–82.

Boo, C., Winton, R.K., Conway, K.M., and Yip, N.Y., 2019. Membrane-less and non-evaporative desalination of hypersaline brines by temperature swing solvent extraction. *Environmental Science & Technology Letters*, 6(6), pp. 359–364.

Brogioli, D., La Mantia, F., and Yip, N.Y., 2019. Energy efficiency analysis of distillation for thermally regenerative salinity gradient power technologies. *Renewable Energy*, 133, pp.1034–1045.

Bundschuh, J., Tomaszewska, B., Ghaffour, N., Hamawand, I., Mahmoudi, H., and Goosen, M., 2018. Coupling geothermal direct heat with agriculture. In J. Bundschuh and B. Tomaszewska (Eds.), *Geothermal Water Management* (pp. 277–300). London: CRC Press. doi:10.1201/9781315734972

Caldera, U., Bogdanov, D., Afanasyeva, S., and Breyer, C., 2018. Role of seawater desalination in the management of an integrated water and 100% renewable energy based power sector in Saudi Arabia. *Water*, 10(1), p.3.

Cox, T., Bywater, J., Heineman, M., Rodrigo, D., and Wood, S., 2019. Forecasting extreme events: Making sense of noisy climate data in support of water resources planning. *H2Open Journal*, 2(1), pp.45–57.

ElBaradei, M., 2020. *The Future of Nuclear Power: Looking Ahead*. Sendai: Address at the Japan Atomic Industrial Forum, 12 April 1999.

Sustainable Development and Future Trends 367

Fantini, R., Mondello, F., Rigallo, A., and Sorbara, D., 2016. Le tecnologie abilitanti per L'IoT (Enabling Technologies for the IoT). Available online: www.telecomitalia. com/tit/it/notiziariotecnico/edizioni-2016/n-3-2016/capitolo-4.html

Fletcher, S., Lickley, M., and Strzepek, K., 2019. Learning about climate change uncertainty enables flexible water infrastructure planning. *Nature Communications*, 10(1), p.1782.

Ghaffour, N., Lattemann, S., Missimer, T.M., Ng, K.C., Sinha, S., and Amy, G., 2014. Renewable energy-driven innovative energy-efficient desalination technologies. *Applied Energy*, 136, pp.1155–1165.

Goosen, M.F., Mahmoudi, H., and Ghaffour, N., 2014. Today's and future challenges in applications of renewable energy technologies for desalination. *Critical Reviews in Environmental Science and Technology*, 44(9), pp.929–999.

Gude, V.G., 2016. Geothermal source potential for water desalination – Current status and future perspective. *Renewable and Sustainable Energy Reviews*, 57, pp.1038–1065.

Kaya, A., Tok, M.E., and Koc, M., 2019. A levelized cost analysis for solar-energy-powered sea water desalination in the emirate of Abu Dhabi. *Sustainability*, 11(6), p.1691.

Lempert, R.J., Groves, D.G., Popper, S.W., and Bankes, S.C., 2006. A general, analytic method for generating robust strategies and narrative scenarios. *Management Science*, 52(4), pp.514–528.

Ling-Chin, J., Bao, H., Ma, Z., Taylor, W., and Roskilly, A.P., 2019. State-of-the-art technologies on low-grade heat recovery and utilization in industry. *Energy Conversion: Current Technologies and Future Trends*, p. 55.

Maiolo, M., Carini, M., Capano, G., Pantusa, D., and Iusi, M., 2019. Trends in metering potable water. *Water Practice and Technology*, 14(1), pp.1–9.

Margulis, S., Hughes, G., Schneider, R., Pandey, K., Narain, U., and Kemeny, T., 2010. *Economics of Adaptation to Climate Change: Synthesis Report* (p.101). Washington, DC: World Bank.

Oussama, N., Bouabdesselam, H., Ghaffour, N., and Abdelkader, L., 2019. Characterization of seawater reverse osmosis fouled membranes from large scale commercial desalination plant. *Chemistry International*, 5(2), pp.158–167.

Pahl-Wostl, C., 2007. Transitions towards adaptive management of water facing climate and global change. *Water Resources Management*, 21(1), pp.49–62.

Petersen, K.L., Heck, N., Reguero, B.G., Potts, D., Hovagimian, A., and Paytan, A., 2019. Biological and physical effects of brine discharge from the Carlsbad desalination plant and implications for future desalination plant constructions. *Water*, 11(2), p.208.

Pramanik, B.K., Shu, L., and Jegatheesan, V., 2017. A review of the management and treatment of brine solutions. *Environmental Science: Water Research & Technology*, 3(4), pp.625–658.

Puig, D., Olhoff, A., Bee, S., Dickson, B., and Alverson, K., (Eds.) 2016. *The Adaptation Finance Gap Report*. Nairobi, Kenya: United Nations Environment Programme.

Qin, M., Deshmukh, A., Epsztein, R., Patel, S.K., Owoseni, O.M., Walker, W.S., and Elimelech, M., 2019. Comparison of energy consumption in desalination by capacitive deionization and reverse osmosis. *Desalination*, 455, pp.100–114.

Ramachandran, A., Oyarzun, D.I., Hawks, S.A., Campbell, P.G., Stadermann, M., and Santiago, J.G., 2019. Comments on "Comparison of energy consumption in desalination by capacitive deionization and reverse osmosis". *Desalination*, 461, 30–36. (LLNL-JRNL-770722).

Saito, O., Kamiyama, C., Hashimoto, S., Matsui, T., Shoyama, K., Kabaya, K., Uetake, T., Taki, H., Ishikawa, Y., Matsushita, K., and Yamane, F., 2019. Co-design of national-scale future scenarios in Japan to predict and assess natural capital and ecosystem services. *Sustainability Science*, 14(1), pp.5–21.

Shahzad, M.W., Burhan, M., and Ng, K.C., 2019b. A standard primary energy approach for comparing desalination processes. *Npj Clean Water*, 2(1), p.1.

Shahzad, M.W., Burhan, M., Ybyraiymkul, D., and Ng, K.C., 2019a. Desalination processes' efficiency and future roadmap. *Entropy*, 21(1), p.84.

Ullah, I., and Rasul, M., 2019. Recent developments in solar thermal desalination technologies: A review. *Energies*, 12(1), p.119.

Wang, L., Dykstra, J.E., and Lin, S., 2019. Energy efficiency of capacitive deionization. *Environmental Science & Technology*, 53(7), pp.3366–3378. doi:10.1021/acs.est.8b04858

Xu, G.R., Xu, J.M., Su, H.C., Liu, X.Y., Zhao, H.L., Feng, H.J., and Das, R., 2019. Two-dimensional (2D) nanoporous membranes with sub-nanopores in reverse osmosis desalination: Latest developments and future directions. *Desalination*, 451, pp.18–34.

Zaragoza, G., Ruiz-Aguirre, A., and Guillén-Burrieza, E., 2014. Efficiency in the use of solar thermal energy of small membrane desalination systems for decentralized water production. *Applied Energy*, 130, pp.491–499.

Terminology

Alkalinity: The buffering capacity of a water body; in other words, a measure of the ability of a water body to neutralize acids and bases and thus maintain a fairly stable pH level. Alkalinity prevents the water pH levels from becoming too basic or too acid and stabilizes its pH levels to around 7. Total alkalinity of water is the sum of all three kinds of alkalinity: carbonate alkalinity, bicarbonate alkalinity, and hydroxide alkalinity. Specifically, the measured alkalinity is set equal to:

$$A_T = \left[HCO_3^-\right]_T + 2\left[CO_3^{2-}\right]_T + \left[B(OH)_4^-\right]_T + \left[OH^-\right]_T + 2\left[PO_4^{3-}\right]_T$$
$$+ \left[HPO_4^{2-}\right]_T + \left[SiO(OH)_3^-\right]_T - \left[H^+\right]_{SWS} - \left[HSO_4^-\right]$$

(Subscript T indicates the total concentration of the species in the solution as measured. This is opposed to the free concentration, which takes into account the significant amount of ion pair interactions that occur in seawater.)

Alkalinity can be measured by titrating a sample with a strong acid until all the buffering capacity of the aforementioned ions above the pH of bicarbonate or carbonate is consumed. This point is functionally set to pH 4.5. At this point, all the bases of interest have been protonated to the zero level species; hence, they no longer cause alkalinity.

Back flushing: A reversed filtration process. In back flushing, the membrane's pores are flushed inside out with permeate. Permeate is always used for a backward flush, because the permeate chamber must always be free of contagion. When the flux has not restored itself sufficiently after back flushing, a chemical cleaning process should be applied.

Biochemical Oxygen Demand (BOD): The amount of oxygen (measured in mg/L) that is required for the decomposition of organic matter by single-cell organisms, under test conditions. It is used to measure the amount of organic pollution in wastewater.

BOD$_5$: The amount of dissolved oxygen consumed in five days by bacteria that perform biological degradation of organic matter.

Brackish water: Water containing low concentrations of soluble salts, usually between 1,000 and 10,000 mg/L.

Brine: Water saturated with, or containing, a high concentration of salts, usually in excess of 36,000 mg/L.

Bubble point pressure: Pressure at which a continuous stream of gas bubbles is pressed through a liquid wet filter.

370   Terminology

Bulk temperature (T_b): Temperature that exists in the bulk phase; in practice, this temperature is equal to the measured temperature.

Chemical Oxygen Demand (COD): The amount of oxygen (measured in mg/L) that is consumed in the oxidation of organic and oxidizable inorganic matter, under test conditions. It is used to measure the total amount of organic and inorganic pollution in wastewater. Contrary to BOD, with COD, practically all compounds are fully oxidized.

Circulation loop: A section of a membrane plant containing one or more circulation pumps ensuring adequate cross-flow velocity of the fluid over the membrane.

Concentration factor (CF): The degree of increasing the concentration of a component in a membrane operation; $CF = C_r/C_f$ (where C_r and C_f the concentrations of the component in the retentate and feed streams, respectively).

Concentration polarization: The concentration gradients at a membrane/solution interface resulted from selective transfer of some species through the membrane under the effect of transmembrane driving forces. Generally, the cause of concentration polarization is the ability of a membrane to transport some species more readily than the other(s); the retained species are concentrated at the upstream membrane surface while the concentration of transported species decreases. In the cases of gas separations, pervaporation, membrane distillation, reverse osmosis, nanofiltration, ultrafiltration, and microfiltration separations, the concentration profile has a higher level of solute nearest to the upstream membrane surface, compared with the more or less well-mixed bulk fluid far from the membrane surface. In the case of dialysis and electrodialysis, the concentrations of selectively transported dissolved species are reduced at the upstream membrane surface, compared to the bulk solution. Concentration polarization strongly affects the performance of the separation process. First, concentration changes in the solution reduce the driving force within the membrane, and hence, the useful flux/rate of separation. In the case of pressure driven processes, this phenomenon causes an increase of the osmotic pressure gradient in the membrane, which reduces the net driving pressure gradient. In the case of dialysis, the driving concentration gradient in the membrane is reduced. In the case of electromembrane processes, the potential drop in the diffusion boundary layers reduces the gradient of electric potential in the membrane. Lower rate of separation under the same external driving force means increased power consumption. Moreover, concentration polarization leads to increased salt leakage through the membrane, as well as increased probability of scale/fouling development. Thus, the selectivity of separation and the membrane lifetime deteriorates. Generally, to reduce concentration polarization, increased flow rates of the solutions between the membranes, as well as spacers promoting turbulence, are applied. This technique results in better mixing of the solution and in reducing the thickness of the diffusion boundary layer, which is defined as

Terminology

the region in the vicinity of an electrode or a membrane in which the concentrations are different from their value in the bulk solution.

Cross-flow velocity (u): The velocity of a fluid flowing parallel to the membrane (also called tangential or parallel velocity). In cross flow operation mode, the feed solution flows parallel to the membrane surface while the permeate flows in a direction perpendicular to the membrane surface.

Deionization (DI): The process of removing ions from water, most commonly through an ion exchange process.

Electrical Conductivity (Water): The ability or power of a substance to conduct or transmit electricity. An electrical current results from the motion of electrically charged particles in response to forces that act on them from an applied electric field. Within most solid materials a current arises from the flow of electrons, which is called electronic conduction. In all conductors, semiconductors, and many insulated materials, only electronic conduction exists, and the electrical conductivity is strongly dependent on the number of electrons available to participate in the conduction process. Most metals are extremely good conductors of electricity, because of the large number of free electrons that can be excited in an empty and available energy state.

In water and ionic materials or fluids, a net motion of charged ions can occur. This phenomenon produces an electric current, and is called ionic conduction. Electrical conductivity of water is directly related to the concentration of dissolved ionized solids in the water. Ions from the dissolved solids in water create the ability for that water to conduct an electric current, which can be measured using a conventional conductivity meter.

Electrical conductivity is defined as the ratio between the current density (J) and the electric field intensity (e) and it is the opposite of the resistivity (r, $[\Omega\,m$ or $W\,m\,A^{-2}]$):

$$s = J/e = 1/r$$

Conductivity is usually measured in micro- or millisiemens per centimeter (μS/cm or mS/cm). It can also be reported in micromhos or millimhos/centimeter (μmhos/cm or mmhos/cm), although these units are less common. One siemen is equal to one mho. Microsiemens per centimeter is the standard unit for freshwater measurements. The relation between the various units is the following:

$$1\,S/cm = 10^3 mS/cm = 10^3 dS/m = 10^6 \mu S/cm = 10^3 EC = 10^4 CF$$
$$= 1\,mho/cm = 0.01\,mho/m$$

Pure water is not a good conductor of electricity. Ordinary distilled water in equilibrium with carbon dioxide of the air has a conductivity of about $10 \times 10^{-6}\,A^2 W^{-1}\cdot m^{-1}$ (10×10^{-6} S/m). Because the electrical current is

372   Terminology

transported by the ions in solution, the conductivity increases as the concentration of ions increases. Thus, conductivity increases as water-dissolved ionic species.

Typical conductivities of waters are: Ultrapure water $5.5 \cdot 10^{-6}$ S/m; drinking water 0.005 – 0.05 S/m, Sea water 5 S/m.

Electrodeionization (EDI): A water treatment process combining an electrodialysis membrane process with an ion exchange resin process to produce high purity, demineralized water.

Electrodialysis: A process that uses electrical currents, applied to permeable membranes to remove minerals from water.

Feed: The fluid entering a membrane module or plant.

Flux (J): Amount of permeate, or of any component in the permeate, that is transported through a membrane per unit of membrane area and per unit of time. The usual units for flux in membrane operations are: volume flux (m^3/m^2s), mass flux (kg/m^2s), and molar flux (mol/m^2s).

Forward flush: A membrane cleaning method where membranes are flushed with feed water or permeate in the forward direction. The feedwater or permeate flows through the system more rapidly than during the production phase, resulting in higher turbulence that causes the particles that are absorbed to the membrane to be released and discharged. Particles that are absorbed inside the membrane's pores are not released in this way. These particles can only be removed through backward flushing.

Forward osmosis: A process for the separation of water from a solution containing unwanted solutes. A *draw* solution of higher osmotic pressure than that of the feed solution is used to induce a net flow of water through a semipermeable membrane, such that the feed solution becomes concentrated as the draw solution becomes dilute. The diluted draw solution may then be used directly (as with an ingestible solute such as glucose), or sent to a secondary separation process for the removal of the draw solute. This secondary separation can be more efficient than a reverse osmosis process would be alone, depending on the draw solute used and the feedwater treated. Forward osmosis is an area of ongoing research, focusing on applications in desalination, water purification, water treatment, food processing, and other areas of study.

Fouling: The deposition of material on the membrane surface and/or in its pores, leading to a change in membrane's performance. Membrane fouling is the main cause of permeate flux decline and loss of product quality in reverse osmosis (RO) Systems, so fouling control dominates RO system design and operation. Sources of fouling can be divided into four principal categories: scale, silt (particular), bacteria (bio fouling, growth of bacteria) and organic fouling (oil,

Terminology

grease). Fouling control involves pretreatment of the feed water to minimize fouling, as well as regular cleaning to handle fouling that still occurs. Fouling by particulates (silt), bacteria, and organics generally affects the first modules in the plant the most. Scaling is related with more concentrated feed solutions. Therefore, the last modules in the plant are most affected, because they are exposed to the most concentrated feed water.

Hardness (water): The amount of dissolved calcium and magnesium in the water. Usually there are two types of water hardness: temporary hardness, which is caused by dissolved calcium or magnesium bicarbonates (which are removed by boiling) and permanent hardness, which is caused by dissolved calcium or magnesium sulfates (which are not removed by boiling). The total water hardness is calculated with the following formula:

$$\text{TOTAL HARDNESS} = \text{CALCIUM HARDNESS} + \text{MAGNESIUM HARDNESS}$$

The calcium and magnesium hardness is the concentration of calcium and magnesium ions expressed as equivalent of calcium carbonate. The molar mass of $CaCO_3$, Ca^{2+} and Mg^{2+} are respectively, 100.1 g/mol, 40.1 g/mol and 24.3 g/mol. The ratios of the molar masses are:

$$\frac{M_{CaCO_3}}{M_{Ca}} = \frac{100.1}{40.1} = 2.5 \text{ and } \frac{M_{CaCO_3}}{M_{Mg}} = \frac{100.1}{24.3} = 4.1 \text{ respectively.}$$

So total water hardness - expressed as equivalent of $CaCO_3$ - can be calculated with the following formula:

$$[CaCO_3] = 2.5[Ca^{2+}] + 4.1[Mg^{2+}]$$

The hardness expressed in ppm or mg/L of $CaCO_3$. Alternatives units are also the French (fH°), German (dH°) and English (eH°) degrees. The correlation among these units is the following:

$$10 \text{ ppm} = 1 \text{ fH}° = 0.56 \text{ dH}° = 0.7 \text{ eH}°$$

Ion exchange: The replacement of undesirable ions with a certain charge by desirable ions of the same charge in a solution, by an ion-permeable absorbent.

Ionic strength: A measure of the concentration of ions in a solution. Ionic compounds, when dissolved in water, dissociate into ions. Ionic strength can be molar (mol/L) or molal (mol/kg water) and to avoid confusion the units should be stated explicitly. The molar ionic strength, I, of a solution is a function of the concentration of all ions present in that solution:

$$I = \frac{1}{2}\sum_{i=1}^{n} C_i z_i^2$$

where $\frac{1}{2}$ is because are included both cations and anions; C_i is the molar concentration of ion i (M, mol/L); z_i is the charge number of that ion, and the sum is taken over all ions in the solution.

IPA bubble point: The IPA (isopropylalcohol) bubble point can be measured according to a standard test method as described in ASTM F-316. This method employs a procedure for determining the maximum pore size and the pore size distribution of a membrane filter by measuring the initial bubble point and gas flow versus pressure through a liquid wet filter. The only difference between the "IPA bubble point method" and ASTM F-316 is that IPA is used as the wetting liquid instead of water.

Langelier Index (LI): An index reflecting the equilibrium pH of water with respect to calcium and alkalinity; used as a measure for control both corrosion and scale deposition.

The formula to calculate the Langelier Saturation Index is:

$$LSI = pH - pH_s$$

where: LSI=Langelier Saturation Index; pH_s=saturation pH; and pH=pH value

The pH of a solution is usually known. Equations (A), (B) and (C) are used to compute the saturation pH (pH_s). Equation (C) is a result of the combination of Equations (A) and (B).

$$\left(HCO_3^-\right)_{aq} \leftrightarrow \left(H^+\right)_{aq} + \left(CO_3^{2-}\right)_{aq} \tag{A}$$

$$\left(Ca^{2+}\right)_{aq} + \left(CO_3^{2-}\right)_{aq} \leftrightarrow \left(CaCO_3\right)_s \tag{B}$$

$$\left(Ca^{2+}\right)_{aq} + \left(HCO_3^-\right)_{aq} \leftrightarrow \left(H^+\right)_{aq} + \left(CaCO_3\right)_s \tag{C}$$

The equilibrium constant K_a of Equation (A) can be calculated with the following formula:

$$K_a = \frac{\gamma_{H^+} \cdot [H^+] \cdot \gamma_{CO_3^{2-}} \cdot \left[CO_3^{2-}\right]}{\gamma_{HCO_3^-} \cdot \left[HCO_3^-\right]}$$

where:

γ_{H^+} = activity coefficient for hydrogen ion;

$\gamma_{CO_3^{2-}}$ = activity coefficient for carbonate;

$\gamma_{HCO_3^-}$ = activity coefficient for bicarbonate;

Terminology

$[H^+]$ = concentration of hydrogen ions;
$[CO_3^{2-}]$ = concentration of carbonates; and
$[HCO_3^-]$ = concentration of bicarbonates.

The equilibrium constant K_{sp} of Equation (B) can be calculated with the following formula:

$$K_{sp} = \gamma_{Ca^{2+}} \cdot [Ca^{2+}] \cdot \gamma_{CO_3^{2-}} \cdot [CO_3^{2-}]$$

where:
K_{sp} = solubility product constant

The equilibrium constant K of Equation (C) can be calculated with the following formula:

$$K = \frac{K_{sp}}{K_a}$$

Substituting in the previously presented equation the expressions for K_{sp} and K_a and rearranging the terms we get the formula of pH_s.

$$K = \frac{\gamma_{Ca^{2+}} \cdot [Ca^{2+}] \cdot \gamma_{HCO_3^-} \cdot [HCO_3^-]}{\gamma_{H^+} \cdot [H^+]} \quad or$$

$$pH_s = -log\left(\frac{K_a \cdot \gamma_{Ca^{2+}} \cdot [Ca^{2+}] \cdot \gamma_{HCO_3^-} \cdot [HCO_3^-]}{\gamma_{H^+} \cdot K_{sp}}\right)$$

For the saturation pH to be determined a water analysis for the calcium and bicarbonate concentrations is necessary, as well as the calculation of the activity coefficients. The activity coefficient can be calculated by the following formula:

$$log(\gamma_i) = \frac{0.5 \cdot (Z_i)^2 \sqrt{I}}{1 + \sqrt{I}}$$

where:
γ_i = activity coefficient of ith ionic species;
I = ionic strength of the solution here water; and
Z_i = charge on ith ionic species.

The ionic strength can be calculated with the following formula:

$$I = 2.5 \cdot 10^{-5} \cdot TDS$$

where:
TDS = Total Dissolved Solids in mg/L or g/m^3

The equilibrium constants K_a and K_{sp} change with temperature. The following values for the equilibrium constants are given in [1]:

Temperature °C	$K_a \cdot 10^{11}$	$K_{sp} \cdot 10^9$
5	2.754	8.128
10	3.236	7.080
15	3.715	6.020
20	4.169	5.248
25	4.477	4.571
40	6.026	3.090

The relation between the constant solubility product K_{sp} of calcium carbonate and temperature is as follows

$$K_{sp} = 9.237 \cdot 10^{-9} e^{-0.0277 \cdot T}$$

The relation between the equilibrium constant K_a of carbonate/bicarbonate (CO_3^{2-}, HCO_3^-) and temperature is as follows:

$$K_a = 9.2 \cdot 10^{-13} \cdot T + 2.3 \cdot 10^{-11}$$

where T is the temperature in degree Celsius in both cases.

REFERENCES:

[1] Metcalf & Eddy, Inc. *Wastewater Engineering Treatment and Reuse*, Fourth Edition. McGraw-Hill, New York, 2003.

Liquid-entry-pressure (LEP): Pressure at which the liquid penetrates into a porous membrane (old term to be replaced by "wetting pressure").

Maximum pore size: The maximum pore size can be calculated by substituting the IPA bubble point pressure into the following formula:

$$\text{pore size} = 4B\gamma/p$$

where: B =pore size morphology constant; γ=surface tension of IPA; and p = bubble point pressure. The pore size morphology constant B is 1 for a circular pore and less than 1 for an elliptical or irregularly shaped pore. Because most pores are not circular, the use of the terms "pore diameter" and "pore radius" is misleading. Unless membranes with circular pores are used, the term "pore size" is recommended.

Terminology   377

Membrane distillation (MD): a thermally driven separation process in which a microporous hydrophobic membrane separates two aqueous solutions at different temperatures. The hydrophobicity of the membrane prevents mass transfer of the liquid, whereby a gas – liquid interface is created. The temperature gradient on the membrane results in a vapor pressure difference, whereby volatile components in the supply mix evaporate through the pores (10 nm – 1 μm) and, via diffusion and/or convection of the compartment with high vapor pressure, are transported to the compartment with low vapor pressure, where they are condensated in the cold liquid/vapour phase. For supply solutions that only contain nonvolatile substances, such as salts, water vapor is transported through the membrane whereby demineralised water is obtained on the distillation side, and a further concentrated salt flow on the supply side.

Module: The smallest practical unit containing one or more membranes and supporting structures (old terms to be replaced are permeator, membrane element).

Osmosis: The spontaneous net movement of solvent molecules through a selectively permeable membrane into a region of higher solute concentration, in the direction that tends to equalize the solute concentrations on the two sides. It may also be used to describe a physical process in which any solvent moves across a selectively permeable membrane (permeable to the solvent, but not the solute), separating two solutions of different concentrations.

Osmotic pressure: is defined as the external pressure required to be applied so that there is no net movement of solvent across the membrane. Osmotic pressure is a colligative property, meaning that the osmotic pressure depends on the molar concentration of the solute but not on its identity. Osmotic pressure is expressed by Van't Hoff's equation:

$$\Pi = i\,C\,R\,T$$

where Π is the osmotic pressure; i is the dimensionless van 't Hoff index; C is the molar concentration of solute; R is the ideal gas constant; and T is the temperature in kelvins. This formula applies when the solute concentration is sufficiently low that the solution can be treated as an ideal solution. Harmon Northrop Morse and Frazer showed that the previous equation applied to more concentrated solutions if the unit of concentration was molal rather than molar. In such a case, it takes the following form

$$\Pi = i\,(v\,m\,M_{H2O}/\,1000\mu_{H2O})\ R\,T$$

where v is the number of ions of an electrolyte; m is the molality (Kmol/kg solvent (water)); M_{H2O} is the molecular weight of water (kg/Kmol); and μ_{H2O} is the partial molecular volume of water (m^3/Kmol).

Permeability coefficient (or simply the permeability): is the transport flux of a material through the membrane per unit, driving force per unit membrane thickness. Its value must be experimentally determined. The most commonly used units are $kmol \cdot m \cdot m^{-2} \cdot s^{-1} \cdot kPa^{-1}$, $m^3 \cdot m \cdot m^{-2} \cdot s^{-1} \cdot kPa^{-1}$, or $kg \cdot m \cdot m^{-2} \cdot s^{-1} \cdot kPa^{-1}$. For pressure driven membrane separations, the driving force is usually the pressure difference across the membrane.

Permeate: The portion of the feed passing through the membrane. Distillate can also be used as a term to describe the *permeate* of membrane distillation, but it is better to use "permeate" because it is commonly used in membrane literature.

Pervaporation: a membrane separation process in which liquid diffuses through a membrane and evaporates at the permeate side of the membrane. The separation characteristics of the pervaporation process are determined by sorption into and diffusion through the membrane. This term should be avoided in membrane distillation processes in which the membrane itself has no influence on the vapor – liquid equilibrium of the liquids to be separated.

Pore size: Openings in a membrane; this term is preferred to "pore diameter" and "pore radius", because all pore shapes can be described by this term.

Porosity: The porosity is defined as the volume of gas that is trapped inside a membrane divided by the total volume of the membrane.

Recovery: The percentage of a component in the feed mixture or solution that is retrieved in permeate. For desalination recovery, recovery expresses the percentage of the feedwater that becomes product water.

$$\% \text{ Recovery } = (\text{Product flow rate } \div \text{ Feed flow rate}) \times 100$$

Retentate: The portion of the feed not passing through the membrane (old term: concentrate).

Retention: The ability of a membrane to hinder a component from passing through it or to retain a component in the fluid.

Retention coefficient (R): The degree of separation of a certain component from the solvent by the membrane under defined operating conditions; $R = 1 - C_p/C_r$ (where C_r and C_p are the concentrations of the component in the retentate and permeate streams, respectively). This term should be used if a solution of a solute (e.g., salt) in a solvent (e.g., water) is treated by membrane distillation.

Reverse osmosis: A separation process that uses pressure to force a solvent through a semipermeable membrane that retains the solute on one side and allows the pure solvent to pass to the other side, forcing it from a region of

Terminology

high solute concentration, through a membrane, to a region of low solute concentration by applying a pressure in excess of the osmotic pressure.

Scaling: The deposition of particles on a membrane, causing it to plug. Without some means of scale inhibition, reverse osmosis (RO) membranes and flow passages within membrane elements scale, owing to precipitation of such soluble compounds such as calcium carbonate, calcium sulfate, barium sulfate, and strontium sulfate. Most natural waters contain relatively high concentrations of calcium, sulfate and bicarbonate ions. In membrane desalination operations at high recovery ratios, the solubility limits of gypsum and calcite exceed saturation levels, leading to crystallization on membrane surfaces. Following an induction period, the surface blockage from the scale results in permeate flux decline, reducing the efficiency of the process and increasing operation costs. The length of this period varies with the type of scale and the degree of super saturation of the sparingly soluble salt. The induction period for calcium carbonate is much shorter than that for sulfate scales, such as calcium sulfate. It is economically preferable to prevent scaling formation, even if there are effective cleaners for scale. Scaling often plugs RO element feed passages, making cleaning difficult and very time consuming. There is also the risk that scaling will damage membrane surface. There are three methods of scale control commonly employed: acidification, ion exchange softening, and antiscalant addiction.

Selectivity (a): This term is to be defined as: $x = \dfrac{(\text{wt.\%A/wt.\%B}) \text{ in permeate}}{(\text{wt.\%A/wt.\%B}) \text{ in feed}}$

This term should be used as both components in the membrane distillation systems are volatile.

Silt Density Index (SDI): A measure of the fouling potential of suspended solids. SDI should not be confused with turbidity, which is a measurement of the amount of suspended solids. SDI and turbidity are not the same, and there is no direct correlation between them. Membranes show very little fouling when the feed water has a turbidity of < 1 NTU or SDI of less than 5. Silt is composed by suspended particulates of all types that accumulate on the membrane surface. Sources of silt are organic colloids, iron corrosion products, precipitated iron hydroxide, algae, and fine particular matter. Silt Density Index testing is a widely accepted method for estimating the rate at which colloidal and particle fouling occur in water purification systems, especially using reverse osmosis (RO) or nanofiltration membranes. The SDI test is used to predict and then prevent the particulate fouling on the membrane surface. Other names for it are the Kolloid-Index (KI) or the Fouling-Index (FI). The test is defined in ASTM Standard D4189, the American Standard for Testing Material. It measures the time required to filter a fixed volume of water through a standard 0.45µm pore size microfiltration membrane with a constant given pressure of 30 psi (2,07 bar). The difference between the initial time and the time of a second measurement, normally after 15 minutes (after silt-build- up), represents the SDI value. The feedwater has to be supplied with a pressure of 2 bars, which is regulated with the pressure regulator and reads off the value on the pressure gauge. The ball valve turns the flow on and off. The filter holder contains the 0.45 µm filter with a diameter of 47mm, which is more likely to clog from colloidal matter than from hard particles such as sand or scale. With this equipment the time required for 500 ml of feed water to flow through the filter. The same measurement is repeated after 15 minutes with the water flowing continues with a pressure of 2 bars through the filter. The SDI is calculated from the following equation:

$$SDI = 100 * (1 - T_i / T_f)/T$$

T_i = Initial time in seconds required to collect the 500 ml sample;
T_f = final time in seconds required to collect the second sample after test time (15 min);
T = total elapsed test time (15 min);

Notes:

a) There is no correction factor or correlation for running the SDI test at pressures other than 30 psi.
b) The SDI may vary as a function of water temperature and values obtained at different temperatures may not necessarily be comparable! The water temperature must remain constant ($\pm$ 1°C) throughout the test. This is necessary as flow rate changes by about 3% per °C.
c) The SDI varies with the membrane filter manufacturer. This, SDI values obtained with filters from different membrane manufacturers, cannot be comparable.

The SDI values give the following indications for reverse osmosis:

SDI < 1	Several years without colloidal fouling
SDI < 3	Several months between cleaning
SDI 3 – 5	Particular fouling likely a problem, frequent cleaning
SDI > 5	Unacceptable, additional pre-treatment is needed

Spiral-wound modules as used in the Nanofiltration test generally require an SDI<5, whereas hollow fine fiber modules are more susceptible to fouling and require an SDI<3. The target SDI after filtration is normally an SDI of 3 – 5 or less. Surface or seawater may have an SDI up to 200, requiring flocculation, coagulation, and deep-bed multimedia filtration before RO treatment.

Total dissolved solids (TDS): is a measure of the dissolved combined content of all inorganic and organic substances present in a liquid in molecular, ionized, or microgranular (colloidal sol) suspended form. Generally, the operational definition is that the solids must be small enough to survive filtration through

Terminology 381

a filter with 2-micrometer (nominal size or smaller) pores. The term usually used to describe the inorganic salts and small amounts of organic matter present in solution in water. The principal constituents are usually calcium, magnesium, sodium, and potassium cations and carbonate, hydrogen carbonate, chloride, sulfate, and nitrate anions. The two principal methods of measuring total dissolved solids are gravimetric analysis and conductivity. Conductivity is the most commonly used method for the determination of TDS in water supplies, and is based on the measurement of specific conductivity with a conductivity probe that detects the presence of ions in water. Conductivity measurements are converted into TDS values by means of a factor that varies with the type of water. The relationship of TDS and specific conductance of groundwater can be approximated by the following equation:

$$TDS = ke * EC$$

where TDS is expressed in mg/L, and EC is the electrical conductivity in microsiemens per centimeter (μS/cm) at 25 °C. The correlation factor ke varies between 0.5 and 0.9. More specifically

$$TDS \ (mg/L) = 0.5 \ x \ EC \ (\mu S/cm) \ or \ 0.5 * 1000 \ x \ EC \ (mS/cm \ or \ dS/m)$$

As the solution becomes more concentrated (TDS > 1000 mg/l, EC > 2000 μs/cm), the proximity of the solution ions to each other depresses their activity and consequently their ability to transmit current, although the physical amount of dissolved solids is not affected. At high TDS values, the ratio TDS/EC increases and the relationship tends toward TDS = 0.9 x EC. In these cases, the aforementioned relationship should not be used and each sample should be characterized separately. For water for agricultural and irrigation purposes, the values for EC and TDS are related to each other and can be converted with an accuracy of about 10% using the following equation:

$$TDS \ (mg/l) = 640 \ x \ EC \ (dS/m \ or \ mmho/cm).$$

Nevertheless, gravimetric methods are the most accurate and involve evaporating the liquid solvent and measuring the mass of residues left, although volatile organic compounds are lost by this method. The main disadvantage of this method is that it is time-consuming. If inorganic salts comprise the great majority of TDS, gravimetric methods are appropriate.

Total solids (TS): The weight of all present solids per unit volume of water. It is usually determined by evaporation. The total weight concerns both dissolved and suspended organic and inorganic matter.

Turbidity: A measure of nontransparency of water owing to the presence of suspended solids that are usually invisible to the naked eye. The more total suspended solids in the water, the murkier it seems and the higher the turbidity.

There are various parameters influencing the cloudiness of the water. Some of these are: phytoplankton, sediments from erosion, resuspended sediments from the bottom (frequently stirred up by bottom feeders like carp), waste discharge, algae growth, and urban runoff. Turbidity is expressed as Nephelometric Turbidity Units (NTU). The WHO (World Health Organization), establishes that the turbidity of drinking water should not be more than 5 NTU, and should ideally be below 1 NTU.

Ultrapure water: Water with a specific resistance higher than 1 megohm-cm.

SOURCES

Water and wastewater treatment technologies (Vol. 1), (2009), EOLSS/UNESCO (ed. Saravanamuthu Vigneswaran)
Environmental Engineering Dictionary and Directory, Thomas M. Pankratz, (2000), CRC Press
US Geological Survey (https://www.usgs.com)
Membranes and Membrane Separation Processes, Heinrich Strathmann, (2000), Wiley Online Library
Lenntech (https://www.lenntech.com)

Index

A

Acid addition, 339
Additives
    Carbon nanotubes (CNTs), 25–28, 67,
        76–88, 219, 221–224
    Graphene oxide (GO), 18–25, 67–76,
        91–121, 221
    Metal salt, 43–47
    Metal-oxides, 41–43
    Nanocellulose, 47–53
    Polyaniline, 53–58
    Silica, 36–40
    Tungsten disulfide, 53–58
    Zeolites, 28–36
Additive Compatibility, 67–88
Archimedes method, 230
Artificial water channels (AWC)
    Aquaporin (AQP) bioassisted membranes,
        212–224, 311
    Nanochannels, 221
    Self-assembled supramolecular channels,
        219–221
    Single molecular channels, 219
Assimilable Organic Carbon (AOC), 333
Atomic Force Microscopy (AFM), 234
Attenuated Total Reflectance – Fourier-Transform
    Infrared (ATR–FTIR), 240

B

Bacterial Growth Potential (BGP), 333
Batch RO systems, 318
Binodal curves, 286
Biodegradable Dissolved Organic Carbon
    (BDOC), 333
Biofilm formation, 331
Biofouling, 145, 331,335
    Potential, 332
Biosolvents, 165, 166
Boltzmann constant (k_b), 294
Brag's Law, 241
Brine management, 349,351

C

Cahn-Hilliard model, 294
Cake filtration, 329
Capacitive deionization (CDI), 358
Capillary flow porometry, 299

Capillary Tube (CT) model, 231
Carbon nanotubes (CNTs), 67, 76–88, 150, 154,
    219, 221–224
Cleaning in place (CIP), 326
Climate change, 347
Closed Circuit Desalination, 318
Composite membranes, 161, 163
Computational Fluid Dynamics (CFD),
    273–275, 293
Concentration factor, 336
Concentration polarization, 133, 145
Concentration polarization coefficient
    (CPC), 133
Conductivity rejection, 246
Contact angle, 147, 241–243
Cost analysis, 354

D

Delaunay triangulation, 275
Demixing, 286
Differential Scanning Calorimetry (DSC), 243
Dispersions, 67–68, 95, 98, 239, 243
Dusty gas model (DGM), 135

E

Electrodialysis, 9–10
Electrospinning, 150, 158, 160, 161
Electrospun nanofiber membranes, 153, 160
Energy Dispersive X-ray Spectroscopy
    (EDX), 237
Energy recovery device (ERD), 307, 320
Euclidean distance maps (EDM), 299
Evaluation of antibacterial properties, 67–88
Evaluation of antifouling properties, 67–88, 110

F

Fouling, 145, 148, 312, 325
    Particulate, 145, 327
Fouling material determination, 241–242
Flory-Huggins
    theory, 294
    free energy density, 297
Flux, 147, 304
Forwards osmosis, 9
Fourier-Transform Infrared (FTIR)
    Microscopy, 236
Functionalization of CNTs, 79–88

383

acids, 85, 88
nitrogen doped, 85
polyamines, 76, 81, 85
Functionalization of GO, 68–76, 95–96, 153, 243
Fe$_3$O$_4$, 95
magnetic chitosan, 96
polyamines, 68–70, 75
silanes, 153, 243
sulfonated nanoparticles, 71
TiO$_2$ nanoparticles, 71, 74

G

Gain Output Ratio (GOR), 140
Glass transition, 148, 149, 150
Grafting, 162
Graphene quantum dots, 234–236, 242
Graphene membranes, 67–76, 91–121, 153, 221
adsorption isotherms, 105–110
applications, 110–120
Gulf Cooperation Council (GCC) countries, 357

H

Heat sources, 350
Heat transfer, 132
Hybrid membrane configuration, 316
Hydrophobicity, 149, 151, 152
Hypertonic solution, 137

I

Inductively coupled radio frequency plasma
technique (RF ICP), 152
Infrastructure planning, 346
Interfacial tension, 147
Ionic strength, 104, 327

J

J, Molar flux, 147
Junction
Mesh, 280
Points, 47

K

Kelvin equation, 232
Knudsen diffusion, 134
Knudsen number (Kn), 134

L

Latent heat of vaporization (Q$_v$), 132
Lipid bilayer, 264

Low temperature hydrothermal (LTH)
technique, 162

M

Membrane capacitive deionization
(MCDI), 361
Membrane crystallization, 126, 136
Membrane crystallizer
Antisolvent configuration, 137
Thermal or osmotic configuration, 137
Membrane distillation, 9–10, 126
Air Gap Membrane Distillation (AGMD),
127, 128
Direct Contact Membrane Distillation
(DCMD), 127, 128
Hollow-fiber multieffect membrane distillation
(HF-MEMD), 129, 130
Material gap membrane distillation
(MGMD), 129, 131
Multieffect membrane distillation
(MEMD), 129
Osmotic membrane distillation
(OMD), 129
Permeate gap membrane distillation (PGMD),
129, 130
Sweeping Gas Membrane Distillation
(SGMD), 127, 128
Vacuum Membrane Distillation (VMD),
127, 129
Vacuum multieffect membrane distillation
(V-MEMD), 129
Melting point, 148, 149, 150
Membrane stability, 126, 148
Membrane thickness, 126,147
Membrane density, 231
Membrane liquid entry pressure of water
(LEPw), 126, 146, 245
Microscopy techniques, 232–238
Mixed matrix membranes (MMMs), 150
Modified Fouling Index (MFI$_{0.45}$), 329
Morphology of membranes, 232–234
Modules
Hollow fiber, 13
Plate and frame, 155
Spiral-wound, 12
Tubular, 12
Molecular dynamics (MD), 252
Molecular simulations, 252
Monoclonal antibodies (mAbs), 141
Multiple effect distillation (MED), 6, 303, 354
Multistage configurations, 313
Multistage flash distillation (MSF), 5,
303, 354
Multiple-pass RO, 314

Index

N

Nanopore 2D, 255
Neutron transparent flow-through cell
        (NTFT-Cell), 288
Nuclear technology, 364
Nusselt number (Nu), 133

O

Omniphobic, 154
Osmotic pressure, 304

P

Particle deposition factor, 331
Permeability, 133
Permeation flux, 245
Phase inversion method, 67–71, 79
Phase separation, 294, 298
    Evaporation induced phase separation
        (EIPS), 159
    Liquid–induced phase separation (LIPS),
        166
    Nonsolvent-induced phase separation
        (NIPS), 150, 159, 298
    Polymerization-induced phase separation
        (PIPS), 295
    Thermally induced phase separation (TIPS),
        150, 159, 166, 294
    vapor induced phase separation (VIPS), 159,
        165, 166
Plasma surface modification, 163
Polyethersulfone (PES) membranes, 68–69,
        74–76, 79, 81, 85
Polyvinylidene fluoride (PVDF) membranes,
        70–74, 80–81, 236
Prandtl number (Pr), 133
Pressure
    Transmembrane, 135, 146, 304
    Osmotic, 304
    Net, 304, 331
Pretreatment, 13–14, 306
Principal component analysis (PCA), 299
Pore size, 231
Porosity of membranes, 132, 147, 230–233
Potential mean force (PMF), 252
Pumps
    High-pressure (HPP), 307
    High flow capacity centrifugal, 320
    Positive displacement, 320

Q

Quantum dots, 234, 236, 242

R

Rankine cycles, 355
REACH regulation, 164
Recovery, 336
Rejection rate, 246, 336
Renewable energy technologies, 354
Reverse osmosis, 8–9, 354
Reynolds number (Re), 133, 134
Roughness of membrane, 234–236

S

Cryogenic Scanning Electron Microscopy
        (Cryo-SEM), 291
Saturation index (SI), 339
Scaling, 145, 336
    Alkaline, 145
    Barium Sulphate, 337
    Calcium Carbonate, 337
    Calcium Phosphate, 338
    Calcium Sulphate, 337
    Indices, 338
    Non alkaline, 145
    Uncharged molecule scale, 145
Salt diffusion coefficient, 305
Salt Rejection, 246
Scanning electron microscopy (SEM), 232
Schmidt (Sc) numbers, 134
Sensible heat (Qc), 132
Sherwood number (Sh), 134
Silt Density Index (SDI, 328
Sintering, 157
Small angle neutron scattering, 288
Spectroscopy techniques, 238–240
Spinodal
    Curve, 296
    Period, 296
    Region, 296
Specific energy consumption (SEC), 303
Split-partial two-pass configuration, 315
Steepest descent, 264
Stretching, 157
Substances of very high concern (SVHC), 165
Supersaturation ratio (S_r), 339
Surface modification, 148

T

Temperature polarization, 132, 145
    Temperature polarization coefficient
        (TPC), 132
Temperature swing solvent extraction, 350
Tensile Strength, 244
Thermal Conductivity, 148

properties, 243–244
Thermal Gravimetric Analysis (TGA), 243
Thermal vapor compression (TVC), 6–7
Thermodynamic ideal minimum SEC_{ideal}, 303
Thin-film composite membranes, 11, 309
Tortuosity factor, 147
Track-etching, 158
Transmembrane flux (J), 135
Transmission Electron Microscopy
    (TEM), 234
Transparent Exopolymer Particles (TEPs), 331

U

Ultrafiltration (UF), 143
Ultrasonic cleaning, 146
Umbrella sampling, 266

V

Volatile organic compounds (VOCs), 165
Voronoi cell, 275
Voronoi decomposition, 275

W

Water permeance Lp, 245
Wilhelmy equation, 242

X

χ (Flory parameter), 294
X-ray
    Diffraction, 241
    Energy dispersive spectroscopy, 238
    Photoelectron spectroscopy, 239
    Tomography, 269

Y

Young Laplace equation, 299
Young's modulus, 37, 69

Z

Zeta potential, 243
Z (coordination number), 294